# Biology of Fishes

# Tertiary Level Biology

A series covering selected areas of biology at advanced undergraduate level. While designed specifically for course options at this level within Universities and Polytechnics, the series will be of great value to specialists and research workers in other fields who require knowledge of the essentials of a subject.

Recent titles in the series:

| | |
|---|---|
| *Social Behaviour in Mammals* | Poole |
| *Seabird Ecology* | Furness and Monaghan |
| *The Biochemistry of Energy Utilization in Plants* | Dennis |
| *The Behavioural Ecology of Ants* | Sudd and Franks |
| *Anaerobic Bacteria* | Holland, Knapp and Shoesmith |
| *Evolutionary Principles* | Calow |
| *Seabird Ecology* | Furness and Monaghan |
| *An Introduction to Marine Science (2nd edn.)* | Meadows and Campbell |
| *Seed Dormancy and Germination* | Bradbeer |
| *Plant Molecular Biology (2nd edn.)* | Grierson and Covey |
| *Polar Ecology* | Stonehouse |
| *The Estuarine Ecosystem (2nd edn.)* | McLusky |
| *Soil Biology* | Wood |
| *Photosynthesis* | Gregory |
| *The Cytoskeleton and Cell Motility* | Preston, King and Hyams |
| *Waterfowl Ecology* | Owen and Black |
| *Tropical Rain Forest Ecology (2nd edn.)* | Mabberley |
| *Fish Ecology* | Wootton |
| *Solute Transport in Plants* | Flowers and Yeo |
| *Human Evolution* | Bilsborough |
| *Principles and Techniques of Contemporary Taxonomy* | Quicke |

*Tertiary Level Biology*

# Biology of Fishes

## Second edition

Q. Bone
Marine Laboratory
Plymouth

N.B. Marshall
Emeritus Professor of Zoology
University of London

and

J.H.S. Blaxter
Scottish Association for Marine Science
Oban

*illustrated by Q. Bone and C. Gore*

## BLACKIE ACADEMIC & PROFESSIONAL
An Imprint of Chapman & Hall

London · Glasgow · Weinheim · New York · Tokyo · Melbourne · Madras

**Published by**
**Blackie Academic and Professional, an imprint of Chapman & Hall,**
**Wester Cleddens Road, Bishopbriggs, Glasgow G64 2NZ**

Chapman & Hall, 2–6 Boundary Row, London SE1 8HN, UK

Blackie Academic & Professional, Wester Cleddens Road, Bishopbriggs, Glasgow G64 2NZ, UK

Chapman & Hall GmbH, Pappelallee 3, 69469 Weinheim, Germany

Chapman & Hall USA, One Penn Plaza, 41st Floor, New York NY 10119, USA

Chapman & Hall Japan, ITP-Japan Kyowa Building, 3F, 2-2-1 Hirakawacho, Chiyoda-ku, Tokyo 102, Japan

DA Book (Aust.) Pty Ltd, 648 Whitehorse Road, Mitcham 3132, Victoria, Australia

Chapman & Hall India, R. Seshadri, 32 Second Main Road, CIT East, Madras 600 035, India

First edition 1982

Second edition 1995

© 1995 Chapman & Hall

Typeset in 10/12pt Times by Cambrian Typesetters, Frimley, Surrey

Printed in Great Britain by St. Edmundsbury Press, Bury St. Edmunds, Suffolk

ISBN 0 7514 0243 5 (HB)
ISBN 0 7514 0022 X (PB)

A catalogue record for this book is available from the British Library

Library of Congress Catalog Card Number: 94–71951

∞ Printed on permanent acid-free text paper, manufactured in accordance with ANSI/NISO Z39.48-1992 (Permanence of Paper).

# Preface

This book, the first edition of which was published in 1982, has been largely rewritten with many new figures, to take account of recent information resulting from the huge rate of publication of scientific papers and books on fishes. As an example, the continuing series "Fish Physiology" (Academic Press) has just reached its 12th volume, covering in two parts only the cardio-vascular systems of fishes.

The original authors, Q. Bone and N.B. Marshall, invited J.H.S. Blaxter to help widen the expertise on fish reproduction, behaviour and exploitation, leading to new chapters on behaviour, fisheries and aquaculture. A chapter on endocrines has been added and earlier chapters have been brought up-to-date. We have chosen those topics which seem to us to be most useful and interesting, inevitably reflecting our own fields of interest. We have, however, tried to make the bibliography sufficiently wide-ranging for the reader to find an introduction to those topics not covered, and to be able to enjoy further forays into those that are.

Fish are the most varied and abundant of vertebrates and the commercial and sport fisheries are of great economic importance. Fish stocks are not vulnerable to drought, as are so many terrestrial sources of protein, but they are highly vulnerable to pollution and overfishing. At least 80% of fish are caught by hunting and this proportion is unlikely to fall; many stocks are shared and lead to political decision-making about management.

We hope that the reader will obtain a quick insight into this fascinating class of vertebrates, with its wide distribution in environments ranging from transient puddles to the abyssal depths of the sea. Life in such varying environments has led to many unique adaptations to deal with the particular problems of body fluid regulation, locomotion and sensory awareness.

We are grateful to Professor Sir Eric Denton and Dr Peter Tytler for reading parts of some chapters in manuscript, and to the Leverhulme Trust which supported Q. Bone and J.H.S. Blaxter with Emeritus Fellowships during part of the writing. We are grateful to Dr R.N. Gibson for allowing us to use the photograph on the front cover.

Q.B.
N.B.M.
J.H.S.B.

*For Susan, Olga and Val*

# Contents

# 1 Diversity of fishes

## 1.1 Introduction

Biologists are fortunate today not only in having a very wide range of kinds of fishes to study, from hagfish to lungfish, but also in being able to examine fishes adapted to every kind of aquatic habitat (and even some which spend most of their time out of water). In consequence of this wide diversity, fishes show us fascinatingly different designs for special modes of life, as well as for solving problems common to them all. Thus for example, we can examine the remarkable ways in which fish eyes are adapted for seeing in different environments, or the different merits of lipid or gas for buoyancy.

Such comparisons are often illuminating, and give us insights into the compromises fishes have to make to satisfy the often conflicting demands of their lives. To understand these demands, we need first to have a clear idea of the kinds of fishes there are, their basic structural features, and how the different types are related to each other.

Over 20000 fishes have been described, and since about 100 new species are described each year, a final total may well exceed 30000 species. This vast array well outnumbers all other kinds of vertebrate: 60% of all vertebrate species are fishes, so on numbers of species alone, a 'typical' vertebrate would be a fish. The great majority of living fishes are bony teleosts, with the cartilaginous elasmobranchs coming a very poor second with around 800 species. Compared with these two, all the other fish groups, containing fish like lampreys, lungfish or sturgeons are insignificant in numbers of species, though of great zoological interest, and also in some instances of gastronomic and commercial importance, like the sturgeons supplying beluga caviar.

How are these different kinds of fishes related? Fortunately for zoologists, all living fishes share a common plan, as in their myotomal design, and such specialised features of the nervous system as the paired giant Mauthner cells of the hindbrain and the neuron patterns of the spinal cord (Chapter 11). These make it clear that they come from a common ancestor (and have retained many of the features of this ancestor). Jefferies has given a seductively well-argued (but ultimately highly implausible) account of the origin of this fish-like ancestor from a group of aberrant echinoderms. More recently, and very surprisingly, the discovery of more-or-less complete Carboniferous fossils of conodonts (long known

as small tooth-like objects of unknown affinity) has shown that they are
fish-like, with notochord, myotomes, caudal fins and anterior large eyes,
and apparently, with bone tissue in the 'teeth'. A recent view of the
relationships of early fossil fishes is shown on page 12. Whenever the fish-
like ancestor prior to known fossils arose, all later fish groups were derived
from it, and the relationships of the surviving groups are now more or less
agreed in broad outline, as seen in Figure 1.1. The names of the larger
categories in Figure 1.1 may seem somewhat arcane, but they are worth
remembering, for they broadly define the groups according to their
outstanding characters. For example, the elasmobranchii are those with
'plate-like' gills; holocephali (all head) are fish with very large heads,
sarcopterygii are the fish with fleshy fins: and actinopterygii are ray-finned
fishes. The reader can (and should!) work out the etymology of the
remainder, for it is much easier to remember names which one understands
and which contain some information about the fishes they include. Readers
of a less advanced age than the authors are not recommended, however, to
follow the distinguished American ichthyologist, David Starr Jordan, who
refused to learn the names of his students on the grounds that if he did so,
he would forget the names of fishes!

Figure 1.1 is somewhat different from those in older texts (and in the
first edition of the present book), and we need to have some idea of why
this should be so. In part it is due to new information about structure (for
example, in the coelacanth *Latimeria*), but it is largely because new
taxonomic approaches have been adopted. As yet, there has been
insufficient molecular sequence data to confirm (or disprove!) the general

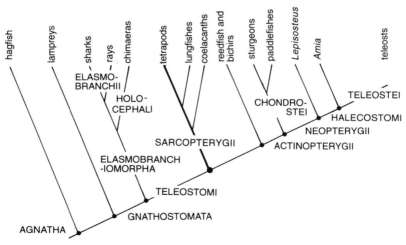

**Figure 1.1** The relationships of living fish groups. Modified from Nelson (1969) *Bull. Amer. Mus. Nat. Hist.*, **141**, 475.

conclusions reached, though more data are steadily being accumulated, and should eventually lead to a definitive phylogeny. There is still ample room for debate, as we shall see later, at a more detailed level, on, say, the relationships of the different teleost groups, or the exact position of the gar pikes and the bowfins, but there is general consensus about the relationships of the larger categories.

As shown in Figure 1.1, the most widely held current view is that hagfishes are separate from the lampreys, the other agnathan group, and from the gnathostome line leading ultimately to ourselves. In other words, hagfishes are (in cladistic terms) the sister group of all other vertebrates. The reader should note, however, that recent molecular data do not support this view (see p. 12).

What is meant by cladistic terms? The naïve reader who has not yet met the cladistic approach will soon find that it plays a significant role in modern fish classification. Although taxonomy has appeared to many biologists as a somewhat dry and unrewarding branch of science, in recent years it has become a fascinating and important battleground between rival workers holding different views of the relationship between evolutionary theory and classifications. What came to be called cladistics (from the Greek κλαδη = *clade* = 'branch') was introduced by the entomologist Hennig in 1965, and it is the taxonomic approach to be found in most recent discussions of fish classification at all levels.

Hennig's phylogenetic method has two components: cladistic analysis with rules for discerning cladistic relations and showing them in cladograms (branching diagrams), and secondly, the use of the cladograms to construct classifications. The cladograms are built from a series of dichotomies, where a parent species gives rise to two daughter species and itself disappears. The two daughter species and all their descendants have equivalent rank in the classification and are sister groups; in Figure 1.1, for example, myxinoids and all other vertebrates are sister groups, as are sharks and rays. Gradual modification along a single line has no place in cladistic taxonomic schemes, which are based entirely on the recognition of branching points where novel characters arise and new sister groups are born. So the 'advanced' or derived characters of a group are more helpful than 'primitive' ones shared by all its members, and the latter are ignored. Hennig and other users of his cladistic method regarded the cladograms as phylogenetic trees, but an influential group of later taxonomists, transformed cladists as they are called, doubt that cladograms have an evolutionary basis, and as a consequence, have sometimes argued that fossils are not helpful in attempting to provide an evolutionary classification of modern forms, although the cladistic technique is used for analysing fossil relationships.

By no means all systematists are cladists or transformed cladists, though at present the great majority are. Evolutionists accept that there is a

systematic relation between a classification and phylogeny, and they
seek natural groups, considering however both advanced and primitive
characters. Scott-Ram gives an interesting and sympathetic account of the
philosophical basis of these different approaches.

Not surprisingly, in view of their vast numbers, and great diversity, it is
the teleosts which have afforded systematists (of whatever school) most
scope to modify the classifications of their predecessors! As assessed by
cladistic methods, the teleosts are generally agreed to fall into four major
monophyletic groups (Figure 1.2). The freshwater Osteoglossomorpha
(bony tongues) contain more than 300 species of mormyrids notable for
their remarkable weak electric systems, and a few large fish like
*Osteoglossum* and *Scleropages*; this group is regarded as the most
primitive. The Elopomorpha, the second group, is about double the size,
and contains the freshwater and marine eels, and marine tarpons (*Elops*);
all have marine leaf-like leptocephalus larvae. The Clupeomorpha (herring-
like fishes) are largely marine and have a unique linkage between the
swimbladder and the inner ear (p. 226). The last group, the Euteleostii are
much the largest group (17000 species in 25 orders and no less than 375

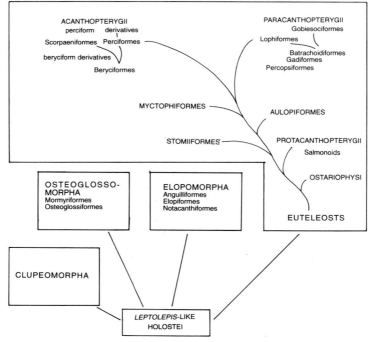

**Figure 1.2** The four major monophyletic teleost groups, the Euteleosts being much the
largest. Partly after Greenwood *et al.* (1966) *Bull. Amer. Mus. Nat. Hist.*, **131**, 339.

families!), and are naturally much the most difficult to organise into a phylogenetic classification. Figure 1.2 shows one possible arrangement of the major groups, but any such arrangement can still only be tentative. At present, for example, both salmon-like and pike-like fishes in the Orders Salmoniformes and Esociformes are uncertainly placed in the Proto-acanthopterygii, near the 6000 or so freshwater carps, characins and catfish of the Ostariophysi, which are distinguished by their Weberian ossicles which link the swimbladder to the inner ear (p. 226).

The remaining euteleost groups make up the Neoteleostei, and share a novel common system of pharyngeal muscles and teeth, enabling them to crush and swallow their food. Most of the important food fish (apart from clupeids) are neoteleosts. Vertebrate life in the oceans is dominated by the perciform acanthopterygians which also contribute importantly to fresh-water faunas, whilst ostariophysans dominate in freshwater. Unfortunately, at present it is far from clear where such large and important neoteleost groups such as the Paracanthopterygii (angler fishes, cod-like fishes) or the spiny-finned Acanthopterygii (around 7000 species of perch-like fishes including tuna, jacks, seabasses and mullets, and some 2000 gobies) fit into the classification. Both of these groups need revision, and much further study before their relationships are better understood. There is general acceptance that the Neoteleostei (Figure 1.3) contain seven groups of fishes, of which the last is by far the largest, containing all the huge array of

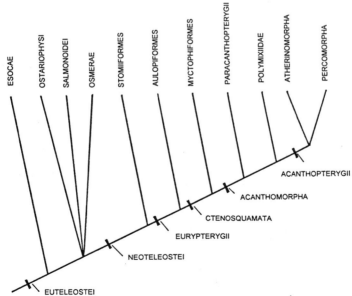

**Figure 1.3** The phylogeny of the Euteleosts. Horizontal names are clades. After Roberts (1993) *Bull. Mar. Sci.*, **52**, 60.

percomorph fishes. Cladistic analyses producing this kind of phylogeny (based on such features as cranial muscle morphology and the structure of the neurocranium), are continuing, and being further refined.

To some extent, the rather dauntingly formidable (even to ichthyologists!) classification of teleost fishes is provisional, and will certainly be amended as new morphological and molecular data become available. As one of the editors of a recent symposium on Percomorph phylogeny justly remarks 'No other vertebrate group, and perhaps no other group of animals, has seen classificatory modifications over the past 25 years equivalent in scope to those in teleost fishes . . .' (Johnson, 1993). These modifications still continue, albeit on a smaller scale. Although it may seem disappointing that a stable classification is yet to be achieved (except in broad outline), the reader should find rather that this is an encouraging sign of continuing activity, and look forward to some interesting surprises such as the recently uncovered relationships between cichlids, wrasses and other groups.

## 1.2 Basic structural features of fishes

### 1.2.1 Body shape, scales and fins

During the evolution of the different fish groups, there have been many changes from the assumed ancestral fusiform (spindle) shape, and as a result, living fishes are so varied that they are hard to categorise and define briefly. The ancient ancestors of fish were possibly something like the conodonts or the present-day protochordate amphioxus (*Branchiostoma*) in general appearance, though we do not know if they were filter feeding like amphioxus, or fed on larger prey as conodonts (apparently) did. Amongst the remarkably preserved middle Cambrian fauna of the Western Canadian Burgess shales, there is a rare chordate-like fossil *Pikaia*, which although as yet not fully described, seems to show the same kind of fusifom body with a series of V-shaped myotomes as in amphioxus, aligned along a dorsal notochord. But we still know very little about the early origins of fishes and further studies of Cambrian faunas may well provide surprises. Many modern fishes have retained the fusiform streamlined body shape, but every other kind of body shape is also found: globular (puffer fish and some angler fish), elongate (eels, lampreys, pipefish), compressed dorso-ventrally (rays), and laterally (sunfish, flatfish, many coral reef fishes).

The most extreme changes in body shape are seen in the remarkable tropical seahorses, which look like animated seaweed, and in many fish larvae which are often quite different from the adults (p. 185). This extraordinarily wide range in body shape is perhaps best appreciated when diving off coral reefs, but similar diversity is found in the deep ocean or in large tropical rivers. The body surface is usually covered with more or less conspicuous tough overlapping scales or denticles, overlain by an outer epithelium. Sometimes, as in eels, the scales are buried under thick epithelial layers so that eels seem to lack scales, but in other fishes like sharks and acanthopterygians or *Latimeria*, although still covered by an epithelial layer, the scales armour the body surface. Scale shape, ornamentation and structure varies in the different groups of fishes (Figure 1.4), and is important in the classification of fossil fishes. Curiously enough, scale structure (seen in ground sections and less well in decalcified material) is still not well known in all groups, for example in the lungfish *Neoceratodus*, new sections had to be ground for Figure 1.4. In bony fishes the scales grow as the fish grows, and fishes living in temperate waters

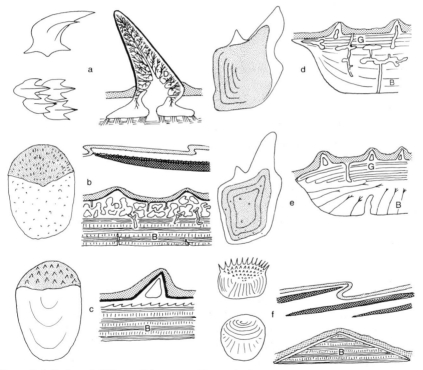

**Figure 1.4** Scales of different fish groups. For each, intact scale on left and section through the scale on right. (a) Elasmobranch; (b) *Neoceratodus* (Dipnoi); (c) *Latimeria* (Coelacanth); (d) *Polypterus* (Branchiopterygii); (e) *Lepisosteus* (Holostei); (f) Teleost (ctenoid scale above, cycloid below). Light stipple = outer layer of dermis; black layer = enamel or enameloid; G = ganoin; D = dentine; B = bone. After Bertin (1958) *Traité de Zoologie*, **13**, fasc2; Kerr (1952, 1955) *Proc. Zool. Soc. Lond.*, **122**, 55 and **125**, 335; Goodrich (1909); Gnadeberg (1926) *Jen. Z. Med. Naturw.*, **42**, 473; and Millot *et al.* (1978) *Anatomie de Latimeria chalumnae*, CNRS, Paris.

where growth rate is seasonal add different amounts to the scales winter and summer. The annual rings so formed, are important in age determination (p. 171). Shark scales (denticles) are intercalated between existing ones as the shark grows, so cannot be used to age sharks (vertebral or spine rings are more useful here). Many of the surface ridges and spines on scales are probably significant in controlling water flow around the fish (see Chapter 3). In *Polypterus* the elasticity of the pattern of interlocking rhomboid scales is used in moving air into the airbladder.

Whatever the kinds of scales and body shape, most fishes have a series of unpaired fins, together with paired pectoral and pelvic fins. Modern lampreys and myxinoids lack the paired fins (though some fossil agnathans had them); those of jawed gnathostome fishes may have arisen from lateral dermal folds of the kind seen on some Upper Silurian agnathan fossils. Once paired fins had appeared, they conferred much greater manoeuvrability, and show a great variety of form according to the mode of life, ranging from lifting foils (sometimes used for flight), to walking legs, suckers, and propellors. Fin structure is also rather varied in different fishes (Figure 1.5). In sharks, the fins are stiffened at their bases by cartilaginous plates and a series of smaller radial cartilages. Stiff bristle-like collagenous ceratotrichia pass outwards from the radials to the edges of the fin. Such fins are fixed, and cannot alter in shape or be furled, but they *can* be tilted by muscles attached to the basal cartilages. In rays, separate muscles attach along each elongate jointed radial of the paired fins which are consequently much more mobile, and flex up and down as the ray wings its way along. Curiously, as well as swimming in this way, many rays can walk slowly along the bottom using the first few radials of the pelvic fin (which are separate from the rest) as legs.

Apart from the extraordinarily mobile thread-like fins of *Protopterus*, and the rotatable and flexible fins of *Latimeria* (see Figure 1.5), fin

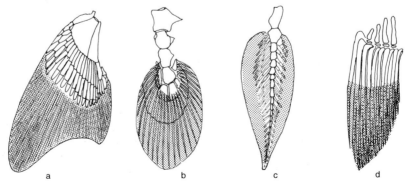

**Figure 1.5** Skeletal structure of pectoral fins in different fish groups. (a) elasmobranch (*Squalus*); (b) *Latimeria*; (c) *Neoceratodus*; (d) teleost. After Goodrich (1909).

musculature is most complex in teleosts, where the fins are built around flexible jointed rays (lepidotrichia), which are moved by basal muscles. This design enables the fish to fold its fins to change their shape, and to make the most delicate movements of astonishing precision. The unpaired dorsal fins of seahorses or knife fishes (gymnotids) which propel their owners slowly about, are exceptionally beautiful examples of motor control, but almost all teleosts owe their manoeuvrability to this advanced fin design. Sometimes, as in sea robins, free pectoral fins rays are used to sample the bottom as the fish 'walks' forwards. Not all teleosts have flexible fins however. For example, the pectoral fin rays of marlin, tunas and swordfish are ossified together to form stiff lifting foils, rather similar to those of large sharks, except that they can be folded back against the body (except in marlins).

## 1.2.2 Internal features

In all fishes, the body is built around an axial notochord, still present in adult hagfish, lampreys, sturgeons and *Latimeria*, but more or less reduced in most adult fish by the development of vertebral elements around it. This provides the incompressible strut flexed by the serial myotomal muscle blocks, and protects the spinal cord which may differ in shape between lampreys, hagfish and other fishes, but is essentially of the same organisation. Anteriorly, some kind of cranium protects the brain, again of basically similar structure in all. All fishes share also gills for gas exchange, a two-chambered heart and circulatory system of essentially the same plan, and a similar complement of viscera. There are, however, striking differences between the internal structure of hagfishes, lampreys and all other fishes. For instance, in hagfish and lampreys the branchial skeleton is attached to the cranium and lies *outside* the gill pouches (in jawed fishes it is free from the cranium and supports the gills directly); there is only a single nasal opening leading to a nasohypophysial sac; a lymphatic system is lacking, and nerve fibres are all non-myelinated.

Much the most important difference is, of course, the development of hinged jaws in the gnathostome fishes, which accounted for their overwhelming success compared with the very few living survivors of the agnathan radiations between the Ordovician and Devonian. The appearance of jaws came together with that of a more streamlined body and paired fins to permit the accurate control of the movements needed to snap up larger and faster prey. Linked with this were stronger skeletal elements to withstand the forces of the more elaborate muscular systems.

The requirement for more accurate control of swimming meant that the brain became larger and more complex; in living lampreys and hagfish the cerebellum is particularly small compared with other fishes (Chapter 10). Once jaws had appeared, a whole series of changes from the agnathan level of organisation became possible, and the new jawed fishes soon became dominant.

The two main gnathostome groups, the sharks, rays and chimaeras (elasmobranchiomorphs), and the different kinds of bony fishes, though less different than agnathans and gnathostomes, each has a distinctive internal structure. The most obvious difference is that the elasmobranchiomorph skeleton is almost entirely cartilaginous. The skull in the adult remains as a cartilaginous neurocranium to which are fused capsules for olfactory organs and ear. The result is a curiously-shaped braincase (especially in chimaeras), to which is attached the palatoquadrate and Meckel's cartilages of the upper and lower jaw (see page 163). The vertebral column is rather simple; circular biconcave centra are separated by intervertebral discs, and bear neural arches dorsally (Figures 1.6 and 1.7). Ventrally there are either short ribs or (in the tail region) haemal arches.

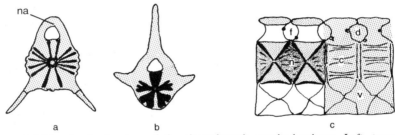

a                              b                              c

**Figure 1.6** The simple structure of the elasmobranch vertebral column. Left, transverse sections through the middle of the centrum of trunk vertebrae in the shark *Lamna* (a) and the ray *Raja* (b) showing radial calcifications (black). Right, lateral view of trunk vertebral column of *Lamna* (in sagittal section on left to show the notochord (n)). c: centrum; f: foramina for dorsal and ventral root nerves; d: interdorsal; v: interventral; na: neural arch. After Goodrich (1909).

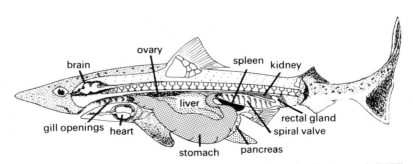

**Figure 1.7** General organisation of the shark *Squalus*. After Lagler *et al.* (1977).

Such a relatively straightforward axial skeleton contrasts with the more complex skull and vertebral column of bony fishes. Although the skull begins in development as a cartilaginous neurocranium similar to that of sharks, it is soon altered by endochondrial bones ossifying in this cartilaginous framework, and by the addition of many dermal (membrane) bones, supposedly derived from a scale layer in the skin. The end result is that the bony fish skull is a composite and complicated structure, and that the homologies of the dermal bones in different groups of bony fishes are by no means easy to sort out. Thus, for example, the large paired dermal bones above the teleost orbit which overlie the pineal were long labelled frontals, but are now accepted to be the equivalent of tetrapod parietals.

In adult bony fish, the jaws are also a mix of endochondrial and membrane bones: separate ossifications in the palatoquadrate cartilage produce the palatine, pterygoid, mesopterygoid and quadrate (some with dermal contributions), and these are covered at the edges of the upper jaw by the dermal membrane premaxilla, maxilla and jugal. The dermal dentary makes up the greater part of the lower jaw, for Meckel's cartilage only ossifies at its hinder end to form the articular. The vertebral column is normally similar to that of elasmobranchiomorphs, with biconcave centra, though the ribs are more complex.

The major differences in the viscera are the presence of a gas-filled swimbladder in most bony fishes (lost in some, and in others like many myctophids and *Latimeria*, filled with lipid), and in the reproductive system. Elasmobranchiomorphs always have internal fertilisation, rare in bony fishes. The spiral valve of the elasmobranchiomorph gut is only found in a few bony fish (*Amia*, *Polypterus* and osteoglossids) where it is less complex. Bony fishes lack the elasmobranchiomorph salt-secreting rectal glands, and high urea content, but curiously, *Latimeria* has both.

### 1.2.3 Distribution and morphology

*Myxinoids.* Until the mid-1980s, hagfishes and lampreys, both lacking jaws, were placed together in the class Agnatha, although it was earlier recognised that they differed in many respects, even if superficially somewhat alike, and were likely derived from different groups of fossil agnathans. Cladistic analyses now suggest, as we have seen, that hagfishes are the sister group to all other living vertebrates, and the two agnathan groups are now generally placed in separate classes. What is more, the most recent studies of fossil agnathans now suggest that instead of myxinoids being derived from one group of fossil agnathans, and lampreys and gnathostomes from another, myxinoids are even more isolated as the sister group of all known fossil agnathans as well as of lampreys and gnathostomes (Figure 1.8). However, recent evidence from 18S ribosomal

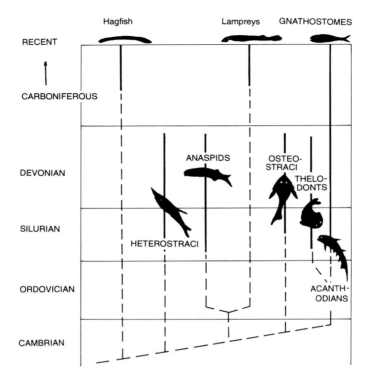

**Figure 1.8** One possible arrangement of fossil and recent agnathans, and the origin of the gnathostomes. Simplified and modified from Forey and Janvier (1993); thelodonts as ancestors to gnathostomes from Turner (1991). In *Early Vertebrates and Related Problems of Evolutionary Biology*. Science Press, Beijing.

RNA sequence analysis suggests that lampreys and hagfish form a natural group (Agnatha), and that this group is the sister group to all other living vertebrates; in other words, the older classification assuming agnathan monophyly was correct! In this book, we accept the conclusion of the cladistic analyses, but the reader is warned to look out for the appearance of further molecular data which should settle the matter.

There are around 50 species of hagfishes, all marine, mostly living in temperate regions down to 2000 m. A single fossil hagfish is known from the upper Carboniferous (230 Myr BP, i.e. 230 million years before the present), and as far as can be seen, is very like living forms. They are naked and eel-like (Figure 1.9), scavenging on dead and dying fishes and invertebrates with a complex toothed radula-like tongue. In several features they are distinct from other vertebrates. There is only a single semicircular canal, the blood is isosmotic with seawater, and vast quantities of slime can be produced from special thread cells opening along the body.

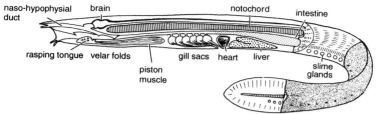

**Figure 1.9** General organisation of the hagfish *Myxine*. After Marinelli and Strenger (1956) *Vergl. Anat. Morphol. der Wirbeltiere*, **2**. F. Deuticke, Wien.

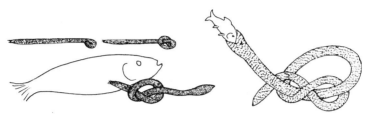

**Figure 1.10** Knotting. Left, in the hagfish. The initial stages of knotting (above), about to pull a mouthful from a dead fish (below). Right, the more complex knot tied by the tropical muraenid eel (*Echidna*) to swallow its prey. After Strahan (1963) *The Biology of Myxine* and Miller (1987) *Copeia*, 1056.

This unique mechanism (a single hagfish can solidify a bucket of water with its slime) is both defensive, and prevents others from sharing hagfish food. It is complemented by the ability of hagfish to tie themselves in a knot, which they slide along the body to wipe off the slime. Knotting (Figure 1.10) is also used to escape capture and (as with the more complicated knots made by moray eels), to tear off food. There are up to 16 pairs of external gill openings, a single olfactory capsule with an exhalent duct opening to the pharynx, and the eyes (which lack eye muscles) are buried under the skin. Hagfish lay large yolky eggs, and unlike lampreys, development is direct; there is no larval stage.

*Lampreys.* Lampreys share with gnathostome vertebrates a more advanced kidney than hagfishes, a photosensitive pineal, more than one semicircular canal (in fact only two), neural and haemal vertebral elements formed around the notochord, a large pancreas, a lateral line, radial muscles in the fins, extrinsic eye muscles and similarities in the histology of the pituitary (Chapter 9). Lampreys are not isosmotic with the surrounding water, those with a marine phase are hypotonic in the sea, and all are hypertonic in freshwater (Chapter 6). For all these reasons, they are now separated from hagfish, and allied with gnathostomes.

Like hagfishes, lampreys are eel-like (Figure 1.11), the largest nearly a metre long, with seven external gill openings, a single opening to the

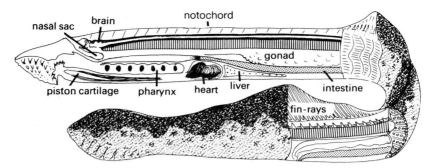

**Figure 1.11** General organisation of lamprey. After Goodrich (1909).

nasohypophyseal sac, and with a spiral valve in the ciliated intestine. Lamprey ancestors must have lived in the Palaeozoic (the fossil *Hardistiella* from the Upper Carboniferous is similar to living forms), but it remains unclear whether they are to be derived from cephalaspid or anaspid fossil agnathans. Living lampreys have an entirely cartilaginous skeleton, with a few cartilaginous 'vertebral' elements around the large unconstricted notochord. Whichever group gave rise to lampreys, it had a heavily mineralised skeleton, and since the cartilage of living lampreys retains the ability to calcify under experimental conditions, their purely cartilaginous skeleton is secondary.

Today, the 40 or so lamprey species are found in rivers north and south of latitudes 30° North and South, and are interestingly divided into roughly equal numbers of parasitic and smaller non-parasitic species. Non-parasitic species probably evolved from their parasitic 'twin' species, perhaps as a result of isolation in parts of river systems where host fishes were rare, and never descend to the sea. There are even examples of some races of the same species being parasitic, others non-parasitic, and evolutionary aspects of lamprey systematics continue to stimulate research. The parasitic species, which may be confined to freshwater or like *Petromyzon marinus* and *Geotria australis*, migrate to the sea as adults (anadromous) rasping the skin of other fishes, and feeding on their blood. *P. marinus*, in the sea, commonly hitches lifts on basking sharks (*Cetorhinus*) and on dolphins, but probably does not feed on the latter. Both kinds lay small eggs in redds excavated in river bed gravel, where after hatching they quickly burrow into the mud at the edge of the rivers. They spend five to seven years as filter-feeding ammocoete larvae (Figure 1.12) before a radical meta-morphosis into adults which then either pass downriver to live for some years in the sea before returning to spawn, or spawn in the river soon after metamorphosis.

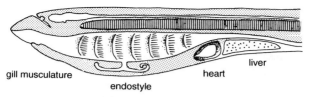

gill musculature

endostyle

liver

heart

**Figure 1.12** Branchial region of lamprey ammocoete larva showing position of endostyle below gill pouches, and in comparison with Figure 1.11, undivided pharynx and absence of teeth and piston cartilage. After de Beer (1928) *Vertebrate Zoology*, Sidgwick and Jackson, London.

*Elasmobranchiomorpha.* The 800 or so sharks and rays and around 30 chimaeroid fishes (Holocephali) are all marine with the exception of one shark which ascends rivers, and a few freshwater stingrays which are (secondarily) confined to freshwater. They share urea conservation for osmoregulation, rectal salt-secreting glands, and internal fertilisation with *Latimeria*, but *Latimeria* certainly acquired both the latter independently. Internal fertilisation in Elasmobranchiomorpha permits a variety of reproductive strategies from laying egg capsules to bearing live young (Chapter 8). There are no very small elasmobranchiomorphs (the little pelagic squaloids *Isistius* and *Squaliolus* (just over 24 cm) are the smallest), possibly because urea retention becomes difficult as surface area increases as body volume decreases. Because their skeletons are almost wholly cartilaginous (see below), and the viscera relatively 'simple' (Figure 1.13), both showing features found in the ontogeny of other fishes, elasmo-branchiomorphs used to be regarded as 'primitive' fish, but their large brains (Chapter 11) and diversity of their feeding mechanisms show that they are far from 'simple'; the cartilaginous skeleton may make them easy to dissect but does not mean that they recapitulate the ancestral fish organisation! In fact, as well as the elasmobranchiomorph skeleton being strengthened at strategic points by granular or prismatic calcifications (see Figure 1.6), in the dogfish *Scyliorhinus*, lamellar bone overlies the vertebral neural arches, and is also found in the bases of the denticles. The X-ray diffraction patterns from shark vertebrae are identical to those from bone, and more work is needed to establish the relations between bone and cartilage in sharks. Strengthening calcifications are less prominent in chimaeras which are rather delicate 'watery' fishes, and unlike sharks and rays have a naked skin, bearing denticles only on the curious claspers. The denticles covering sharks and rays have a plate-like base where true bone tissue may occur, which supports the main body of the denticle composed of dentine with an enameloid capping (see Figure 1.4). Denticles are variously modified to form fin spines, flattened scales and filtering combs (on the gills of basking sharks), and on the jaws, they form teeth (Figure 1.14). Shark teeth are replaced from behind forwards, only one or two rows normally being in use at one time. These teeth are in some species

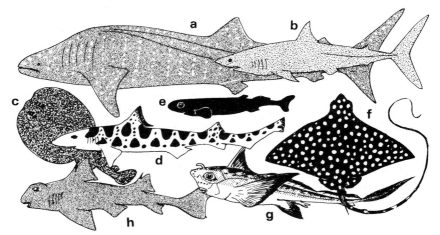

**Figure 1.13** Various elasmobranchiomorphs (not to same scale). (a) whale shark (*Rhincodon*); (b) mako (*Isurus*); (c) electric ray (*Torpedo*); (d) cat shark (*Triakis*); (e) cookie cutter (*Squaliolus*); (f) eagle ray (*Aetiobatis*); (g) rabbitfish (*Chimaera*); (h) Port Jackson shark (*Heterodontus*). After Dean (1906) *Chimaeroid Fishes and their Development*, Carnegie Institute, Washington, publ. 32; Daniel (1922) *The Elasmobranch Fishes*, California University Press; Lineaweaver and Backus (1970) *The Natural History of Sharks*, Lippencott, Philadephia; and Marshall (1971).

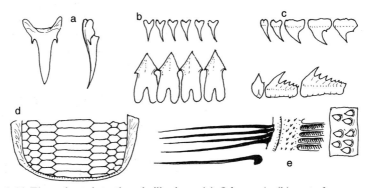

**Figure 1.14** Elasmobranch teeth and gill rakers. (a) *Odontaspis*; (b) part of upper and lower set of the cookie-cutter shark (*Isistius*); (c) same for the six-gilled shark (*Hexanchus*); (d) battery of crushing plates from lower jaw of the eagle ray (*Myliobatis*); (e) filtering gill rakers (left), and minute teeth (right) from the basking shark (*Cetorhinus*). After Bigelow and Schroeder (1948, 1953) *Fishes of the W.N. Atlantic*, Mem. Sears Fndn., **1**, parts 1 and 2.

changed quite frequently; every 9–12 days in sandbar sharks (*Ginglymostoma*, though only two to four times a year in the blue shark (*Prionotus*). In the tiny pelagic cookie-cutter sharks (*Squaliolus*) which bite circular lumps out of marine mammals and larger fish, the upper and lower sets of teeth are replaced as units, and swallowed, but this economy is not found in other sharks where the teeth simply drop out to the ocean floor. Since teeth are almost the only parts of sharks known as fossils and are very abundant, interesting calculations have been based on the numbers of fossil teeth on the sea bed, and replacement rates of modern teeth, in an attempt to work out the abundance of sharks in ancient seas. Rays have batteries of crushing teeth to grind molluscs and crustaceans, whilst holocephalans deal with their varied diet using curious longer-lasting toothplates reminiscent of rabbit incisors.

All elasmobranchiomorpha have large numbers of fantastically sensitive tubular electroreceptors (the ampullae of Lorenzini) (Chapter 10) and rays (but no sharks) can generate electrical pulses with weak electric organs in the tail (probably for intraspecific communication). Some, like *Torpedo*, use much more powerful pectoral organs to stun their prey.

Just over half of living elasmobranchiomorphs are rays, dorsoventrally flattened, with ventral gill openings and pectorals fused to the head. Most of this group have large pectorals like *Raja*, the main skate genus, but sawfishes (*Pristis*) and guitar fishes (*Rhinobatis*) have much smaller pectorals and look much like sharks. The flattened shark, *Rhina*, looks rather like a sawfish without the saw, but its gill openings are lateral rather than ventral. Rays swim by undulating or flapping their pectorals, and the tail may be much reduced or virtually absent, in sting rays being whip-like and bearing venomous spines. Hammerhead sharks (*Sphyrna*) which prey on stingrays, often contain large numbers of such spines. In sawfish and guitar fish, which swim by oscillating the tail rather than by flapping their pectorals, the tail is similar to that of sharks. Although most rays are bottom feeders, some are pelagic (like the butterfly rays *Gymnura*), and the huge manta (*Mobula*) is a filter-feeder.

Most sharks are galeomorphs, a large superorder containing smooth dogfishes (triakids); some 200 carcharinids like tiger (*Galeocerdo*) and grey (*Carcharinus*) sharks, as well as the advanced fast-swimming isurids with warm red muscles (see page 63), culminating in the great white shark *Carcharodon*. *Carcharodon* has an immense oil-filled liver for buoyancy, like the deep-sea squaloids (Chapter 5); in the largest specimen reliably recorded, 6.5 m long and weighing 3300 kg, the liver weighed no less than 456 kg. The larger basking shark (*Cetorhinus*) also has an immense liver for buoyancy, and cruises slowly around filtering copepods with its gill rakers, the teeth being minute. The peculiar megamouth (*Megachasma*), first caught in 1976, also has minute teeth and filter-feeds on small crustaceans and jellyfish. It has an apparently inefficient filtering system with a different type of gill raker, but the suggestion that plankton is attracted into its mouth by a bioluminescent pharyngeal lining seems unlikely, and its method of feeding remains mysterious. The biggest shark, indeed the biggest of all fishes, the whale shark (*Rhincodon*) exceeds 15 m, and feeds on plankton and shoals of small fishes. Whale sharks are grouped with the carpet and nurse sharks (orectolobids) of shallow warm seas.

Squaliform sharks include the six- and seven-gilled hexanchids, and pristiophorid saw-sharks (remarkably like the pristid sawfishes which are rays), but the largest group of squaliforms are the squaloids like the spur dogfish, *Squalus*. Squaloids have radiated widely in the deep sea, where new species turn up regularly. These are dark brown or black, with tapeta at the back of the eye reflecting green light. When hooked on deep-sea long lines and brought on deck in daylight, however, the attractive feature of their green eyes is somewhat discounted by their extreme corpulence, resulting from their large oil-filled livers which enable them to achieve neutral buoyancy and to cruise above the seabed. In older zoology texts, it is stated that the heterocercal tail in sharks (with its flexible lower lobe) provides lift as it sweeps across, but the tails of deep-sea sharks are markedly heterocercal although they are neutrally buoyant and no lift is required; heterocercal tails can provide lift, but their intrinsic musculature can be adjusted so that they do not do so, or indeed, may generate downthrust!

*Sarcopterygii.*
*Lungfish (Dipnoi).* Six species of lungfish survive today (Figure 1.15) from a group that appeared in the Devonian and was widespread in the Palaeozoic. Modern lungfish are in several ways specialised and simplified versions of their more completely ossified ancestors. The Australian *Neoceratodus* is the most heavily ossified of living forms and although there are differences in the skull and toothplates, resembles what is known of the Triassic *Ceratodus*. It is covered with large overlapping scales, the notochord is large and unconstricted and the girdles and fin skeletons are cartilaginous. The paired fins are lobe-like (sarcopterygian), contrasting with the elongate fins of the other living genera. Young *Neoceratodus* and

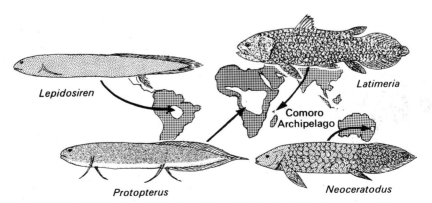

**Figure 1.15** The distribution of living lungfish and *Latimeria*. After Norman and Greenwood (1975) *A History of Fishes*, 3rd edn. Ernest Benn, London; and Nelson (1976).

*Protopterus* 'walk' along the bottom with their paired fins very like urodele amphibia. Unpaired fins are absent, but it seems that the symmetrical 'caudal' may have arisen from the union of dorsal and anal fins, since there is no trace in the development of living lungfish of the heterocercal elasmobranch-like tail seen in fossil lungfish.

Lungfishes eat small invertebrates and large amounts of plant material, which they grind and crush with paired upper and lower tooth plates. The gut is simple and ciliated, there is no stomach or hepatic caecum, but a spiral valve is present. Both inhalent and exhalent openings to the nasal chamber lie in the roof of the mouth, and although the exhalent opening used to be supposed to be equivalent to the internal naris of tetrapods, more recent detailed embryological studies have shown that it is simply the homologue of the actinopterygian posterior nasal opening secondarily displaced into the mouth. The airbladder is septate and lung-like (Chapter 5), paired in *Protopterus* and *Lepidosiren*, but single in *Neoceratodus* which normally only breathes air when stressed and very active. Like most other fish groups, lungfish have electroreceptors, but recent studies of the snout of *Neoceratodus* have shown that the interpretations of the cranial tubular systems of fossil lungfish as a complex electroreceptor system are probably mistaken, and that the curious vascular loops in *Neoceratodus* skin, like the fossil tubular canal systems are likely to be related to the modifications of the dermal bones as the fish grows. A remarkable feature of *Protopterus* and *Lepidosiren* (but not *Neoceratodus*) is that when the pools in which they live dry up, they make burrows and aestivate (p. 116). In fossil lungfishes the well-formed opercular apparatus indicates water breathing was more important than lungs in the Palaeozoic, but fossil burrows (*Griphognathus*) show that air breathing was early acquired and was widespread in the group. Aestivation is not, however, peculiar to lungfish (see page 116).

A simple valve in the truncus arteriosus of the heart allows a partial separation of the pulmonary and systemic circulations. In this, and the structure of the brain and reproductive system, and in the interesting feature that the skin of the embryo and early hatchling is ciliated (and in *Neoceratodus* propagates action potentials), lungfishes resemble amphibians. *Neoceratodus* simply attaches its large amphibian-like eggs to water plants and to the submerged roots of riverside trees, and then abandons them, but the other living genera make nests.

*Crossopterygii.* The living *Latimeria chalumnae* (first caught off southern Africa in 1938) is known from over 170 later specimens all caught from the volcanic islands of the Comoro archipelago, apart from the latest and largest, a female trawled off Mozambique in August 1991. *Latimeria* is the sole survivor of a specialised crossopterygian group first known from the mid-Devonian. It is much larger (the latest and largest was 1.6 m long) than its forebears, but otherwise very similar to them. It is covered with large overlapping bony spinous scales, and has the typical coelacanth trilobed tail and lobate fins.

Recent expeditions to the Comoro Islands have successfully used manned submersibles to film *Latimera* in its benthic habitat around 200 m, and have obtained very striking film of the remarkable individual sculling actions of the paired and unpaired fins. Being close to neutral buoyancy (the swimbladder is lipid-filled), *Latimeria* can drift slowly around near the bottom, sometimes adopting a head-down attitude. Despite the lack of obvious intromittent organs in the male, *Latimeria* bears live young. The large female caught in 1991 gave birth to 26 young whilst in the trawl net. Fossil coelacanth embryos with yolk sacs are known from the Carboniferous (see p. 183).

*Actinopterygii.*

*Chondrostei.*   The surviving 25 species of chondrostean fishes (Figure 1.16) are divided between the bichirs and reedfishes (*Polypterus*, *Erpetoichthys*) in the Brachiopterygii, and the acipenseriform sturgeons and paddlefishes (*Acipenser*, *Polyodon* and *Psephurus*). Both groups share primitive characters like spiracle and intestinal spiral valve, but *Polypterus*, with its closely-set rhomboid scales covered with shiny ganoin, is much closer to its Devonian palaeoniscid ancestors. Indeed, the only speculation that E.S. Goodrich, the most distinguished of all comparative anatomists, ever permitted himself was to entitle a paper '*Polypterus*, a palaeoniscid?'. The paired lung-like septated swimbladder, connected to the oesophagus

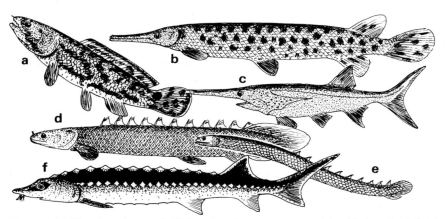

**Figure 1.16** Chondrosteans and Holosteans (not to same scale). (a) bowfin (*Amia*); (b) garpike (*Lepisosteus*); (c) paddle fish (*Polyodon*); (d) bichir (*Polypterus*); (e) reedfish (*Calamoichthys*); (f) sturgeon (*Acipenser*). After Goodrich (1909) and Marshall (1971).

by a ventral glottis, probably shows the ancestral condition. Bichirs and reedfish are air breathers, and die if denied access to the surface to gulp air, which they do in a curious way, using the elastic recoil of the scales to refill the bladder emptied by intrinsic muscles. In *Polypterus*, the swimbladder is also used to produce moans and thump-like sounds.

Sturgeons have a mainly cartilaginous skeleton, with an enormous unconstricted notochord, but some dermal elements in the skull and a line of lateral scutes along the body are ossified. The scales of the tail region have lost the ganoin layer and consist of concentric layers of bone, as in teleosts. The swimbladder is not respiratory. There are four barbels in front of the mouth, and sturgeons use these and ampullary electroreceptors on the snout to find their (mainly invertebrate) prey, rooting them out with the snout and sucking them in with the toothless protrusible mouth. Some sturgeons are large (5 m and 1000 kg) and may include bottom fishes in their diet. Most sturgeons live in the oceans or in large inland seas and lakes, ascending rivers only to breed where they attach their sticky eggs to the bottom after external fertilisation. The spiny larvae which hatch spend a year or more in the river after metamorphosis and only then pass down out of the river. Sturgeons are of economic importance for their roe (caviar) and isinglass extracted from the swimbladder is used in fining beer.

The Mississippi paddle fish (*Polyodon*) filters plankton from the water with a gill raker sieve, but the Chinese *Psephurus* feeds on fish and crabs, which it catches with its protrusible jaws. Both, like sturgeons, are equipped with many electroreceptors on their flattened snouts, perhaps using them to detect the muscle action potentials of swimming prey.

*Holostei.*    Even fewer holosteans than chondrosteans survive today; the bowfin *Amia* of rivers in eastern North America, and seven species of garpikes (*Lepisosteus* and *Altractosteus*) from fresh and brackish waters of North and Central America are all that remain. Garpikes are more primitive than *Amia*, and as shown in Figure 1.1, within the Neopterygii, are the sister group to *Amia* and teleosts. However, both are much more like teleosts than chondrosteans. The skeleton is strongly ossified, the fins are more flexible, though with fewer fin rays than chondrosteans, and there is no spiracle. The heart has a large conus, and there is a reduced spiral valve in the intestine, but these primitive characters are over-shadowed by such teleost features as the development of an eye muscle canal (myodome) in the floor of the skull, the loss of the clavicle from the pelvic girdle, and the freeing of the maxillary from the pre-operculum strapping together the bones supporting the jaws. *Lepisosteus* retains the thick rhomboidal ganoin-plated scales which make the body relatively inflexible (and reduce its fast-start performance); it is an ambush predator, hiding amongst weeds in shallow water, catching its prey with a sideways snap of the jaws. To achieve neutral buoyancy with this thick dermal armour, the airbladder is large (12% of body volume) and septated, and is used also for respiration. Since the giant tropical alligator gar (*L. tristoechus*) grows to 3.5 m, it is a formidable predator.

*Amia* is covered with thin teleost-like bony scales, and is more teleost-like than *Lepisosteus* except in its reproductive system where the ovary is not continuous with the oviduct, and like elasmobranchiomorphs and chondrosteans the small eggs are shed into the body cavity before entering the oviduct. Cleavage is total, whilst in *Lepisosteus* it is partial only (meroblastic) as in teleosts.

*Teleostei.* The numbers of teleost species far exceed those of any other fish group, or any other kind of vertebrate, for 96% of living fishes are teleosts. This has been the most successful fish group since they began to radiate in the Cretaceous, and it is immensely diverse. Teleosts live in freezing Antarctic waters and in hot springs (up to 44°C), in alkaline lakes and acid streams, in the deep sea and largely out of water on land, or even up trees. They range in length from the smallest of all vertebrates, the tiny marine goby *Trimmatom nanus* (8–10 mm) and the freshwater *Pandaka pygmaea* (10–11 cm), to the elongate giant freshwater European catfish, *Silurus glanis*, where the record specimen is no less than 5 m long, though only two-thirds the weight of large specimens of the slightly smaller swordfish (*Xiphias*) and large tunas. This wide range of body size enables them to fit into many different habitats. About 10% of teleosts are 10 cm or less long, whilst at least 80% are between 10 cm and 1 m. The adaptive advantages of these small- and medium-sized teleosts are easily seen in such environments as coral reefs which provide food, shelter and spawning territories suitable for gobioids a few centimetres long to lutjanids and lethrinids up to 50 cm long. In the reef channels, larger predatory carangids (and off the reef edges) scombroids prey on the medium-sized fishes.

In comparison with holosteans, teleosts are generally more active rapid swimmers and more lightly built. The skeletal elements are normally well calcified, but are lightened by being built from a scaffolding of struts unlike the dense cancellous bone and thick scales of holosteans. The skull is similar to the holostean, but the lower jaw has been simplified to only three components: dentary, angular and articular.

There are (as we might expect) exceptions to the generalisation that teleosts have a well-calcified but nonetheless lightly built skeleton. For example, the ostracodont trunkfishes are enclosed in a heavy bony cuirass, and (in the opposite direction) many deep-sea teleosts have much reduced skeletal calcification to save weight, as also in the immense sunfish *Mola* which is hardly calcified at all (it feeds on a watery diet of jellyfish). Teleost fins are usually more flexible than those of holosteans, with fewer fin rays, and the caudal fin is usually symmetrical. Internally (Figure 1.17), the

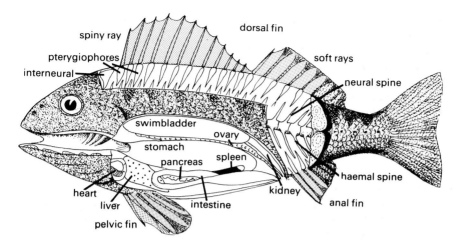

**Figure 1.17** General features of a teleost (*Perca*). After Dean (1895).

teleost heart has an elastic bulbus instead of the holostean contractile conus, there are usually complex gut diverticula (pyloric caeca), and only rarely (e.g. osteoglossids) a spiral valve.

The evolution of buoyant eggs produced by many coastal and deep-sea species was important for marine teleosts, since despite high mortality in the plankton, the wide distribution of eggs and larvae by currents has enabled these species to exploit large areas of suitable adult habitats.

## Bibliography

Alldredge, R.J., Briggs, D.E.G., Smith, M.P., Clarkson, E.N.K. and Clark, N.D.L. (1993) The anatomy of conodonts. *Philosophical Transactions of the Royal Society of London, Series B*, **340**, 405–421.

Carpenter, K.E. (1993) Optimal cladistic and quantitative evolutionary classifications as illustrated by fusilier fishes (Teleostei:Caesionidae). *Systematic Biology*, **42**, 142–154.

Compagno, L.J.V. (1977) Phyletic relationships of living sharks and rays. *American Zoologist*, **17**, 303–322.

Dean, B. (1895) *Fishes Living and Fossil*, New York.

Echelle, A.A. and Kornfield, I. (eds) (1984) *Evolution of fish species flocks*, Univ. Maine Press.

Fink, S.V. and Fink, W.L. (1981) Interrelationships of the ostariophysan fishes (Teleostei). *Journal of the Linnaean Society of London, (Zool.)*, **72**, 297–353.

Fink, S.V. and Weitzmann, S.H. (1982) Relationships of the stomiiform fishes (Teleostei) with a description of *Diplophos*. *Bulletin of the Museum of Comparative Zoology Harvard*, **150**, 31–93.

Forey, P.L. (1988) Golden jubiliee for the coelacanth *Latimeria chalumnae*. *Nature*, **336**, 727–732.

Forey, P.L. and Janvier, P. (1993) Agnathans and the origin of jawed vertebrates. *Nature*, **361**, 129–134.

Goodrich, E.S. (1928). *Polypterus* a Paleoniscid? *Palaeobiologica*, **1**, 87–92.

Grobben, K. and Kühn, A. (1932) *Lehrbuch der Zoologie*. Springer-Verlag, Berlin.

Jefferies, R.P.S. (1986) *The Ancestry of the Vertebrates*. British Museum (Natural History), London.

Johnson, G.D. (1993) Percomorph phylogeny: progress and problems. *Bull. Mar. Sci.*, **52**, 3–28.

Quicke, D.L.J. (1993) Principles and Techniques of Contemporary Taxonomy. Blackie Academic & Professional, Glasgow.

Scott-Ram, N.R. (1990) *Transformed Cladistics, Taxonomy and Evolution*. Cambridge University Press.

Stock, D.W. and Whitt, G.S. (1992) Evidence from 18S ribosomal RNA sequences that lampreys and hagfish form a natural group. *Science*, **257**, 787–789.

# 2 Fishes and their habitats

## 2.1 Introduction

The habitats of fishes vary greatly not only in physical features such as pH, salinity, temperature, oxygen content and light level, but also differ immensely in the space available. Some fishes have extraordinarily (and dangerously) restricted distributions, like the relict populations of 15–20 species of desert pupfishes (*Cyprinodon* spp.) endemic to isolated small spring systems in the desert regions of south-west USA and Mexico (some species can withstand temperatures up to 44.6°C!), or the equally isolated populations of cave fishes, such as *Lucifuga* in Cuba and the Bahamas. In the marine environment, living spaces are usually greater, though even here, some fish may be restricted to the reefs around single atolls, or like the bithytid vent fish (probably *Diplacanthapoma*), to the thermal vents of the Galapagos rift at 2400 m. By contrast, other marine fishes, like the pelagic blue shark (*Prionotus*), range over all the oceans, whilst the minnow-sized bathypelagic stomiatoids *Cyclothone microdon* and *C. acclinidens* are found below 100 m worldwide, and must comprise many billions of individuals.

## 2.2 Marine habitats

### 2.2.1 The open ocean

*Epipelagic fishes.* The open ocean beyond the continental shelf covers nearly two-thirds of the surface of the Earth, and some 2500 species are found there, about half being benthic, half pelagic. Near the surface is the euphotic zone, where light drives (phytoplankton) photosynthesis year-round in the tropics and sub-tropics, and for the warmer part of the year in cold and temperate waters (Figure 2.1). This zone of primary production (down to 100 m in the clearest waters), supports an epipelagic fish fauna of some 250 species, as well as many larvae of fish from deeper levels (Figure 2.2). Sharks, flying fish, scombroids such as tunas and billfish, halfbeaks, garfish, the large sunfish *Mola* and stromateoids are typical of this zone, and are usually coloured dark blue above and lighter below. Floating objects attract both smaller epipelagic fishes such as the stromateoid

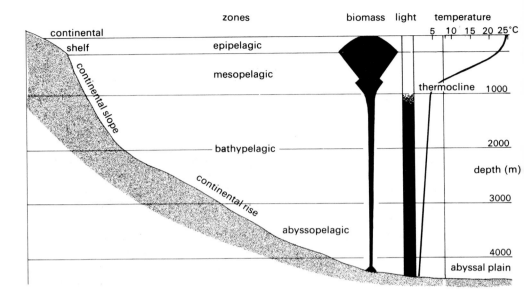

**Figure 2.1** Zones of the ocean—note that plankton biomass increases close to the ocean floor. Modified from Marshall (1971).

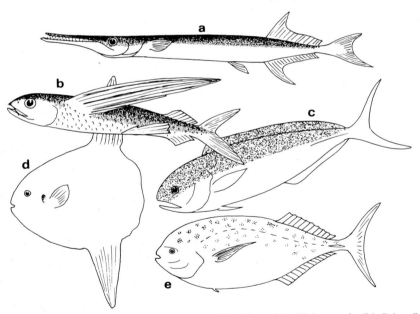

**Figure 2.2** Epipelagic fishes (not to same scale). (a) garfish (*Tylosaurus*); (b) flying fish (*Exoceoetus*); (c) dolphin (*Coryphaena*); (d) sunfish (*Mola*); (e) louvar (*Louvarus*). After Fitch and Lavenburg (1971) *Marine Food and Game Fishes of California*. California University Press; and Herre (1928) *Philippine J. Sci.*, **36**, 215.

driftfish (*Nomeus*) and medusafish *Schedophilus*, which hide under medusae for protection, as well as larger scombroids preying on the smaller fishes. The perciform wreckfish (*Polyprion*) is named from its habit of living under floating wreckage, old teacases seemingly being favourite lairs. The epipelagic fauna is richest in warmer regions, but some species, like the 'warm' isurid sharks and blue-fin tuna *Thunnus thynnus* (see page 64) migrate to colder waters in the productive season.

*Mesopelagic fishes.*    The 900 or so mesopelagic fish species (Figure 2.3) live above the thermocline in a zone where daylight still penetrates; many migrate upwards at night towards the surface, sinking again before dawn, following the migrations of their zooplankton food. Not all vertical migrators travel upwards far enough to reach the surface, and differences in the amplitude and timing of such migrations effectively partition different feeding levels, so reducing competition. For example, in the Rockall trough, as judged by the species composition of the copepods they feed on, the hatchet fish *Argyropelecus olfersi* feeds at lower depth horizons than *A. hemigymnus*, whilst the third common sternoptychid *Maurolicus muelleri* feeds closest to the surface.

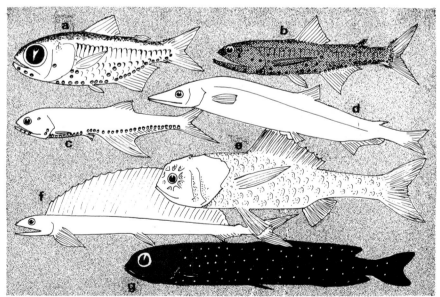

**Figure 2.3** Mesopelagic fishes (not to same scale). (a) myctophid (*Electrona*); (b) myctophid (*Lampanyctus*); (c) stomiatoid (*Bonapartia*); (d) adult *Paralepis*—these are rarely caught, the postlarval and larval stages are elongate and thinner; (e) *Melamphaes*; (f) lancet fish (*Alepisaurus*); (g) alepocephalid (*Xenodermichthys*). After Grey (1964); Rosen (1966); Gibbs and Wilimovsky (1966); Ebeling and Weed (1973) *Mem. Sears Fndn. Mar. Res.*, **1**, 397; and Marshall (1965) *La Vie des Poissons*, Éditions Rencontre, Lausanne.

The very rare accidental capture of a large mesopelagic fish like the megamouth shark (*Megachiasmodon*) shows us that there are some large fishes in this zone. All biologists who have fished in the open ocean with mid-water trawls are well aware that these devices never catch the large squid which are known to be abundant (from the stomach contents of marine mammals, much more efficient sampling engines), and so it is possible that other large mesopelagic fish remain to be discovered. Those we can catch are almost all smaller than 30 cm, the myctophid lantern fishes (ca. 250 species) mostly being 10 cm or smaller. Myctophids and several stomiatoids, feeding on copepods and small crustaceans, undertake vertical migrations of 400 m or so up and down each night. Larger fishes which feed on fishes and other larger prey, are non-migrators, like the alepisauroids (lancet fishes) up to 2 m long, and chiasmodonts (giant swallowers) with larger jaws and distensible gut and body walls. Many mesopelagic fishes have relatively large eyes and are covered with silvery scales and light organs (for intraspecific communication and camouflage, see p. 249), though others like the alepocephalids are dark brown or black.

**Figure 2.4** Bathypelagic fishes (not to same scale). (a) gulper (*Eurypharynx*); (b) *Cyclothone braueri*; (c) gulper eel (*Saccopharynx*); (d) whalefish (*Cetomimus*); (e) angler (*Linophryne*); (f) snipe eel (*Cyema*). After Böhlke (1966) *Mem. Sears Fndn. Mar. Res.*, **1**, part 5, 168; Marshall (1971); and Fitch and Lavenburg (1971) *Marine Food and Game Fishes of California*, California University Press.

*Bathypelagic fishes.*    Compared to the more robust mesopelagic fishes, bathypelagic fishes (Figure 2.4) are 'economy' designs, reducing the calcification of their skeletons except for the all-important jaws, and with watery muscles. The water content of such fish is high, e.g. 95% in the angler fish (*Melanocetus*) and 94% in the gulper eel (*Eurypharynx*). Even without gas-filled swimbladders, they are near neutral buoyancy. Their hearts are very small, they have very little red muscle and their low haematocrits (packed red blood-cell volumes) average 8%, compared with the mackerel with a haematocrit of about 50%! Most of the 150 or so species of bathypelagic fishes are ceratioid angler fishes (about 100 species), but the dominant forms in numbers of individuals are black species of the stomiatoid genus *Cyclothone*. Like angler fishes, *Cyclothone* species have smaller males than females, but they are not so much reduced and do not fuse with the females as do ceratioid males. Many bathypelagic fishes must live a rather sedentary life, hanging in the water waiting opportunistically for the occasional meal to come within range of the jaws. *Cyclothone* species live on a diet of copepods and small fish, as do angler fishes, some of which (like *Melanocetus*) can swallow fish two to three times their own length, attracting them within range of the jaws with luminous lures. Although no daylight penetrates to the deep-sea, many benthopelagic fishes have normal-sized eyes often with specialisations to increase their sensitivity (see p. 243), and the fitful flashes and glows of bioluminescence must obviously be significant in their lives.

*Benthopelagic fishes.*    Rather surprisingly, the great majority of bottom-living fishes from upper slope levels at 200 m or so to around 8000 m in the deep ocean (Figure 2.5) are neutrally buoyant and live not on the bottom, but just off it, like the deep-sea squaloid sharks (floating by means of their large oil-filled livers, p. 81). Although shallow-water rays are dense fish and rest on the bottom, their deep-sea relatives resemble the squaloids in having very large livers and are also close to neutral buoyancy, presumably also being bathypelagic. Amongst the teleosts, cusk-eels (ophidioids) and rat-tails (macrourids) dominate this cosmopolitan fauna, whose biomass may be considerable, and like the deep-sea cods (morids), deep-sea eels, notacanths and halosaurs, all have gas-filled swimbladders (Figure 4.4, p. 87). They feed on other fishes, on benthopelagic zooplankton, and on benthic invertebrates. Apart from those around thermal vents, the benthopelagic and benthic invertebrates ultimately depend on organic matter like faecal material raining down from surface waters, and it is understandable that the biomass of the plankton decreases rapidly with depth. For example, at 1000 m it is only 1% of that at the surface, and at 5000 m only about 0.01%.

Around thermal vents and methane seeps, however, food for bentho-pelagic fishes is richer, symbiotic bacteria supplying the energy source for worms, crustaceans and molluscs.

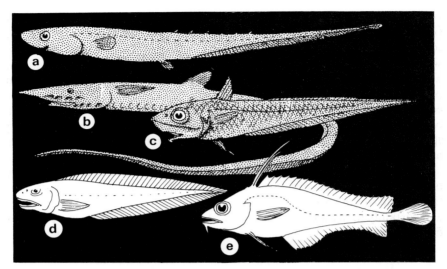

**Figure 2.5** Benthopelagic fishes (not to same scale). (a) notacanth (*Polycanthonotus*); (b) halosaur (*Aldrovandia*); (c) macrourid (*Coelorhynchus*); (d) brotulid (*Bassogigas*); (e) gadoid (*Lepidion*). After McDowell (1973) *Mem. Sears Fndn. Mar. Res.*, **1**, part 6, 1; and Ebeling and Weed (1973) *Mem. Sears Fndn. Mar. Res.*, **1**, part 6, 397.

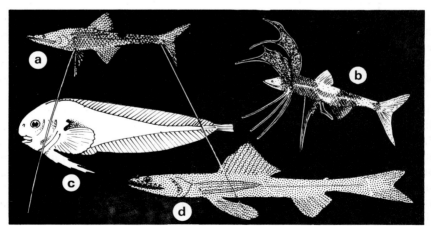

**Figure 2.6** Deep benthic fishes (not to same scale). (a) tripod fish (*Bathypterois grallator*); (b) tripod fish (*B. bigelowi*) in fishing attitude; (c) seasnail (*Careproctus*); (d) lizard fish (*Bathysaurus*). After Marshall (1979) *Developments in deep-sea biology*, Blandford Press; Mead (1966) *Mem. Sears Fndn. Mar. Res.*, **1**, part 5, 19, and Marshall (1971).

*Benthic fishes.* In contrast to the benthopelagic fishes, benthic fishes (Figure 2.6) lack swimbladders and are dense, resting on the bottom, sometimes, like tripod fishes (*Bathypterois* spp.) on stiff elongate fin-rays, sitting aligned into the currents that bring zooplankton to their mouths. Tripod-fish and green-eyes (both chloropthalmids) and lizard-fishes (synodontids) are the dominant forms, and in temperate and polar waters, eel-pouts (zoarcids) and seasnails (liparids) are also important. As with the benthopelagic fishes, benthic fishes of these kinds often show interesting specialisation of the eyes (see page 237), and some, like *Chloropthalmus* have luminous organs (see page 250).

Because of the lack of cues like light and temperature change, it was once thought that there was no seasonality in the deep sea. This is not so— the fall-out of detritus depends on seasonal production cycles at the surface and certainly in higher latitudes, there are annual growth cycles in deep sea fish species.

### 2.2.2 Shallow seas and coastal regions

*Warm water fishes.* By far the greatest number (80%) of the 10 000 or so species of fishes in shallow seas live in warm temperate or tropical waters, most associated with coral reefs and atolls (Figure 2.7) in waters where mean temperatures during the coldest part of the year do not fall below 18°C. Coral reefs are widespread in the Indian and western Pacific oceans between latitudes 30°C North and 30°C South, and there are also large reefs in the Caribbean and around the West Indies. There is a striking difference in the number of species of coral fishes in different regions (Figures 2.8 and 2.9), from the richest central Indo-West Pacific reefs of the Philippines, New Guinea, and the Australian Great Barrier Reef to the less rich reefs around Florida where only 500–750 species live. The cline in species number seen in Figure 2.9 may reflect the Indo-West Pacific origin of the global coral reef fish fauna, divided into four main regions: Indo-West Pacific, Pacific American (Panamanian), West Indian and West African (Figure 2.10). The Eastern Pacific oceanic barrier (an east-west distance of some 3000 miles), between the Indo-West Pacific and Panamanian faunas has been crossed by very few species. Some of these are large active swimmers like the tiger shark (*Galeocerdo*), an important predator on seasnakes, and the spotted eagle ray (*Aetiobatis*); others have made the crossing as the long-lived pelagic leptocephalus larva stage, like the bonefish (*Albula*) and six species of moray eels. On the whole of the Great Barrier reef 1300 species are known, but at its southern end (the Capricorn–Bunker group, which has been very well sampled, and is the best-known region of the reef) only about two-thirds of these (859 species) occur. This impoverishment, however, seems primarily due to lowered

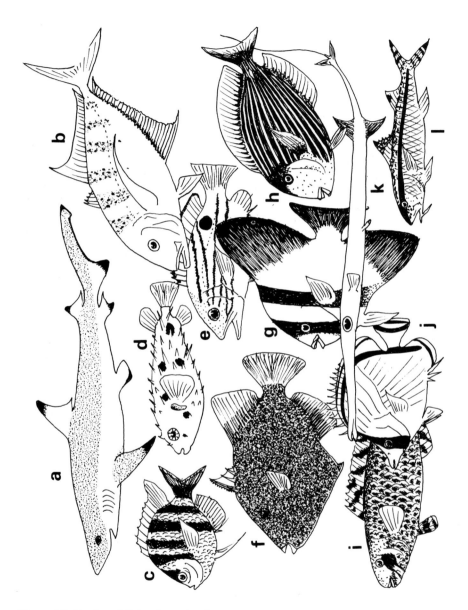

**Figure 2.7** Coral reef fishes from the Great Barrier Reef (not to same scale). (a) blacktip shark (*C. melanopterus*); (b) jack (*Caranx*); (c) damselfish (*Abudefduf*); (d) puffer (*Tragulichthys*); (e) snapper (*Lutjanus*); (f) trigger fish (*Cathidermis*); (g) batfish (*Platax*); (h) sturgeon fish (*Ctenochaetus*); (i) parrotfish (*Leptoscarus*); (j) butterfly fish (*Chaetodon*); (k) cornetfish (*Fistularia*); (l) goatfish (*Upeneus*). After Marshall (1965) *Fishes of the Great Barrier Reef and Coastal Waters of Queensland*, Wynnewood, Livingstone.

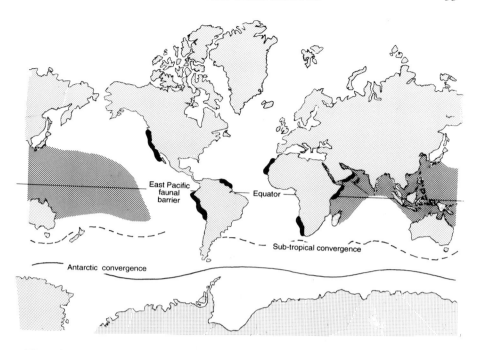

**Figure 2.8** The world oceans, showing areas of coral reefs (stippled), coastal upwelling zones of high productivity (black), and the Antarctic and sub-tropical convergences. Modified from Marshall (1965) *La Vie des Poissons*, Éditions Rencontre, Lausanne.

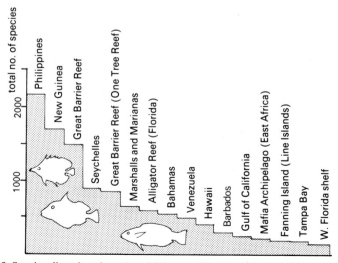

**Figure 2.9** Species diversity of coral reef fishes at different reef locations. After Sale (1980).

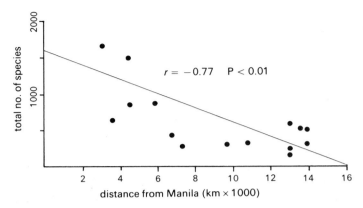

**Figure 2.10** Numbers of species of coral reef fishes from each of the 16 sites in Figure 2.9,
showing geographical cline in diversity. After Sale (1980).

habitat diversity (in the north, there are outer barrier reefs and inshore
coastal reefs), and not to within-habitat diversity.

Nearly all coral fishes are acanthopterygians, and many families like
gobies (Gobiidae), wrasses (Labridae), damsel fishes (Pomacentridae),
butterfly fishes (Chaetodontidae) and squirrel fishes (Holocentridae) are
represented globally at all coral reefs, though in each faunal area, the
species are largely different. On the Great Barrier Reef, 43% of all fishes
are gobies, whilst wrasses and damselfishes each comprise 23%, and
butterfly fishes 8%.

Coral fishes make their living in many specialised ways. Some are
herbivores, cropping algae, or like parrotfishes (Scaridae) scraping and
biting off coral to obtain the algal symbionts. Others like puffer fishes
(Tetraodontidae), boxfish (Ostraciontidae), gobies and some damselfishes,
eat invertebrates, and these include the filefishes (Monacanthidae) and
butterfly fishes which eat the coral polyps themselves, picking their food
with their forceps-like mouths. Yet others prey on fishes, like the trumpet
fishes (Aulostomatidae) which stalk small fishes by swimming close to other
non-predatory fishes. Manoeuvrability is important in the reef habitat, and
as a result, many coral reef fishes have abandoned normal oscillatory
swimming except in emergencies and instead flap their pectoral fins
(wrasse, parrotfish and surgeonfishes (Acanthuridae)) or the dorsal and
ventral unpaired fins (triggerfishes (Balistidae)), or undulate unpaired fins
(seahorses, pipefishes and trumpetfishes).

Most coral fishes are dazzlingly brightly coloured, and change colour
during courtship as well as by day and night. During the night, large-eyed
nocturnal feeders like squirrelfishes and the luminescent pempherids
emerge from their daytime hiding places, whilst day-feeding parrotfishes

retire to sleep in mucous cocoons. As well as the brightly-coloured and patterned coral fishes, there are also cryptically camouflaged ambush predators like carpet sharks (orectolobids) and frogfish (the Western Australian *Batrachomoeus rubricephalus*, as well as looking rather like a frog, croaks like one!)

*Fishes of temperate and cold waters.* The coastal and shallow sea fish fauna (Figure 2.11) of temperate regions is much less rich in species than that of warm waters, since there are only around 1000 species in the temperate North Pacific, and many fewer in the temperate North Atlantic. Both regions contain members of the same families, like the scorpion fishes (Scorpaenidae), kelpfishes (Clinidae) and eel-blennies (Lumpenidae), but in the North Pacific these families are more diverse, and this region also has some endemic families like the surf perches (Embiotocidae) and the greenlings (Hexagrammidae).

**Figure 2.11** Coastal and shallow sea fishes of temperate and cold waters (not to same scale). (a) bass (*Morone*); (b) mackerel (*Scomber*); (c) herring (*Clupea*); (d) cod (*Gadus*); (e) orange roughy (*Hoplostethus*); (f) John Dory (*Zeus*); (g) turbot (*Scophthalmus*); (h) ice fish (*Notot) (*Notothenia*). After Marshall (1972) *Le vie des Océans*; Bigelow and Welsh (1925) *Bull. U.S. Bur. Fish.*, **40**, part 1, 1; Hayward, P.J. and Ryland, J.S. (eds) (1990) *The Marine Fauna of the British Isles and North-West Europe* **2**, Oxford Science Publications; Herald (1961); and Whitehead, P.J.P., Bauchot, M-L., Hureau, J-C., Nielsen, J. and Tortonese, E. (eds) *Fishes of the North-Eastern Atlantic and the Mediterranean* **I–III**, Unesco 1984–1986.

Although there are fewer species than in warmer waters, temperate and cold waters contain the most important food fishes like the cod, pollack and haddock (Gadidae), the herrings and their relatives (Clupeidae), and the pleuronectid flatfishes like the plaice, soles and flounders (p. 318). Upwelling currents along coasts and mid-ocean convergences bring nutrient-rich deep water to the surface, supporting phytoplankton blooms nourishing the zooplankton on which fish feed. For example, off the Peruvian coast, upwelling is the basis for the fishery for the Peruvian anchoveta (*Engraulis ringens*), and oceanographic changes which reduce upwelling have catastrophic consequences for the fishery and for the seabirds which feed on the anchoveta. On the continental shelf, deeper nutrient-rich waters interact with surface water at fronts, which are regions of high productivity.

Temperate seas vary seasonally in temperature, much more than warm or polar seas which have a nearly constant temperature year-round. In consequence, phytoplankton and zooplankton abundance is seasonal. Thus, for example, Icelandic waters can vary from 0°C in February to 10°C in August, whilst, the more temperate sub-Antarctic waters of the Southern Ocean are in the range 5–15°C. These waters lie between the Antarctic and Subtropical Convergences, and the former, where temperatures rapidly drop by 2–3°C, has long formed a barrier to exchange in sub-Antarctic and Antarctic regions. Once Antarctica had become fully separated from the other parts of Gondwanaland in the late Cenozoic (30–23 Myr BP), the Antarctic circumpolar current decoupled sub-tropical waters from Antarctica, and this isolation permitted the wide adaptive radiation of notothenioid fishes (over 100 species). Notothenioids are found also on the southern New Zealand shelf and on the Patagonian shelf, perhaps relics of the severe cooling of the Southern Ocean in the Miocene and Pliocene which advanced the Convergence 3–7° north of its present mean position.

Apart from notothenioids, eel-pouts (Zoarcidae) and seasnails (Liparidae) also occur as coastal fishes in Antarctica, where there are different species to their nearest relatives across the Convergence in the coastal waters of the sub-Antarctic Southern Ocean. These mainly lie over the Chilean and Patagonian shelves. The commercially important gadoids, flatfishes (pleuronectids) and clupeids, such a conspicuous part of the fauna of north temperate seas, are poorly represented here, whilst such fish as the orange roughy (*Hoplostethus*) are presently commercially important, though a catastrophic collapse of the New Zealand and Australian stocks seems inevitable due to overfishing.

*Estuarine fishes.* Like other estuarine animals, the fishes found in estuaries and river mouths (Figure 2.12) are mainly euryhaline (salinity-tolerant) forms which can live in unstable surroundings, where salinity is

**Figure 2.12** Estuarine fishes. (a) shad (*Alosa*); (b) sciaenid (*Cynoscion*); (c) puffer (*Tetraodon*); (d) stickleback (*Gasterosteus*); (e) flounder (*Platichthys*); (f) tarpon (*Elops*); (g) threadfin (*Polynemus*). After Bigelow and Welsh (1925) *Bull. U.S. Bur. Fish.*, **40**, part 1, 1.

variable and the waters are often turbulent and muddy, edged in the tropics by mangroves. In temperate estuaries, typical fishes are grey mullets (Mugilidae) flatfishes like flounders (*Pleuronectes flesus*) and shads (*Alosa*), whilst in warm waters, many species of marine origin such as lutjanids, pomadasyids, catfishes, sciaenids and threadfins (Polynemidae) are found in estuaries, supporting valuable fisheries such as that of the Niger estuary. Many estuarine fishes are migrants, entering estuaries seasonally, like herring and sprat, or on passage to and from the sea, like eels and salmonids. Fishes of muddy estuaries may look rather like deep-sea fishes, with their long fin-rays and small eyes. The Bombay duck (*Harpadon*), of Indian estuaries, not only looks like a deep-sea fish, but is related to the deep-sea lizard fishes.

*Intertidal fishes.*   The intertidal zone is a demanding environment, where fishes are alternately buffeted by waves and isolated in pools or on mudflats (Figure 2.13). Some intertidal fishes, such as the mudskippers (Periophthalmidae) and leaping blennies like *Alticus kirki* of the Red Sea, are truly amphibious, emerging from the water to graze on algal films on mud or rock in or above the splash zone. These fishes have remarkable behavioural and physiological adaptations to avoid (or withstand) desiccation and to regulate nitrogen excretion (Chapter 6).

Most intertidal fishes, however, remain in the water and to avoid being

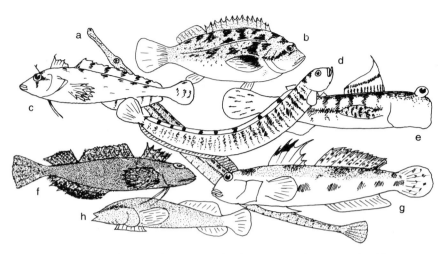

**Figure 2.13** Intertidal fishes (not to scale). (a) pipefish (*Siphostoma*); (b) priest fish (*Sebastodes*); (c) clingfish (*Heteroclinus*); (d) butterfish (*Pholis*); (e) mudskipper (*Periophthalmus*); (f) triple fin (*Tripterygion*); (g) goby (*Gobius*); (h) sucker fish (*Lepadichthys*). After Bigelow and Welsh (1925) *Bull. U.S. Bur. Fish.*, **40**, part 1, 1; Ayling and Cox (1900) *Collins Guide to the Sea Fishes of New Zealand*, Collins, London and Auckland; and Marshall, N.B. (1965) *Fishes of the Great Barrier Reef and Coastal Waters of Queensland*, Livingstone, Wynnewood.

washed away, generally are dense, small fishes (less than 20 cm), thin or flattened to hide in holes and crevices, or may have the pelvic fins modified into suckers. An interesting and obviously necessary behavioural feature of many intertidal fishes is their pronounced homing ability, particularly striking in pool-dwelling blennies which leap into the air to view the surrounding terrain. Intertidal areas are very important as nursery areas for young fishes, for example for young flatfishes in the Wadden Sea, and to avoid stranding, they move up and down as the tide ebbs and flows.

## 2.3 Freshwater fishes

Most of the 8000 or so freshwater fishes (Figure 2.14) known (more turn up each year) live in lakes and rivers out of reach of the sea. Of these, an estimate in 1970 of a total of 6650 species (now more probably over 7000) are primary freshwater fishes which evolved in freshwater and cannot tolerate seawater. By far the majority (over 93%) of primary freshwater fishes are ostariophysan catfish, carps and characins, the remainder including the weakly electric mormyrids in Africa, and osteoglossids in Australia, South-East Asia, South America and Africa. Although it is probable that the little Western Australian newt-like *Lepidogalaxias*

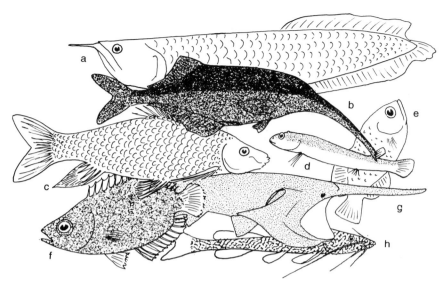

**Figure 2.14** Freshwater fish (not to scale). (a) arapaima (*Osteoglossum*); (b) elephant snout fish, a mormyrid (*Campylomormyrus*); (c) (*Labeo*); (d) (*Lepidogalaxias salamandroides*); (e) archer fish (*Toxotes*); (f) climbing perch (*Anabas*); (g) paddlefish (*Polyodon*) feeding on plankton; (h) catfish (*Mystus*). After Herald (1961); Merrick and Schmida (1962) *Australian Freshwater Fishes*, and Inger and Kong (1962) *Fieldiana: Zool.*, **45**, 1.

*salamandroides* is a primary freshwater fish, the only two certainly primary freshwater fishes of Australia are the lungfish *Neoceratodus*, and the osteoglossid *Scleropages*. These are large fishes, common in their restricted distributions, but smaller southern native freshwater fishes, like the New Zealand galaxeid *Galaxias brevipinnis*, have been largely exterminated where they were once abundant by the introduced rainbow trout (*Oncorhynchus mykiss*), and by cross-transfer of the native smelt *Retropinna*.

Several marine families, like clupeids, puffer fishes (Tetraodontidae) and drum fishes (sciaenids) have freshwater representatives, for example the small ctenothrissid clupeids of West African rivers or the Australian clupeid *Nematalosa erebi*. There are even a few entirely freshwater stingrays (Dasyatidae and Potamotrygonidae). These are all secondary freshwater fishes which entered from the sea in the past. They mostly now live exclusively in freshwater, and even though the main groups (cichlids, cyprinodonts and poeciliids) have species which have some salinity tolerance and can live in brackish water, they must spawn in freshwater. Only a very few such fishes have adapted to spawning as well as living in brackish water, like the cyprinids in the Caspian and Aral seas, whose eggs can develop in salinities up to 8–10‰. Diadromous fishes migrate between fresh and salt waters. Some, like the 15 species of freshwater eels

(*Anguilla*) leave freshwater to spawn in the sea (catadromous species), making very interesting anticipatory changes in serum ion content, body colour and visual pigments before and during their passage downriver (p. 244). As well as the eels, one or more pleuronectids, scorpaenids and mullets (Mugilidae) are catadromous. Many more (anadromous species) make the reverse migration to spawn in freshwater after periods up to four or five years growing and maturing in the sea. Parasitic lampreys, many salmonids, shads (*Alosa*, *Ilisha*) and alewives (*Pomolobus*) and smelts (Osmeridae) are all anadromous. A third category of diadromous fishes, amphidromous species, migrate into or out of the sea but do not do so for breeding purposes. For example, the newly-hatched larvae of the ayu (*Plecoglossus altivelis*), a fish much prized as a delicacy in Japan, are swept downstream to the sea whence they return as small juveniles to grow and feed for several years before breeding. A similar situation is seen in the southern salmoniform galaxeids. McDowell has examined the systematic distribution of diadromy in teleosts, concluding that it is an ancient life-history style, even though for a few species, a recent marine ancestry seems plausible. Figure 2.15 shows possible steps in the evolution of diadromy in fishes.

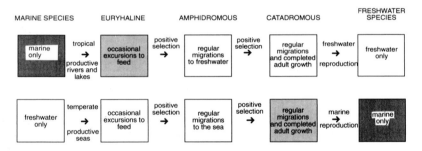

**Figure 2.15** Possible steps in the evolution of diadromy in fishes. After Gross (1987).

Ostariophysan fishes dominate the freshwater fish fauna throughout the globe, both Africa and South America having more than 2500 species. In South America, where cyprinoids are absent, the (possibly) more primitive characoids have radiated widely, evolving herring-like, mullet-like, gobie-like, minnow-like and salmonid-like forms. There are even characoid flying hatchetfishes (Gastropelecidae) which are reputed to flap their elongated pectorals at high frequency to remain airborne! Catfishes (siluroids) have also radiated in South America, where there are 14 endemic families and 1100 species, almost as many as characoids (15 endemic families, over 1500 species). In Africa, the different ostariophysan groups occupy different broad habitat types: characoids almost always inhabit sluggish lowland streams, cyprinids higher level flowing streams and rivers, whilst catfish occupy most habitats, presumably side-stepping competition from these other diurnal fishes by being active only at night. The 600 or so acanthopterygian cichlids are mainly found in the extraordinary species flocks of the African rift lakes like Lake Malawi.

## 2.4 Ostariophysan success

The reader may well enquire why it is that the ostariophysi have been so successful in freshwaters, though few (only 140 out of over 2200 catfish species) have become marine. It may seem improbable that the acuteness of their hearing, paralleled in the sea by the clupeids, is the most important basis of their success. However, mormyrids testify to the importance of hearing in freshwaters, for in addition to their sophisticated electrolocation and signalling systems, they also have gas-filled swimbladder remnants linked to the inner ear (Chapter 10).

The predator-warning pheromone signalling system is a second notable specialisation (found also in a single cichlid) (see p. 261). For schooling fishes, as most characins and cyprinids are, this system must be especially valuable; it is even likely that the evolution of the Weberian ossicle system (see p. 226) was linked to a schooling habit, as excellent hearing seems particularly useful to help fish in a school to monitor the movements of their neighbours.

## 2.5 The variety and origin of some freshwater fish faunas

The variety of freshwater fishes is striking, for they show radiations into different habitats with some of the most extraordinary adaptations of all fishes. The African mormyrids and the South American gymnotids, for example, have independently evolved amazingly specialised electrical signalling systems for use in crowded and turbid waters, and many freshwater fishes have developed different accessory respiratory devices to enable them to live in swamps where the oxygen content of the water is low. Something of this variety is seen in Figure 2.14. The radiations of the cichlid species flocks of the African lakes provide what are probably the best examples of evolution in progress at the species level.

Equally fascinating are the distributions of freshwater fishes, which are closely linked to the geological history of the regions they inhabit, and may indeed, offer useful clues to the complex history of the changes in the Earth's surface. For example, one scenario for the dominant freshwater group, the ostariophysi, following a relatively simple continental drift scheme, is that ostariophysans arose in Gondwanaland in the early Cretaceous before the South American and African plates separated.

What seem to be the most primitive ostariophysans now live in South America, hence the ostariophysi arose in the South American portion of Gondwanaland, and the modern African characoids and siluroids came from ancestors which reached West Africa before the two continents had finally separated. Antarctica, Australia, India and Madagascar had already separated, hence ostariophysans in these regions came later, like the two secondarily freshwater Australian catfishes. The Laurasian plate also separated early, but the African plate contacted the Laurasian at the end of the Cretaceous, and ostariophysans were able to enter Laurasia. Cyprinoids then evolved from this ostariophysan stock and spread throughout Laurasia and most of Africa, though they did not manage to reach South America (Figure 2.16). Today, South-East Asia has the greatest diversity of cyprinoids.

It will hardly come as a surprise to the reader (who should by now be accustomed to the notoriously revisionist tendencies of fish taxonomists!) that this scenario has been challenged on various grounds. Thus, the 'classical' idea that characoids gave rise to cyprinoids has been inverted by Fink and Fink, to make cyprinoids the basal ostariophysan group, and the South American portion of Gondwanaland has been rejected as the site of origin of the group. Much more work on the ostariophysan lineages is needed to decide between these alternatives, though the Finks' scheme now seems the more likely, even if it does not explain why cyprinoids are absent from South America! The simpler distribution of living lungfishes (see Figure 1.15) shows (even if we had no fossil lungfish from Antarctica) that they evolved earlier than the ostariophysi, before the separation of Australia from Gondwanaland.

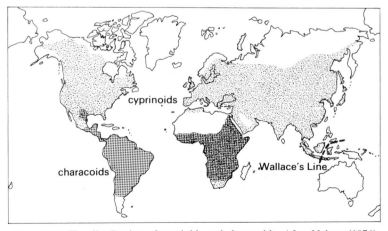

**Figure 2.16** The distribution of cyprinids and characoids. After Nelson (1976).

# Bibliography

Bemis, W.E. and Northcutt, R.G. (1992) Skin and blood vessels of the snout of the Australian lungfish, *Neoceratodus forsteri*, and their significance for interpreting the cosmine of Devonian lungfishes. *Acta Zoologica (Stockholm)*, **73**, 115–139.

Dadswell, M.J., Klauda, R.J., Moffitt, C.M., Saunders, R.L., Rulifson, R.A. and Cooper, J.E. (eds) (1987) Common strategies of Anadromous and Catadromous fishes. *American Fisheries Society Symposium*, **1**.

Echelle, A.A. and Echelle, A.F. (1993) Allozyme perspective on mitochondrial DNA variation and evolution of the Death Valley pupfishes (Cyprinodontidae: *Cyprinodon*). *Copeia*, 275–287.

Fink, S.V. and Fink, W.L. (1981) Interrelationships of the ostariophysan fishes (Teleostei). *J. Linnean Soc. London (Zool.)*, **72**, 297–353.

Gross, M.R. (1987) Evolution of diadromy in fishes. In *Common Strategies of Anadromous and Catadromous Fishes*, pp. 14–25.

Herald, E.S. (1961) *Living Fishes of the World*. World of Nature Series No. 6. Hamish Hamilton, London. 304 pp.

Johnson, G.D. (1982) Monophyly of the Euteleostean Clades-Neoteleostei Eurypterygii and Ctenosquamata. *Copeia 1992*, 8–25.

McDowall, R.M. (1993) A recent marine ancestry for diadromous fishes? Sometimes yes, but mostly no. *Env. Biol. Fishes*, **37**, 329–335.

McDowell, S.B. (1973) Order Heteromi (Notacanthiformes). Family Halosauridae. Family Notacanthidae. Family Lipogenyidae. In: *Fishes of the western North Atlantic. Memoirs of the Sears Foundation for Marine Research*, **1**, 1–228.

Marshall, N.B. (1971) *Explorations in the Life of Fishes* Harvard University Press, Cambridge, Mass.

Nelson, J.S. (1976) *Fishes of the World*. John Wiley and Sons, New York.

Sale, P.F. (1980) The ecology of fishes on coral reefs. *Oceanography and Marine Biology Annual Reviews*, **18**, 367–421.

Stiassny, M.L.J. (1986) The limits and relationships of the acanthomorph teleosts. *Journal of Zoology, London*, **1**(2), 411–460.

# 3 Locomotion

Some fishes can fly, and others move around on land, even in some cases, climbing trees, or burrowing 2 m or so into the soil, but the great majority of fishes swim in water. Most fishes swim by oscillating their bodies, which is what the greater part of this chapter is about; this is not an easy system to analyse.

## 3.1 The problem of analysis

The solid of least resistance that the great pioneer aerodynamicist Cayley drew from the body of the trout over a century ago (Figure 3.1), looks very like the shape of a submarine hull. The density and viscosity of the medium dictate the same streamlined form for submarines and fishes, but there is an important difference between the two. Submarines do not change shape as they move along, and thrust is provided by an aft propellor. Consequently, we can make models, test them in wind or water tunnels, and using well-established formulae, calculate the thrust required from the propeller (and the power output of the engine) to overcome the drag of the hull and drive it at different speeds. Some fishes swim slowly using their paired and unpaired fins to propel themselves forwards without body movements; like submarines they are more or less rigid bodies with attached propellors, and so the same kind of formulae can be applied to calculate the thrust that

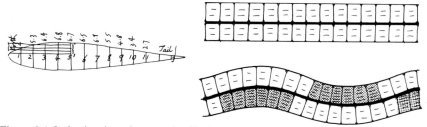

**Figure 3.1** Left: drawing of a trout by Sir George Cayley (1773–1857), the 'Father of the Aeroplane', who designed and flew in 1852 the first glider to carry a man (his coachman). Right: schematic diagram illustrating bending of a body with a flexible but incompressible backbone (thick black line) by segmented muscle blocks (contracted myotomes stippled). Partly after Gibbs-Smith (1962) *Sir George Cayley's Aeronautics*, Science Museum Handbook.

their fins have to provide, and the power needed from the fin muscles. The fin movements themselves may well alter the flow pattern around the body (and hence the drag it incurs) even if it remains more or less rigid, but the calculations are much the same for such fish as for submarines.

However, the great majority of fishes swim differently, for as they move forwards, their bodies flex and change shape. Even fishes that keep their bodies rigid at slow speeds, such as the trigger fishes (Balistidae), which flap dorsal and ventral unpaired fins from side to side like birds turned on their sides, or tropical wrasse and parrot fishes which scull along with their pectorals, all switch to swimming with changes in body shape when swimming faster. Plainly, formulae for rigid bodies are inappropriate, and it is very striking that the challenging problems of calculating the drag, power output, and efficiency of fishes swimming in different ways have now largely been solved. With smaller fish at least, reasonable estimates can be made for what we would like to know for any self-propelled machine: how fast it can go, the power required and the efficiency of the process.

This chapter first examines the muscles fish employ to move their bodies, then how these movements generate propulsive thrust, how fast fish swim and how efficient swimming is, and then how fish may reduce the costs of swimming.

## 3.2 The myotomal muscles

### 3.2.1 Myotomal structure

Oscillation of the body and of the tail fin results from the contraction of the segmented axial musculature, divided into myotomes by the myoseptal connective-tissue partitions on which the muscle fibres insert. Because the supporting central notochord or vertebral column is incompressible, the body bends laterally as the myotomes on one side contract and shorten (Figure 3.1). A simple model of such a system may be made by tying string along either side of a hacksaw blade, and pulling one or other of the strings (to simulate contraction) when the blade bends but does not shorten. But (unsurprisingly) such a simple model gives no idea of the mechanical subtleties of the myotomal layout, which after all has been refined since its origin in the Pre-Cambrian. The notochord or vertebral column lies dorsally (with the viscera below) and presumably must always have done so since the origin of the chordate body shape. The consequence of this is that the greater bulk of the myotomes lies below the notochord, and to avoid simply bending the body downwards at either end when they contracted, the ancestral myotomes were V-shaped with the apex of the V at notochordal level and a longer backward pointing arm below. Although

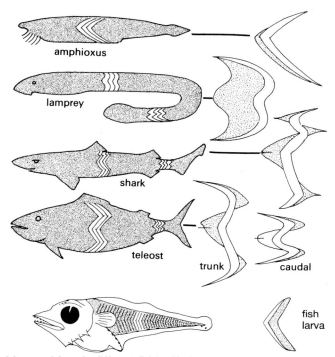

**Figure 3.2** Myotomal form in different fishes. Single myotomes at right, their outer surfaces stippled. After Nursall (1956), *Proc. Zool. Soc. Lond.*, **126**, 127, and Klawe (1961) *Pacific Science*, **15**, 487.

the fascinating *Pikaia* from the lower Cambrian Burgess shales still awaits detailed description, it is clear that its myotomes were of this form, as were those of conodonts. Similar V-shaped myotomes are seen today in amphioxus and in early fish larvae (Figure 3.2). Once the fish larva begins to ossify the vertebral column, and so restrict its movements to the lateral plane, increased overlap and folding of the myotomes begins, leading to the complex structures of living adult fishes.

Fish myotomes are indeed most complex in shape, and apart from amphioxus, in all other living adult fishes the outer borders of the myotomes are folded into W-shapes (Figure 3.2), after beginning in ontogeny as simpler V-shapes. The folding and overlap of the myotomes to make a set of nesting cones means that in a transverse section (Figure 3.3) portions of several myotomes are cut across, separated by horizontal septa (extensions of the apices of the myosepta), so the muscle fibres of the myotomes lie in a series of horizontal tubes of connective tissues running along the length of the fish.

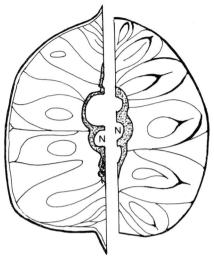

**Figure 3.3** Transverse hemi-sections across caudal region of shark (right), and trunk region of elver stage of an eel (left) showing sections of the nesting myotomal cones. N: notochord. After Willemse (1972) *Z. morphol. Tiere*, **72**, 221, and Egginton and Johnston (1982), *Cell and Tissue Res.*, **222**, 563.

Since it is far from easy to make two-dimensional diagrams of such three-dimensional structures, figures like Figures 3.2 and 3.3 are no substitute for examining them for oneself. Probably the best approach is to take a fair-sized fish and lightly boil or steam it to separate the myotomes and then try and model them from plasticine.

Shape is not the only complex feature of the myotomes. The arrangement of the muscle fibres that they contain is also complicated. We might expect that all the myotomal muscle fibres would lie parallel to the long axis of the fish, which is the arrangement found in amphioxus, where the muscle fibres are not more or less circular in section, but are large, very thin flat plates extending from the outer to the inner edges of the myotomes. In higher fishes, like teleosts and elasmobranchs, only the most superficial fibres just under the skin lie along the long axis; deeper fibres spiral in the myotomes at quite large angles (up to 30°) to the axis (Figure 3.4). In advanced teleosts, the trajectories of these fibres in adjacent myotomes form segments of a helix (Figure 3.4); in sharks they are ordered less regularly. This may seem at first sight a curious arrangement, but it has two important consequences. First, it permits these deeper fibres all to contract to a similar extent for a given amount of bending of the body. Secondly, whilst calculations based on the angles of the fibres to the longitudinal axis indicate that the longitudinally-running superficial fibres shorten during swimming by about 10% of their resting length, the deeper fibres

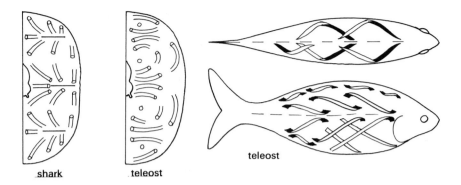

**Figure 3.4** Left: thick transverse slices across a shark and a teleost showing orientation of myotomal muscle fibres. Note tendons in shark (horizontal lines.) Right: dorsal and lateral views of typical teleost showing course of muscle fibres in successive myotomes along the body. The helices shown were obtained by taking the origin of one muscle fibre from the point at which the muscle fibre in the myotome next anterior inserts onto the common myoseptum, and so on along the fish. After Alexander (1969).

shorten less (only 2–3% in advanced teleosts) so contraction is near-isometric, the least amount of mechanical work is performed, and the fibres operate at the most advantageous point of the force–velocity curve. The painstaking analyses of microscope sections which revealed these arrangements were notable, for they explained a puzzling feature of myotomal design, and hinted at the reason for the complex folding of the myotomes, probably arranged to allow optimum fibre packing and orientation in a wide myotome.

### 3.2.2 Myosepta

The myosepta onto which the muscle fibres attach consist of a meshwork of inextensible overlapping collagen fibres. As a result, the myosepta are like loosely-woven pieces of cloth, which can be deformed but not elongated; in many teleosts they are often stiffened (to limit this deformation) by ribs and by small intermuscular bones, irritatingly obvious when eating clupeids. Just how these stiffening elements function is not entirely clear, nor is it obvious why they should be lacking in other fish (all sharks for example), but it seems probable that they act to permit only lateral movements when the myotomal fibres contract, and they serve also to reduce the general flexibility of the body. The inner borders of the myosepta attach to the vertebral column, the outer to the connective-tissue layer of the skin, which consists, like the myotomes, of overlapping collagen fibres.

How is the force produced by myotomal muscle contraction transmitted to bend the vertebral column? When a fibre contracts, the tension produced is transmitted to the collagen fibres of the myoseptal cones, and thence backwards and inwards to the vertebral column via the horizontal and median septa. This avoids the transmission of force from active muscle fibres to inactive compliant fibres in other myotomes, and (because the myosepta are also attached to the connective-tissue layer of the skin), means that the connective-tissue fibres of the skin become involved in the process of bending the body. The skin connective-tissue fibres are wound around in a helical lattice, making angles of 50–70° to the long axis. Direct measurements in small sharks have shown that intramyotomal pressures rise in active myotomes (in consequence of increase in diameter of the myotomal muscle fibres as they contract and shorten) to as much as 20 times resting values (up to 2.8 MPa) during burst swimming. The pressure rise imposes a circumferential stress on the skin, shed diagonally (which contributes to the shortening of the skin brought about by the shortening of the underlying myotomes), and at the same time the contractile force is transmitted to the vertebral column at head and tail.

So the skin plays a role in the bending of the fish, and it is possible (though not yet proven experimentally) that it may also store elastic energy to aid in accelerating unbending as the myotomes of the opposite side of the fish contract. Calculations based on biaxial stress tests of shark skin suggest that such energy storage is only likely to be of consequence during burst swimming. In fast-swimming fish like oceanic scombroids, the posterior myosepta are very oblique, and produced posteriorly into long tendons, giving rise to a tendinous caudal peduncle (Figure 3.5). This means that the mass and inertia of the caudal peduncle is kept as low as possible to allow increased frequency of tail beat, and its width can also be reduced for hydrodynamic reasons. Additional benefits of this arrangement are that the inertia of the anterior part of the body is increased, so reducing the tendency of the tail fin to cause yaw of the head as it sweeps across, and that there is likely to be significant energy storage in these caudal tendons.

### 3.2.3 Myotomal muscle fibres

In almost all fishes, the muscle fibres of the myotomes are of two quite different kinds, easily distinguished by their colour like wines. There are red and white fibres (and in some fish pink or rosé ones too!). In fish, this distinction between fibres by colour (actually myoglobin content) is convenient, for the two main types are not intermingled, as can easily be seen by cutting across the post-anal region of any fresh common food fish like a clupeid or gadoid. A thin layer of red fibres lies just below the skin and covers the main part of the myotome which consists of white fibres. In fixed material or even in smoked fish like kippers, the red fibres are brown,

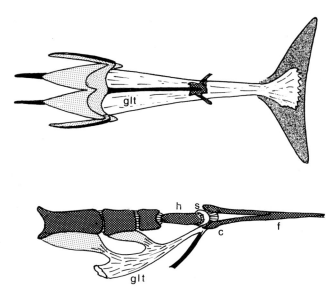

**Figure 3.5** Caudal tendons of scombroid fishes. Above: side view of caudal region of generalised scombroid, showing single myotome on left, with smaller tendons (black), and the great lateral tendon (glt) inserting on the rays of the caudal fin. Below: median horizontal section through tail of *Scomber japonicus*, showing two posterior myotomes contributing to the great lateral tendon (glt) inserting on the middle fin ray (f). The anterior end of the fin ray articulates with the hypural (h) via cartilages (c) enclosing a synovial cavity (s). After Fierstine and Walters (1968) *Mem. S. Calif. Acad. Sci.*, **6**, 1.

but still easily distinguishable from the underlying paler fibres. In higher vertebrates, different fibre types *are* intermingled, and colour differences are not always obvious; nor if they do occur, are they necessarily related to other features of the fibres (Table 3.1, Figure 3.6). For example, some fast muscle fibres in mammals are red, whereas in fish, fast fibres are always white, slow fibres always red. In fish therefore, though not in mammals, fibre colour unambiguously distinguishes the two main fibre types. However, to anticipate the experimental results, it is probably best to distinguish them by contraction speed rather than colour, as slow (red) and fast (white) fibres.

   If we examine the red and white portions of the myotome more closely, we see that the muscle fibres in each are quite different from each other (Table 3.1, Figure 3.6), and it is evident that fish have two very different motor systems in their myotomes. Table 3.1 suggests that the well-vascularised red muscle strip is designed for aerobic operation, whilst the much larger mass of poorly vascularised larger fast fibres, poor in mitochondria, seem specialised for anaerobic operation. We might infer that these two motor systems are used for different kinds of swimming, and

**Table 3.1** A comparison between the fast and slow muscle fibres in fish myotomes (Bone, 1978)

| *Slow* | *Fast* |
| --- | --- |
| Smaller diameter (20–50% of fast fibres) | Large diameter (may be more than 300 μm) |
| Well vascularized | Poorly vascularized |
| Usually abundant myoglobin, red | No myoglobin, usually white |
| Abundant large mitochondria | Few smaller mitochondria |
| Oxidative enzyme systems | Enzymes of anaerobic glycolysis |
| Lower activity of $Ca^{2+}$-activated myosin ATPase | High activity of enzyme |
| Little low molecular wt. $Ca^{2+}$-binding protein | Rich in low molecular wt. $Ca^{2+}$-binding protein |
| Lipid and glycogen stores | Glycogen store, usually little lipid |
| Sarcotubular system lower volume than in fast fibres | Relatively larger sarcotubular system |
| Distributed cholinergic innervation | Focal or distributed cholinergic innervation |
| No propagated muscle action potentials | Propagated action potentials; may not always occur in multiply-innervated fibres |
| Long-lasting contractions evoked by depolarizing agents | Brief contractions evoked by depolarizing agents |

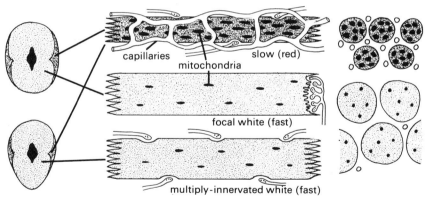

**Figure 3.6** Summary diagram showing structure and innervation pattern of red and white myotomal muscle fibres in fishes. Note that multiply-innervated white fibres are found only in higher teleosts (see text).

knowing that aerobic metabolism is more efficient in fuel use than anaerobic, make the reasonable further inference that the slow fibres of the red muscle strip are used for cruising, whilst the main white mass of fast fibres is used for bursts of speed.

Both inferences are correct, as has been shown directly by extracellular recording from the myotomes of swimming fish. Since the different kinds

of muscle fibres are not intermingled, it is not hard to record electromyograms (EMG) from the different regions of the myotome containing
each type of fibre. By placing simple thin insulated copper wire electrodes
(bared at the tips) in different positions in the myotomes, we can record
the electrical activity of the zone of red or white muscle fibres around the
electrode tips as they contract. Spinal dogfish are particularly suitable for
this kind of experiment, since if appropriately set up they continue to swim
steadily after the brain has been destroyed.

When swimming slowly, regular bursts of electrical activity are seen
from the electrodes placed amongst the slow fibres, whilst those in the
white portion of the myotome are silent. Only if the dogfish is stimulated to
swim rapidly are large bursts of activity seen from the fast fibres. Such
experiments (Figure 3.7) clearly show that the red portion of the myotome
is used during cruise swimming, and the white portion only during bursts of
activity. These results have been confirmed in other fishes such as herring
where the fast fibres are innervated only at one or both ends (focally) as
they are in dogfish (Figure 3.6). In other fishes where the fast fibres are
innervated along their length by many end-plates (Figure 3.6), they play
some part in slow speed cruising as well as in burst swimming (Section
3.2.5), but it remains true that even in these fishes, the slow and fast fibres
are basically two separate specialised motor systems. Studies of the

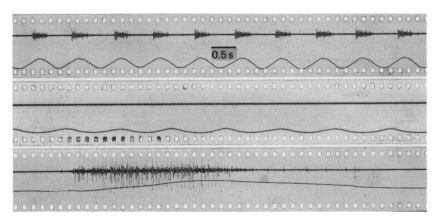

**Figure 3.7** Electromyographic records from swimming spinal dogfish. Upper line in each case
represents electrical activity from muscle fibres, lower line the swimming movements of the
fish. Top: electrode tips in red muscle; middle: electrode tips in white muscle; bottom, same
as middle, but white muscle activated by pinching tail of fish. Bottom trace at higher
recording speed than the others, to show separate muscle action potentials from white fibres.
Such electromyograms are nowadays recorded either with high-speed pen recorders or with
on-line computers, but these first paper records (1965) from fish myotomes were made with a
motorised camera attached to an oscilloscope; the movements of the fish being monitored
with a liquid potentiometer.

depletion of fuel stores in the different zones of the myotome after different types of swimming reinforce this view.

So fishes have two motor systems in their myotomes, used under different swimming conditions. The arrangement is analogous to the jet engines of military aircraft, which run reasonably economically in the cruise condition, but can vastly increase power output (and fuel consumption!) by using reheat for short emergency bursts. However, such aircraft do not have to meet the later penalty fish suffer of repaying the oxygen debt incurred during bursts of rapid swimming.

It may seem paradoxical to the reader aware of motor unit rotation in our own locomotor musculature, that fishes are able to set aside (as it were) the greater part of the mass of their locomotor musculature, to be used only during emergency escape and attack manoeuvres. But fish suffer much less from the effects of gravity than terrestrial animals do, since they are buoyed up by the dense medium, so there is little weight penalty for carrying around a mass of fast fibres of only occasional use. Evidently, the fish arrangement would be quite unsuitable on land; terrestrial animals not only have relatively less muscle in their bodies than fish (muscle in common domestic animals like sheep, goats and cattle is around 30% of body weight, whilst in scombroid fishes like tunas it is around 60%), but also that their muscle fibres are arranged differently and all fibre types are more similar to one another.

Operating aerobically, the slow fibres of the myotome can contract for long periods. Indeed some fish such as mackerel or pelagic sharks cruise continually all their lives, but the anaerobic fast fibres have a very limited duration before they exhaust their fuel stores, probably as in dogfish or herring, around 1–2 minutes only. This is not such a severe drawback as might appear, since it is very rare for any fish to swim at its maximum speed for more than a few strokes of its tail before it glides to rest or into slow cruise swimming. Calculations based on the glycogen content of rested dogfish white fibres show that continuous operation for 2 minutes would allow the fish to travel about 600 m before this system was exhausted and the fish would then no longer be able to do other than swim slowly using the red fibre system.

## 3.2.4 Development of the two systems

In some fish, such as sharks, which do not pass through a larval stage, the young are born or hatch from the egg case, as small adults and like the adults, have a similar dual motor system. In the early development of sharks, all fibres begin as myotubes containing relatively large amounts of mitochondria, and the mitochondria-poor fast white fibres of the adult are transformed from these, attaining a fixed number before the baby shark begins its free-swimming life. Further growth simply takes place by increase in fibre diameter and length. In large adult sharks, such as the basking shark (*Cetorhinus*), the fast myotomal muscle fibres are some 300 μm in diameter and ⩾13 cm long! The development of the dual motor system is different in teleosts. Some, with large eggs, like the salmonids, hatch at an advanced stage of development, but most teleost larvae are only a few millimetres long. Their transparent bodies have serial V-shaped myotomes (Figure 3.2), and as development proceeds, new myotomes are added posteriorly and all become more complex in shape, interdigitating with those adjacent. In zebrafish (*Brachydanio*), herring (*Clupea*), plaice (*Pleuronectes*) and northern anchovy (*Engraulis*) (and probably in many other species) the slow (red) muscle at hatching consists of a thin cylindrical sheath around the larger mass of fast (white) fibres (see Figure 3.8). The slow fibres contain large amounts of mitochondria, particularly in active species like herring. In larvae which are more advanced at hatching, like those of the whitefish (*Coregonus*), the slow muscle extends as a thin strip dorsally and ventrally from the flank lateral line. In most fish larvae

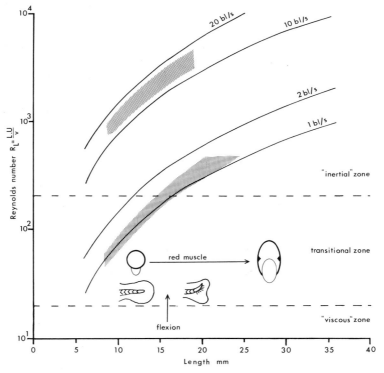

**Figure 3.8** Graph and diagrams showing the change in Reynolds number for cruising and burst speeds during the growth of a fish larva. Note the change in distribution of slow (red) muscle and the length where flexion of the caudal fin occurs.

studied, the slow muscle eventually concentrates along the mid-lateral line of the flank, but in plaice, the slow muscle monolayer is still present after metamorphosis. Unlike elasmobranchs, however, teleost myotomal muscle fibres not only increase in size after the end of the yolk sac stage, but also increase in number throughout life.

In addition, there are changes in the contractile proteins of the muscle fibres during development. Herring, seabass (*Morone*), barbel (*Barbus*) and plaice, and, no doubt, all fish larvae, have muscle fibre types with embryonic or larval isoforms of myosin and other contractile proteins. These fibres arise from pre-myoblast cells in the embryo somites well before hatching, but their number and diameter depend on hatching temperature. As the larvae age and metamorphose, the isoforms of the myofibrillar proteins change with age. Presumably contraction speed (apparently linked to the nature of the myosin heavy-chain isoforms) changes as the fish increases in size and its muscle fibres elongate and increase in diameter.

As development of the larva proceeds, other changes occur in relation to locomotion. The primordial fin fold is gradually replaced by median dorsal, anal and caudal fins, although the lateral fins, especially the pectorals are likely to be present at an early stage to give stability. A major event in development is the flexion of the caudal fin (Figure 3.8) when the tip of the notochord turns up and hypural finrays form the base of the caudal fin. At this time (certainly in the longer, slimmer larvae), locomotion changes from an eel-like mode (p. 67), where locomotor waves pass along the whole body, to oscillations of the caudal region alone, and swimming

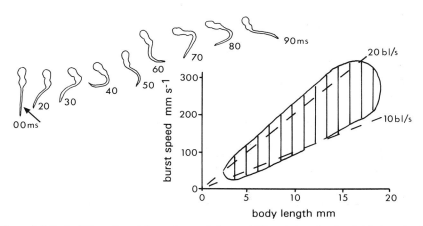

**Figure 3.9** Left: Silhouettes of 5-mm long sole larva at 10-ms intervals after initiation of a C-start (from high-speed video). Right: The relationship between body length and burst swimming speed of fish larvae; the shaded area includes a number of relationships for different species.

performance greatly improves. As the larvae grow and swim faster, so operating at higher Reynolds numbers, inertial rather than viscous forces predominate (see p. 72), and beat-and-glide swimming becomes energetically more efficient.

Larvae of zebra fish, herring, cod (*Gadus*), and other species are equipped with rapid 'hard-wired' escape responses from predators at hatching. These are triggered by tactile stimuli (and later, in herring larvae, by light flashes and sounds), and take the form of a C-start, where the body flexes to a C-shape, before the larva swims off at high speed. The C-start is driven by large neurons in the hindbrain which include the Mauthner pair (p. 283) whose axons decussate before descending the cord, and so drive the myotomal motoneurons on the contralateral side to that receiving the stimulus. Activation of the Mauthner system thus leads to a turn to the non-stimulated side, a sensible response to a predator. After the initial flexion of the C-start, the larvae only swim a short distance, but during this, they achieve speeds of up to 20 body lengths per second, a performance otherwise only seen in adult scombroids (see Figure 3.9).

The general strategy for predator evasion seems to be as follows. Very small and transparent larvae only respond with a C-start when touched, presumably relying on their transparency to make them inconspicuous: as they become older and less transparent (the reader may care to consider why increasing age leads to reduction in transparency) they respond to an attack as do adult fish (see Figure 7.9, p. 167). When the predator is a short distance away, they dart swiftly aside with a C-start at a point when it is too late for the predator to reprogramme its attack, just as a matador steps aside at the last moment from the charge of the bull.

### 3.2.5 Overlap between the two myotomal motor systems

We have seen that the experimental results from dogfish and herring clearly show that only the red portion of the myotome is active during cruise swimming, and the white mass of fast fibres only during bursts.

If we look in more detail at the histochemistry of the myotomal muscle fibres (Figure 3.10) we find that there is a gradient in enzyme activity within the red zone. It seems probable that future less crude experiments than those shown in Figure 3.9, will show that successive layers of slow fibres are brought into use as the fish increases its cruising speed from the minimum value, in other words, that the differences between the slow fibres mean that there is functional partitioning of the red zone of the myotome.

In actinopterygians, fast fibres are multiply innervated (like the slow fibres of all fishes), and there is good evidence that they may be active during cruise swimming in some species (e.g. carp and goldfish). These fish do not accumulate an oxygen debt during sustained swimming, and there

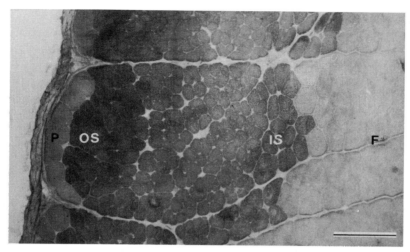

**Figure 3.10** Cryostat section of outer region of dogfish myotome stained for malate dehydrogenase activity, showing graded enzyme activity within the slow zone (S), and in the outer border of the fast zone (F). The superficial single layer (P) of fibres (which are absent in most fishes) probably play a postural role rather than being involved in swimming. Scale bar: 1 μm.

are certainly more mitochondria in the often 'rosé' fast fibres next to the inner border of the red zone, which are more 'aerobic' than those next to the inner border of the myotomes. Again, it is likely that there is functional partitioning so that it is these fibres which are involved with the slow fibres in cruise swimming.

The multiple innervation of fast fibres in the Euteleostei suggests that it must be advantageous over the focal innervation pattern seen in elasmobranchiomorphs, sarcopterygians, elopomorpha and some ostario-physans; perhaps it simply means that some use can be made of the otherwise inactive fast fibres during sustained cruising.

## 3.3 The cost of speed in water

As they swim along, fish have to overcome drag opposing their motion. Most of the drag results from the viscosity of the water which causes water particles to attach to the skin surface and to be carried along with it in a boundary layer, giving rise to skin friction drag ($D_{sf}$). This and the other kinds of drag involved are considered in Section 3.4; for the present, all we need to bear in mind is that for fishes larger than 3–5 cm long $D_{sf}$ is proportional to the *square* of the swimming speed. Since at any constant speed $V$, drag must equal thrust or the fish would speed up or slow down, the thrust will also be proportional to $V^2$. The work performed being thrust × distance, the power (i.e. the rate of doing work), required from the myotomal muscles will be proportional to $V^3$. The power required will not be exactly proportional to $V^3$ (it will more probably be proportional to $V^{2.8}$) for several reasons which the acute reader may ponder, but it remains true that any increase in swimming speed is thus extremely costly in terms of the power needed from the muscles, just as it is (for the same reason) for all machines like submarines, or cars and aeroplanes which travel through a fluid.

It is small wonder that the white portion of the myotome used for maximum speed burst swimming is much larger than the red portion, and that the fast fibres are highly specialised for power production. Ultimately, the power produced by a given mass of muscle depends on the interaction between the actin and myosin filaments, thus modifications allowing more myofilaments in a given volume take precedence over other considerations. So the white muscle fibres are large (up to 300 μm diameter in sharks and rays), few mitochondria interrupt the filament array, and there are few muscle capillaries taking space that could be filled by muscle fibres.

Obviously, such specialisations for maximum power output imply anaerobic operation, and as is seen both by metabolite depletion and enzyme profile studies they operate by anaerobic glycolysis. This only provides 3 moles of ATP per glycosan unit, to drive the actin–myosin interaction and the fibre ion pumps, so it is a relatively costly process in terms of the amount of fuel used, and the glycolytic pathway shows adaptations for high ATP turnover. For example, the rate-limiting enzymes phosphorylase and phosphofructokinase are at high levels. Interestingly, whilst in mammals, the phosphorylase is activated hormonally by catecholamines circulating in the blood, in dogfish hormonal activation does not occur, the activity of the enzyme is directly regulated by calcium release from the sarcoplasmic reticulum triggered by motor nerve activity. Since there is a very poor vascular bed in the white muscle, an 'anticipatory' increase in ATP turnover under catecholamine control would be unsuitable for the white muscle system. The poor vascularity of the white muscle has two important consequences. First, the glycogen stores of the fibres are rapidly depleted by a short burst of activity (in dogfish 1–2 minutes of vigorous activity reduce glycogen stores to 5% of resting levels) and they cannot quickly be restored since activities of the enzymes of gluconeogenesis are low. Secondly, the large amounts of lactate produced by anaerobic glycolysis can only pass slowly into the circulation (presumably to be oxidised to pyruvate at the gills and liver), so there has to be a high buffering capacity in the white fibres. Blood levels of lactate remain high for hours after white muscle activity.

The slow muscle fibres are specialised in an entirely different direction. They operate by aerobic glycolysis, which yields 38 moles of ATP per glycosan unit, or by aerobic lipolysis (over 100 moles of ATP per mole of

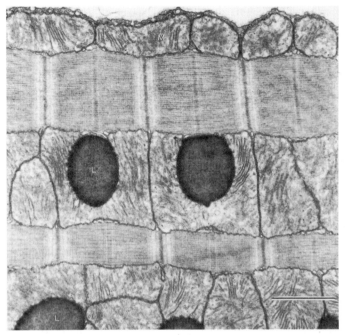

**Figure 3.11** Electron micrograph of longitudinal section of the edge of a slow fibre from *Scomber* showing mitochondrial and lipid droplets (black). After Bone (1978) In: *The Physiological Ecology of Tunas* Sharp and Dixon eds, p. 183. Academic Press. Scale bar: 500 μm.

lipid); plainly this is a much more efficient use of fuel. These processes require oxygen, so there is a rich vascular bed, and myoglobin for internal oxygen transport, to support the high levels of oxidative mitochondrial enzymes such as succinic and lactate dehydrogenases. In some scombroids, no less than 50% of fibre cross-sectional area is taken up by mitochondria, and the myofilament array may also be interrupted by many lipid droplets (Figure 3.11). The power produced per unit of muscle volume will be much less than for the fast fibres, but for cruising, what is needed is efficiency and economy of operation, rather than maximum power output.

## 3.4 Swimming speeds

The two separate motor systems in the myotomes mean that most fish can swim for long periods at cruising speeds (when the fish is using its aerobic musculature), and can also make short bursts at a much higher speed. Since fishes range in size from larvae (and some adult teleosts) only a few millimetres long to the 12-m whale shark (*Rhinocodon*), they operate over a very wide Reynolds number range (see p. 72 and Figures 3.8 and 3.23) and swim at very different cruising speeds. For comparison therefore, speed measurements are usually given in body lengths per second (BL $s^{-1}$) as well as in centimetres or metres per second. Most trout- or herring-sized fishes cruise at around two body lengths per second.

### 3.4.1 Cruising speed and slow muscle

The proportion of slow fibres in the myotomes gives a good indication of the importance of sustained cruising swimming in the life of the fish, so that, for example, spur dogs (*Squalus*) have relatively more slow fibres than the sluggish *Scyliorhinus* or carpet sharks (orectolobids) and salmonids more than tench (*Tinca*). In most fish, the slow fibres make up around 5–10% of the myotomes, and such fish can cruise at one to two body lengths per second for long periods. Fish which cruise more rapidly have a higher proportion of slow fibres in the myotomes, but there is a limit to what can be done to increase cruising speed by simply increasing the proportion of slow muscle. It needs an abundant oxygen supply which has to be acquired by the gills, and unless there are special adaptations for oxygen acquisition and transport (as in scombroids), around 15% of the total myotomal mass seems to be the limit for red muscle. Maximum sustained cruising speed is limited by the oxygen requirement of the slow muscle, but few fish cruise at this maximum. Instead, as we should expect, pelagic fish (like many sharks and scombroids) cruise at speeds which accord well with the expectations based on mimumum energy expenditure for distance travelled, given by $V = 0.5L^{0.43}$. For example, a 9 m basking shark tracked whilst swimming 8 m below the surface swam at 94 cm $s^{-1}$, very close to the expected 1 m $s^{-1}$, whilst 2 m sharks of two carcharinid species swam close to the expected 68 cm $s^{-1}$, and migrating salmon also swim close to predicted speeds, around 50 cm $s^{-1}$.

Minimum energy expenditure for distance travelled is, however, not the only criterion by which cruising speeds are determined. Recent direct measurements of swimming speeds in blue marlin (*Makaira nigricans*), a large scombroid, show that these fish (1.5–2.0 m long) cruise at relatively low speeds (15–25 cm $s^{-1}$) when within 10 m of the surface, whilst at depths greater than 50 m, they cruise at higher speeds, up to 120 cm $s^{-1}$. Short bursts up to 2.25 m $s^{-1}$ were recorded, usually when changing depth. In schools, yellowfin tuna 48–79 cm long, swam between 1.6 and 5.5 m $s^{-1}$, well above the predicted speed, but these fish were cruising to catch food and hence swam faster than for minimum energy expenditure.

### 3.4.2 Maximum speeds of fishes

Maximum burst speeds are hard to measure, for burst speeds are naturally only sustained for brief periods of a minute or less. Various techniques have been employed to estimate burst speeds, including measuring the height to which fish of known weight jump, or their ability to catch bait trolled at known speeds. For example, the exit velocity of a jumping mako shark (*Isurus*) (when it was presumably making a maximum effort), was calculated to be 10 m s$^{-1}$ or 36 km h$^{-1}$ (around 5 BL s$^{-1}$). There have also been direct measurements of line run out from hooked fish, from videos of fish attacking prey, and most recently, by telemetry from fish released with speed transducers and by conditioning experiments in large tanks where fish are trained to swim between alternately flashing lights. A quite different and novel approach, by Wardle and his colleagues (1989), has been to measure the twitch contraction time of the fast muscle fibres, since it is this which will limit the tail beat frequency (which increases linearly with speed). Larger fish beat their tails more slowly than small ones, but if account is taken of this by including fish size (Figure 3.12), we see that the fish moves forwards for the same fraction of body length for each tail beat. In most fishes, this distance (the 'stride') is around 0.7 × BL. To swim faster then, a fish increases its tail beat frequency, so increasing the

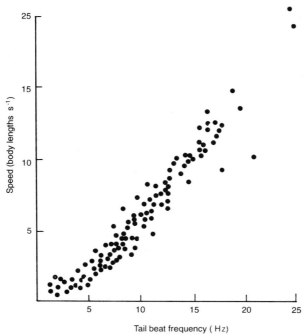

**Figure 3.12** Relation between tail beat frequency and relative swimming speed in dace (*Leuciscus*) of different sizes. Above a tail beat frequency around 5 Hz the relative swimming speed is directly proportional to tail beat frequency. After Bainbridge (1958) *J. Exp. Biol.*, **35**, 109.

number of 'strides' in a given time. Wardle reasoned that the upper limit to
the tail beat frequency must be set by the time that was required for the
white muscle to contract, that is, the duration of the muscle twitch. By
stimulating isolated muscle bundles therefore, it should be possible to
estimate the upper limit of tail beat frequency, and therefore, the
maximum swimming speed of the fish, from the simple equation:

$$M = AL/2T$$

where $M$ = maximum swimming speed, $A$ = 'stride' length, $L$ = fish
length and $T$ = muscle contraction time. Very good agreement was given
between the maximum speeds estimated from this formula, and those
obtained from video records of fish swimming at burst speeds, though
scombroids differed from other fishes in having values of $A$ close to unity at
burst speeds instead of around 0.6–0.8. For example, from video records of
mackerel (*Scomber scomber*) 30.5 cm long, in a large enclosure (Figure
3.13), maximum burst speed was 5.5 m s$^{-1}$, i.e. 18 BL s$^{-1}$ (at 12°C). Stride
length $A$ was around 1.0 rather than 0.7, and substituting this value for $A$ in
Equation (3.1) and measuring minimum muscle contraction time (0.026 s)
gave maximum burst speed as 19 BL s$^{-1}$, in excellent agreement with the
measured burst swimming speed. Scombroids are certainly the fastest
fishes, and the maximum burst speed recorded for any fish is 20.7 m s$^{-1}$
(74.5 km h$^{-1}$) for small yellowfin tuna (*Thunnus albacares*) measured after
being hooked (with a tape recorder on the reel). This was almost
18 BL s$^{-1}$, and presumably was biased low by the additional drag the fish
had to overcome from the line. Wardle and his colleagues (1989)
investigated fast muscle twitch contraction time in the much larger giant
bluefin tuna (*Thunnus thynnus*), concluding that the 50 ms contraction
time set an upper limit of 10 Hz for tail beat frequency in a bluefin 2.26 m
long. If (as in the mackerel) stride length is then around 1.0, at burst
speeds, this means that such a large fish could achieve the astonishing
maximum speed of 22.6 m s$^{-1}$ (81.4 km h$^{-1}$) or 10 BL s$^{-1}$!

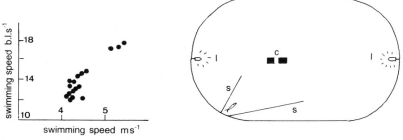

**Figure 3.13** Swimming speeds of mackerel *Scomber* as measured by ciné-photography and
video recording in the tank shown at right. l: Lights; c: cameras; s: screens. After Wardle and
He (1988).

It is still not entirely clear why scombroids should have a different burst stride length to other fishes, but it is because they do so that they are able to achieve such high swimming speeds.

## 3.5 Warm red muscle

Almost all fish operate with body (and muscle) temperatures very close to that of the ambient water, very few are able to run their red muscles above ambient. There is no problem in generating heat (an inevitable byproduct of muscular contraction, which is why we shiver when chilled), but the high heat capacity of water and its contact with the whole body surface, means that any heat generated will normally be lost immediately to the water. What is more, the blood circulating around the body has to be intimately in contact at the gills with the water from which it acquires oxygen (see Chapter 5); the gills are rather like radiators. Surprisingly perhaps, only about 30% of the heat generated in the body is lost at the gills; much more is lost from the body surface.

The more advanced scombroids and isurid sharks like the great white shark (*Carcharodon*), are able to run their red muscles above ambient water temperature. To do so, they have evolved remarkable convergent special modifications of the vascular system. By making the warm blood leaving the red muscle portion of the myotomes pass close to the entering cool oxygenated blood, in special parallel networks of arteries and veins (the retia mirabilia), heat is exchanged to the incoming blood and the muscle can maintain an elevated temperature. Measurements on albacore (*Thunnus alalunga*) indicate heat transfer efficiency in the retia around 98%! The arrangement of these ingenious countercurrent retia is seen in Figure 3.14. Very similar retia are found at other sites in the body, to maintain elevated brain and gut temperatures (p. 276), and lamnid sharks like the porbeagle (*Lamna*) have suprahepatic retia to keep the viscera warm. Retia are also found in the teleost swimbladder gas-generating system (p. 90). At all these sites, the added vascular resistance is outweighed by the advantages of heat and gas retention.

Curiously enough, although Humphry Davy's brother John (on a voyage to India in 1816) found that the red myotomal muscles of bonito were 10°C above seawater temperature, and shark warm red muscles have been known since the late 1960s, it is still not entirely clear (despite much further work) *why* these fish have evolved such complex vascular arrangements to retain heat. One possibility is suggested by the fact that although cruising speeds of similarly sized warm skipjack tuna (*Euthynnus pelamis*) and cold bonito (*Sarda chiliensis*) are similar, the skipjack's burst speed is nearly double that of the bonito. Perhaps the red muscle zone (the site of heat production) is kept warm not only to increase its own power output during

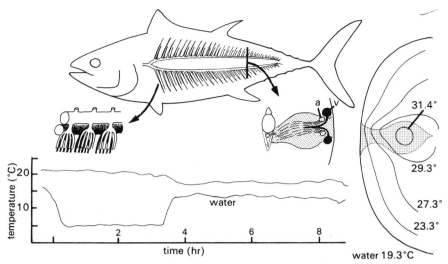

**Figure 3.14** Organisation and function of the retial thermoregulation system in the bluefin tuna (*Thunnus thynnus*). Upper: bluefin showing lateral vessels (details of branches to left, position of lateral vessels and retial system in red muscle to right. a, artery; v. vein). Right: thermal profile obtained with thermistor probes, red muscle stippled. Bottom: records of water temperature (lower line) and stomach temperature (upper) obtained by telemetry from free-swimming bluefin, showing independence of body temperature from changes in water temperature. After Carey *et al.* (1971) *Amer. Zool.*, **11**, 137, and Gibbs and Collette (1967) *Fish and Wildlife ser. Fish. Bull.*, **66**, 65.

cruising, but also so that it can warm the white fast fibres of the myotome and hence increase their power output during burst swimming. Another possible advantage, in the more advanced blue-fin funa (*T. thynnus*) which can thermoregulate, is obviously that the muscle enzyme systems can always operate at their thermal optima when the fish migrates to waters of different temperature. Once a fish has evolved a countercurrent heat exchanger to retain metabolic heat, it faces the problem of possible overheating, but this is avoiding by vascular shunts which bypass the heat exchanger when necessary.

### 3.6 The generation of thrust

Fishes generate forward thrust by transferring momentum from the musculature of the body or fins to the water. They do this in two main ways: (1) by a lift-based (or vorticity) method, and (2) by a drag-based (or added mass) method. The majority of fishes use the second technique, involving the passage of transverse body waves along the body towards the tail, but it is easiest to consider first the more advanced faster swimming

**Figure 3.15** Successive dorsal views of tuna (*Euthynnus affinis*) at intervals of 10 ms, swimming at 8.2 body lengths s$^{-1}$ (approximately 4 m s$^{-1}$) in an experimental tank. After Fierstine and Walters (1968) *Mem. S. Calif. Acad. Sci.*, **6**, 1.

fishes like the large scombroids which use an oscillating lifting caudal foil and are much closer to our own rigid machines.

### 3.6.1 Caudal fin oscillations

Tuna are essentially rigid well-streamlined masses of myotomal muscle attached to a rigid lunate (crescentic) foil by a narrow neck (Figure 3.15). They have virtually separated the propulsive myotomal musculature from the thrust-generating caudal propellor system. The myosepta are prolonged into caudal tendons which pass across a flexible joint in this narrow caudal peduncle, so that as the muscles contract, the foil is oscillated from side to side across the joint, acting much like a propellor. Because the foil oscillates in the vertical plane (rather than rotating like boat or aircraft propellors), it operates with the same side alternately forming the leading and trailing surface and so is symmetrical in section. Since it comes to a halt, as it were, and reverses at the end of each stroke, the thrust produced is oscillatory, though smoothed by the high frequency (up to 10 Hz!) of tail beat.

How is thrust generated by this oscillating foil, and how is it specially adapted for this purpose? Here we are on firm ground, for the same process generates lift by aeroplane wings and thrust by rotating propellors. In fact, the hydrodynamic analysis of fish swimming with lunate tails has been most successfully approached by using oscillating aerofoil theory.

### 3.6.2 Circulation, lift and thrust

The forces acting on a foil as it passes through a fluid arise from the displacement of fluid by the foil: they are of two kinds. Frictional forces (arising from the viscosity of the fluid) may be neglected for the present as they are only indirectly concerned. Lift or thrust forces can only result from a net difference in the pressure of the fluid on either side of the foil. As an aeroplane wing passes through the air, it is both sucked upwards from above, and pushed up from below, for lower pressures are found above the wing, and higher pressures below it (Figure 3.16). How do these pressure differences arise?

The relationship between pressure and velocity in the flow of an incompressible fluid is given by:

$$P + \tfrac{1}{2}fV^2 = \text{constant}$$

where $P$ = static pressure, $f$ = density and $V$ = velocity. The quantity $\tfrac{1}{2}fV^2$ (often abbreviated to $q$), is the *dynamic pressure*, the force experienced if one puts one's hand out of the window of a moving car, and it is this which is the fundamental source of aerodynamic or hydrodynamic forces. As we shall see, $q$ will appear in formulae for calculating thrust, lift and drag. At the speeds fishes swim, fluid density remains the same, so the differences in pressure above and below a lifting foil in a fluid (like an aeroplane wing, or a shark pectoral fin) must mean that there are differences in the velocity of flow on the two surfaces; the velocity of flow must be higher above a lifting foil than below it. Lanchester's circulation theory (see below) showed how these velocity differences arise. When a lifting foil is moved through water, it is possible to visualise the displacement of the water caused by the passage of the foil using a high-speed camera or video system to track small particles like polystyrene beads distributed in the water. Of course, the more convenient (and equivalent) experimental arrangement is to fix the foil and flow water past it using a water tunnel. If we monitor the displacement of two particles (A

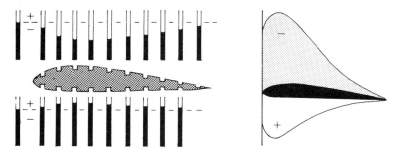

**Figure 3.16** Pressure distribution around a foil at a positive angle of attack in a wind or water tunnel. Left: manometers connected to openings in upper and lower surfaces show pressures above and below ambient (dotted line). Right: typical pressure distribution around such a lifting foil.

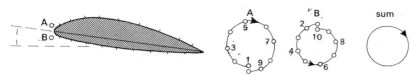

**Figure 3.17** Circulation and lift. Two particles A and B are caused to change position as the foil passes through the fluid. Right: the successive positions of the two particles (numbering as on left). The sum of their movement is a net clockwise circulation producing lift.

and B, Figure 3.17) from their initial positions before the water reaches the leading edge of the foil, until their final positions after they have passed it, we find that they have followed curved paths of opposite sense. That is, they are forced to circulate as they pass over and under the foil. The displacement of A above the foil is greater in the same time than that of B under it, so the algebraic sum of the two results in a net clockwise rotation of circulation. Because it is in this sense, the relative velocity of flow of the fluid above the foil will be greater than that below it, hence lift will be generated. Evidently, the stronger the circulation, the greater the lift.

Although we have been considering circulation and the origin of lift for a lifting foil, the generation of thrust by propellors (whether rotating or oscillating) of course has the same basis. This circulation theory of lift was first described at the end of the last century by one of the greatest English engineers, F.W. Lanchester (known also for his radical early motor car designs) and whose seminal paper was rejected by *Physical Reviews*. He was obliged to publish it himself in book form!

### 3.6.3 Body waves and drag-based thrust generation

Eels swim in a way quite different to tunas: instead of separating the propulsive muscles from the thrust-generating high aspect ratio (AR) caudal foil as tuna do, they have no caudal foil at all, and the movements of the body themselves produce thrust. As Gray (1933) showed, when eels swim, they pass transverse waves of increasing amplitude down the body (Figure 3.18). These pass backwards faster than the fish swims forward. Eel swimming ('anguilliform locomotion') can be seen in terms of the transverse motions of short segments of the body (e.g. those shown on Figure 3.18) as the propulsive waves pass backwards. This approach calculates the resistive forces acting on the segments which depend on the instantaneous velocity of the fish relative to the water. Reactive (inertial) forces are neglected, and except for animals much smaller than almost all fish, these are important.

More recently, a version of a reactive slender-body theory has proven a more fruitful and appropriate approach, which has been used not only to calculate drag and thrust requirements in steady swimming, but also during

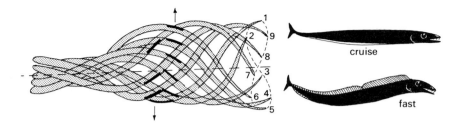

**Figure 3.18** Left: drawings from frames (1–9) of film of young *Anguilla*, superimposed to show short sections of the body (thick lines) as planes inclined to axis of progression when the eel moves forwards. The oblique dashed line is the axis of progression, the tail tip describes a figure-eight pattern. After Gray (1933) *J. Exp. Biol.*, **10**, 88. Right: the two swimming patterns of the trichiurid *Aphanopus*.

accelerations, and manoeuvring. In essence, this approach (mainly due to Lighthill) deals with the forces met by the water masses next to the body surface. These result from the inertia of the water next to the fish, and are proportional to the rate of change of relative velocity of the surface of the fish, with respect to the water next to it. As the locomotor wave passes backwards along the fish, it increases the momentum of water passing backwards (as it is 'pushed' by the inclined planes seen in Figure 3.18); the rate of shedding of this momentum into the wake is proportional to the thrust.

Although eels travel huge distances to spawn, they are relatively inefficient swimmers, because, although long, the body is rounded in section, and so a relatively small mass of water is affected by the passage of the eel through it. Deepening the body to give a deep flat cross-section like many carangids or the elopomorph leptocephali larvae, much increases the virtual mass of water given momentum by the swimming movements, so producing greater thrust. Many fishes which swim by passing waves down their bodies (even if less than a whole wavelength is formed along the body) deepen the body section and have median fins of various kinds to extend body depth. These not only stabilise the fish in roll, but are important in thrust generation. It is not necessary to have a continuous fin, for provided the spacing of the fins along the body is suitable, as it is in many sharks or in gadoids, momentum shed from the anterior fin is 'passed' to the next posterior. Interrupting the series of fins is in fact an advantage, for it incurs less drag (see Section 3.5).

If we consider what fishes need to do in their lives, we see that almost all have to compromise between a body shape suitable for low- or high-speed cruising, and a shape suitable for rapid acceleration and manoeuvre. Some actually adopt different shapes for the two requirements! The deep-sea scabbard fish (*Aphanopus carbo*), for example, has an elongate body ending in a very small caudal fin of fairly high aspect ratio. It swims slowly by oscillating this foil at the end of the rigid body. To accelerate rapidly, it unfurls the long median dorsal fin to deepen the body section, and passes locomotor waves backwards along the body (Figure 3.18). Amputation experiments on salmonids and gadoids have shown that caudal fin removal has little effect on steady swimming speed. Most of the thrust generated must be provided by the locomotor waves passing back along the body. Amputation of the fins, especially the caudal fin, however has a marked effect on acceleration, so it seems that the wide deep caudal fins of many fishes are adaptations for rapid acceleration as are the deep bodies of fish like carp (*Cyprinus*). Since increasing the depth of the body increases wetted surface area, drag is increased (see below) and so rapidly swimming fishes like garfish (e.g. *Belone*), or cypsilurid flying fishes have elongate thin bodies.

## 3.7 Drag

There are two kinds of drag forces which retard the forward motion of fishes or indeed any other object in a flow. The different kinds of pressure drag result from the pressure at the nose of the fish being higher than at the tail, owing to vorticity in the wake. Skin friction drag results from the viscosity of the water making it stick as a boundary layer to the fish surface. Total drag therefore is the sum of pressure drag (form drag + drag associated with circulation and lift) and skin friction drag.

How do these different kinds of drag arise, and how can they be minimised? Any drag-reduction achievable will be greatly advantageous since it will mean that for a given power output, a fish can travel faster, or for a given speed need less power.

### 3.7.1 Pressure drag

Unstreamlined objects (such as bricks) passing through water leave a large turbulent wake behind them because what boundary layer can form near the leading edge soon separates to give rise to the wake. A large wake means a large pressure difference between nose and tail and so a large form drag. This can be very greatly reduced by streamlining to avoid abrupt changes of contour and awkward excrescences. We have seen that trout approximate to the solid of least resistance, and scombroids like mackerel have transparent eye fairings. Drag tests on dead mackerel show that the drag incurred is almost exactly that of a flat plate of similar wetted area; this is because form drag is virtually absent, so excellent is the streamlining. When the mackerel swims, however, some part of the vortices trailing in its wake arise from the generation of thrust and of lift, so that this second vortex drag component is an inevitable consequence of these processes.

### 3.7.2 Vortex, induced, or lift (thrust)-associated drag and circulation

The brief account of circulation theory in Section 3.3 dealt with the lift generated by circulation around the main part of a foil. So what happens where the foil ends, at the tip? This is an important question, for whilst circulation around the body of the foil generates lift (or thrust), at the tip it gives rise to drag. The circulation of the fluid over the main part of the foil continues downstream at the tip in a series of tip vortices, giving rise to upwash just behind the tip, and downwash across the span, decreasing to the root of the foil (Figure 3.19). Vortex production requires energy; the energy expended to maintain the tip vortices is one component of the pressure drag incurred as the foil passes through the fluid, called variously induced, lift-associated, or vortex drag ($D_v$) (proportional to $1/V^2$).

Since the tip vortices depend upon the circulation, the stronger the circulation, the greater $D_v$ incurred. Like aeroplanes, fishes take advantage of two possibilities to reduce this component. First, since the same lift or thrust can be produced by a weak circulation over a long span as by a strong circulation over a short span, doubling the span halves the $D_v$. This is why the caudal fins of fast-swimming fishes like tuna or lamnid sharks are long and thin, as are the wings of efficient sailplanes and the lifting

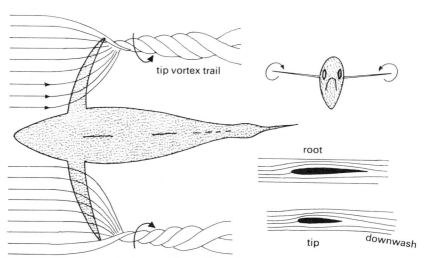

**Figure 3.19** Flow patterns around a gliding skipjack (*Katsuwonus*). Tip vortices cause vortex of lift-associated drag. Lower right: flow over root and near tip of pectoral fins showing downwash near tips.

pectorals of dense fast-swimming fishes. The aspect ratio, AR (span$^2$/area), of tuna caudal fins is up to six, much lower than that of sailplane wings, but as high as can be arranged without sacrificing the structural strength needed to cope with the forces involved in transmitting thrust.

Secondly, it can be shown that for a given AR, $D_v$ is least if the spanwise lift distribution is elliptical; this can be conveniently achieved by using an elliptical planform, as R.G. Mitchell did for his hard-to-manufacture but efficient Spitfire wing. All fishes which normally cruise at high speed, like scombroids or fast sharks, have lunate elliptical caudal thrust-generating foils and lifting pectoral foils (Figure 3.20).

High AR elliptical caudal and pectoral fins are designed to minimise $D_v$ in fast-cruising fishes. But such fins are not suitable for rapid acceleration or for agile manoeuvring, and fishes like salmonids which need these capabilities have much broader lower AR fins.

Why are high AR fins unsuitable? Symmetrical section foils (almost all fish fins are like this) have to be operated at an angle of attack to their axis of movement, or circulation will be equal above and below the foil, and no lift or thrust will result. At small angles of attack, the fluid will flow around the foil in an ordered way, but if the angle of attack is increased above a certain limit, the flow above the foil will separate from it: it will stall with loss of lift and greatly increased drag. A long thin foil of high AR is much more prone to stall as the angle of attack is increased than is a short wide low AR foil, and it is when accelerating or manoeuvring that the angle of attack of the fins is greatest. So the high AR design of tuna fins or albatross wings is not suitable for trout or pheasants which need rapid acceleration from rest and both have low AR, short, broad designs. The convergent designs of the lifting foils of birds, and the lifting and thrust-generating foils

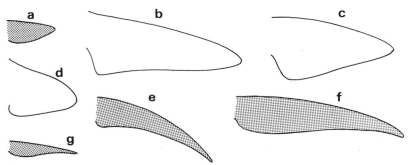

**Figure 3.20** The lifting pectoral foils of different fishes (teleosts stippled). (a) Mackerel (*Scomber*); (b) *Carcharhinus longimanus*; (c) basking shark (*Cetorhinus*); (d) dogfish (*Scyliorhinus*); (e) carangid (*Trachurus*); (f) longfin tuna (*Thunnus alalunga*); (g) swordfish (*Xiphias*).

of tunas and similar fast-cruising fishes are not the only convergences
between fishes and birds and aeroplanes, as we shall see later.

### 3.7.3 Skin friction drag, boundary layers and Reynolds number

The most important drag component incurred by fish results from the
viscosity of the water. Water tends to stick to the surface of the fish as it
moves forward, forming a thin boundary layer in which there is a steep
velocity gradient between the still water carried along by the fish, to the
water going past in the free stream outside the boundary layer (Figure
3.21). This velocity gradient means that there is a large shear stress in the
boundary layer, resulting in skin friction drag, $D_{sf}$. The properties of the
boundary layer, including its thickness and the $D_{sf}$ incurred, depend on the
ratio between the viscous and inertial forces acting on the fish. These are
given by the useful Reynolds number, Re (introduced by Osborne
Reynolds who experimented on flow in tubes). For general flow conditions
this number is defined as

$$Re = \frac{\text{Length } (L) \times \text{Velocity } (V) \times \text{Density of medium } (\delta)}{\text{Viscosity of medium } (v)}$$

The kinematic viscosity of the medium, $v$, is viscosity/density (for water,
constant at any given temperature, approximately 0.001), so Re is usually
given as $Re = LV/v$, which in dimensional symbols is:

$$Re = \frac{L \times LT^{-1}}{ML^{-1} T^{-1}/ML^{-3}}$$

where L = length, T = time and M = mass.

Re is thus dimensionless and so whatever their size, if two similarly
shaped objects pass through a fluid at different speeds so that Re is the
same, the flow regime around them will be the same, and flows can be
tested with models. If we know Re, we have at once a good idea of what
the flow pattern around an object (such as a fish) will be. In practice, at Re

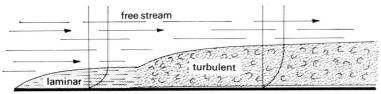

**Figure 3.21** Development of boundary layer on a flat plate in a fluid flowing from left to right.
The anterior thinner laminar layer has a less steep velocity gradient than the posterior
turbulent layer.

below $5 \times 10^5$ viscous forces predominate and the boundary layer on a flat plate will be laminar. At higher Re, over $5 \times 10^6$, inertial forces are more important, and flow within the boundary layer will be turbulent. The two kinds of boundary layer flow are seen in Figure 3.21, whilst Figure 3.22 shows the relation between Re and fish swimming speeds. Evidently, fish larvae will operate at low Reynolds numbers, and as they grow in length and in the speed at which they swim, Re will increase.

On a flat plate of sufficient length placed in a water tunnel, the boundary layer will be laminar near its upstream (front) edge but it will thicken as it passes downstream (aft) and change to a turbulent boundary layer. The point of transition depends much on the smoothness of the surface, for the laminar layer is rather sensitive to any adverse pressure gradients produced by surface irregularities, because energy to maintain it can only diffuse across the layers within it.

Skin friction drag on a flat plate is given by:

$$D_{sf} = \frac{1}{2}fV^2AC_f$$

where $f$ = density, $V$ = velocity of flow, $A$ = wetted area and $C_f$ is a drag coefficient which depends on the type of boundary layer.

We saw that attention to streamlining, and to the planform and AR of

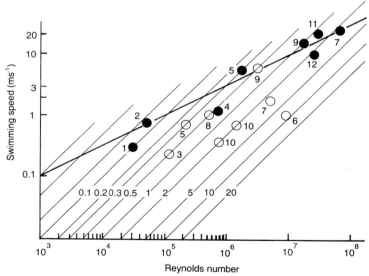

**Figure 3.22** Relation between Reynolds number and swimming speeds (open circles: cruising speeds, filled circles: burst speeds). Oblique lines: fish lengths (m). (1) *Pholis*; (2) *Clupea*; (3) *Scyliorhinus*; (4) *Anguilla*; (5) *Scomber*; (6) *Cetorhinus*; (7) *Thunnus thynnus*; (8) *Salmo*; (9) *Thunnus albacares*; (10) *Makaira*; (11) *Acanthocybium*; (12) *Isurus*. Note that the bluefin tuna (*T. thynnus*) when sprinting operate at a Reynolds number of $6.5 \times 10^7$. Partly after Webb (1975) *Bull. Fish. Res. Bd. Canada*, **190**, 1.

lifting and thrust generating foils, will minimise pressure drag. What can be done to minimise skin friction drag, almost always much more important than pressure drag?

### 3.7.4 Mechanisms for reducing skin friction drag

*Reduction of wetted area.* Anything which reduces wetted area ($A$) is worth doing; it is for this reason that fast-swimming fishes like scombroids have fins which retract into slots when not required for manoeuvring, and instead of continuous fins, body depth in fish making lateral body movements is increased by a discontinuous series of fins. Caudal fins are often scooped out at the rear to reduce wetted area.

*Reduction of lateral movements.* Fishes which make substantial lateral body movements as they swim incur greatly increased $D_{sf}$ compared to that for a flat plate of equivalent wetted area, or to the same fish gliding without body flexures. This is because the transverse motion of the body thins the boundary layer on its leading side, which increases the velocity gradient and hence the stress across it. Calculations of thrust and oxygen consumption measurements suggest that this process increases drag over the flat plate value by around five times. For fast-swimming fishes, lateral movements of the body must be avoided, and the fastest all use lift-based caudal fin propulsion; only slower swimmers have large amplitude body movements.

*Boundary layer control mechanisms.* The great majority of fishes are small enough so that even at burst speeds they operate in a flow regime where Re does not exceed $5 \times 10^5$ and boundary layer flow will be laminar. But larger and faster fishes operate at Re above this, where inertial forces are more important and where the boundary layer is expected to be largely turbulent. Since the drag coefficient for a laminar boundary layer is much smaller than that for a turbulent bundary layer, anything that the fish can do to maintain a laminar layer, delaying transition to a turbulent layer, is worth while. Sharks have denticles with low sharp-edged ridges parallel to the direction of motion (Figure 1.4, p. 7). Such ridges have been shown experimentally to reduce drag on flat plates by some 10–15%, and ridging of this kind has been used on yacht hulls. The ridges are believed to delay transition by reducing microturbulence in a laminar boundary layer. Recollecting that the laminar sublayer is sensitive to adverse pressure gradients, another possibility is to move these as far aft as possible, and some (not all) large scombroids do this and have body sections similar to our own laminar flow aerofoil section profiles (Figures 3.15 and 3.19).

Sometimes (paradoxically at first sight), fishes seem to have adopted boundary-layer control devices which, rather than making the boundary layer laminar as far as possible, *induce* transition to the higher-drag turbulent layer. This is because the turbulent layer is less sensitive to disturbance than the laminar layer, and so less prone to separate from the body surface. Separation is to be avoided at all costs, as it produces a vortex-laden wake and the severe penalty of a large pressure drag component. So in some cases it is advantageous to induce transition to the higher drag turbulent layer rather than risk separation.

The large mesopelagic castor oil fish (*Ruvettus*), so-called because its tissues, and especially the bones, contain a purgative oil, has a remarkably specialised integument, with two separate mechanisms for maintaining a turbulent boundary layer. The very sharp-pointed spines of the ctenoid (comb-like) scales all over the body project about 1.0 mm above the body surface and act as vortex generators, entraining fluid from the free stream as seen in Figure 3.23. This input to the boundary layer helps to stabilise it and prevents separation during the accelerations and bursts of speed of this big predator. A less complex vortex generator array occurs on the upper surface of the wings in some airliners, at the point on the chord where separation (stall) needs to be avoided during landing and take-off. Vortex generators seem such a simple idea, and perhaps all teleost fishes with

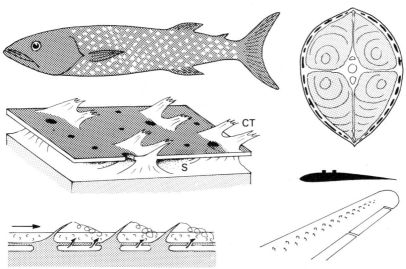

**Figure 3.23** The specialised integument of the castor oil fish *Ruvettus*. Upper left: ctenoid scales (white dots) form a regular pattern over the surface. Upper right: transverse section of body showing sub-dermal spaces (black). Middle: stereogram of integument showing ctenoid scales (CT) and sub-dermal spaces (S). Anterior to left. Lower left: assumed operation of integument injecting momentum into boundary layer (see text). Lower right: vortex generators on aircraft wing. After Bone (1972) *Copeia*, 78.

ctenoid scales use them as vortex generators, but this remains to be proved. They would not be suitable for fishes where cruising economy is important, since they increase drag at low speeds.

*Ruvettus* also injects fluid into the boundary layer. The bases of the ctenoid scales form pillars supporting the outer skin away from the connective-tissue layer over the muscles; the sub-dermal spaces resulting communicate with the boundary layer by backward-facing openings (Figure 3.23). As the fish flexes its body, seawater will alternately be sucked into and squeezed out of these openings, so energising the boundary layer and keeping it attached. This is certainly the most complex integument of any fish yet examined, but analogous simpler injection systems are seen in some stromateoids, like the salp-eating *Tetragonurus*. Fluid injection into the boundary layer became common aircraft practice with 'blown' flaps and control surfaces, once these could easily be fed from jet engines.

*Drag-reducing behaviours?*   Fish denser than water could save energy as many small birds do; by alternating propulsive body movements with periods of gliding with the body rigid when $D_{sf}$ is less. An advantage of 20% or more could result, but we do not know whether dense fish (like many scombroids or the pelagic blue shark *Prionotus*) actually swim in such a glide-power cycle. Small whales and dolphins reduce the energy costs of swimming by 'porpoising' in and out of the water, and it seems that the pop-eye mullet of Northern Australia which 'porpoises' as it enters estuaries on the incoming tide, does so to reduce $D_{sf}$.

## 3.8 Efficiency

Efficiency is essentially measured by the ratio of input to output; there are various measures of efficiency according to the input:output properties considered. We could compare the efficiency of the swimming fish to that of our own machines such as motor cars, which generate power (as does the swimming musculature) by converting chemical to mechanical energy, and then transmit this power to the road to drive themselves along. For such machines, engineers define two principal kinds of efficiency.

$$1. \quad \text{Mechanical efficiency } (\eta_m) = \frac{\text{Brake horsepower} \times 100}{\text{Indicated horsepower}}$$

This is the ratio of the power developed at the pistons vs. the power available at the output shaft after frictional losses in bearings, gear trains, etc. In practice, the mechanical efficiency of a motor car is usually around 80–85%.

2. $$\text{Thermal efficiency } (\eta_t) = \frac{\text{Work done} \times 100}{\text{Heat supplied}}$$

This is the ratio of one kind of energy, heat, supplied during the combustion of fuel in the cylinders to the useful energy obtained in the form of work done by the engine driving the wheels. Diesel engines have thermal efficiencies around 40%, petrol engines rather less.

Overall efficiency, $N$, is the product of the two, so $N$ of a well-designed diesel car will be something like 30%. How do fishes compare with this? Compared with land animals, fishes have efficient locomotion, as measured by the energy used per unit of body mass required to travel a given distance, for as Schmidt-Nielsen pointed out, in fishes this 'cost of transport' is about 10 times less than that land animals require. In practice, cost of transport is determined by measuring oxygen consumption for fish swimming under controlled conditions in water tunnels or flumes, and is expressed as calories $g^{-1}$ $km^{-1}$. In the white crappie (*Pomoxis annularis*), for example, swimming at its most efficient speed of 20–25 cm $s^{-1}$ (1.1–1.5 BL $s^{-1}$), the total cost of transport was found to be 0.78–0.72 cal $g^{-1}$ $km^{-1}$, whereas in the much larger hammerhead shark (*Sphyrna tiburo*) which cruises continuously, the cost of transport for small specimens (body length 34–95 cm) cruising at 29–67 cm $s^{-1}$ was significantly lower (0.4–0.67 cal $g^{-1}$ $km^{-1}$).

$N$ may be considered as the product of the efficiency of transfer of momentum $m$ from the fish body and fins to the water (usually called $\eta_p$), and that of the conversion of chemical to mechanical energy in the muscles ($\eta_m$). Values for $N$ obtained from oxygen consumption measurements during sustained aerobic cruising by salmonids work out between 5% at low speeds to around 20–22% at the maximum sustainable speeds. Knowing $N$, we could obtain $\eta_m$ or $\eta_p$, if we knew one of these. Unfortunately, few measurements exist of $\eta_m$ for fish muscle. Thus $\eta_p$ has to be estimated using values for $\eta_m$ known for other animals such as frogs. Fish muscle is not very different in structure to frog muscle, so this is not an unreasonable assumption to make, and if we do so, estimates of $\eta_p$ range up to 75% in trout swimming at 2 BL $s^{-1}$. It seems highly probable that tunas and other lunate-tail swimmers will have a higher $\eta_p$ than this (perhaps 85%), and that $N$ will be higher than for trout, possibly greater than 30%. So fish compare quite favourably with motor cars!

## Bibliography

Alexander, R.McN. (1969) The orientation of muscle fibres in the myomeres of fishes. *J. Mar. Biol. Ass. UK*, **49**, 263–290.

Blake, R.W. (1983) *Fish Locomotion*. Cambridge University Press.

Bone, Q. (1978) Locomotor muscle. pp. 361–424. In *Fish Physiology*, 7 (Hoar, W.S. and Randall, D.J. eds) Academic Press.

Bone, Q. (1988) Muscles and Locomotion. In *Physiology of Elasmobranch Fishes* (Shuttleworth, T.J. ed), 99–142. Springer-Verlag, Berlin.

Crockford, T.C. and Johnston, I.A. (1993) Development changes in the composition of myofibrillar proteins in the swimming muscles of Atlantic herring (*Clupea harengus* L) *Mar. Biol.*, 115, 15–22.

Daniel, T.L. (1988) Forward flapping flight from flexing fins. *Canadian Journal of Zoology*, 66, 630–638.

Daniel, T.W., Jordan, C. and Grunbaum, D. (1992) Hydromechanics of swimming. In *Advances in Comparative and Environmental Physiology*, 11, Mechanics of Animal Locomotion, pp. 17–49.

Eaton, R.C. and DiDomenico, R. (1980) Role of the teleost escape response during development. *Trans. Amer. Fish. Soc.*, 115, 128–142.

Fricke, H. and Hissmann, K. (1992) Fin coordination and body form of the living coelacanth *Latimeria chalumnae*. *Environmental Biology of Fishes*, 34, 329–356.

Gray, J. (1933) Studies in animal locomotion. I. The movement of fish with special reference to the eel. *J. Exp. Biol.*, 10, 88–104.

Holst, R.J. and Bone, Q. (1993) On bipedalism in rays. *Philosophical Transactions of the Royal Society of London*, B, 339, 105–108.

Jayne, B.C., and Lauder, G.V. (1993) Red and white muscle activity and kinematics of the escape response of the bluegill sunfish during swimming. *J. Comp. Physiol. A*, 173, 495–508.

Lighthill, J. (1971) Large-amplitude elongated body theory of fish locomotion. *Proceedings of the Royal Society of London*, B, 179, 125–138.

Lighthill, J. (1973) Aquatic animal locomotion. In *Applied Mechanics. Proceedings of the Thirteenth International Congress of Theoretical and Applied Mechanics.* (ed. Becker, E. and Mikhailov, G.K.) Springer-Verlag, Berlin.

Lighthill, J. and Blake, R. (1990) Biofluid dynamics of balistiform and gymnotiform locomotion. Pt. 1. Biological background and analysis by elongated-body theory. *Journal of Fluid Mechanics*, 212, 183–207.

Parsons, G.R., and Sylvester, J.L. (1992) Swimming efficiency of the white crappie, *Pomoxis annularis*. *Copeia 1992*, 1033–1038.

Videler, J.J. (1993) *Fish swimming.* Fish and Fisheries series (T.J. Pitcher ed), 10, Chapman & Hall, London.

Schmidt-Nielsen, K. (1971) Locomotion: energy cost of swimming, flying and running. *Science NY*, 177, 222–228.

Wardle, C.S. (1977) Effects of size on the swimming speeds of fish. In *Scale Effects in Animal Locomotion.* (Pedley, T.J. ed). Academic Press, New York. pp. 299–313.

Wardle, C.S. and He, P. (1988) Burst swimming speeds of mackerel, *Scomber scombrus* L. *Journal of Fish Biology*, 32, 471–478.

Wardle, C.S. and Videler, J.J. (1993) The timing of the electromyogram in the lateral myotomes of mackerel and saithe at different swimming speeds. *Journal of Fish Biology*, 42, 347–359.

Wardle, C.S., Videler, J.J., Arimoto, T., Franco, J.M. and He, P. (1989) The muscle twitch and the maximum swimming speed of giant bluefin tuna, *Thunnus thynnus*. *Journal of Fish Biology*, 335, 129–137.

Webb, P.W. (1975) Hydrodynamics and energetics of fish propulsion. *Bulletin of the Fisheries Research Board of Canada*, 190, 1–159.

Webb, P.W. and Weihs, D. (1986) Functional locomotor morphology of early life history stages of fish. *Transactions of the American Fisheries Society.* 115, 115–127.

Webb, P.W. and Weihs, D. eds. (1983) *Fish Biomechanics.* Praeger, New York.

# 4 Buoyancy

## 4.1 Dynamic lift

We saw in Chapter 3 that some fishes are denser than the water in which they swim, and have to generate dynamic lift by using their outspread pectorals as lifting foils. This process inevitably generates drag, which makes a significant contribution to the total drag and energy requirement of such fishes. For example, a mackerel (*Scomber*) weighing around 25 g in seawater has to generate some $1.2 \times 10^4$ N of dynamic lift during level swimming—mackerel effectively are climbing a 1 in 15 hill all their lives! But increased energy expenditure is not the only drawback to keeping aloft in the water by generating dynamic lift. Dense fishes must maintain a minimum forward cruising speed to generate enough lift to prevent themselves sinking, so they cannot hover or swim backwards. The swimming rhythms of spinal sharks (35–40 tail beats $min^{-1}$ in spinal *Scyliorhinus*) are apparently related to this minimum swimming speed. For benthic fishes that rest on the bottom and only swim occasionally, dynamic lift generation is appropriate, but for other fishes, a better solution would seem to be to store light materials to provide static lift (as do airships and submarines) and thus avoid the drawbacks of generating dynamic lift.

## 4.2 Static lift

Most of the materials making up a fish are denser than water. For example much of the body consists of locomotor muscle (density 1050–1060 g $litre^{-1}$ in common marine fishes). Skeletal tissues loaded with heavy mineral salts are correspondingly denser (2040 g $litre^{-1}$ for typical teleost bones). Nevertheless, water is so much denser than air that even without special stores of low-density materials, fishes are not so much denser than the water in which they swim. They have the option (denied to birds) of storing sufficient low-density material that they can make themselves the same density as the water; they can thus achieve neutral buoyancy and need expend no muscular energy to keep station in the water.

Fishes use two quite different materials to provide static lift. Gas is efficient in giving lift, since its density is very low, and most teleosts possess gas-filled swimbladders. Usually the swimbladder is about 5% of the body volume of marine fishes, and about 7% in freshwater fishes, and provides enough lift for neutral buoyancy. Fish swimbladders cannot significantly resist changing in volume as the fish swims up and down in the water and the ambient pressure changes; indeed, the swimbladders of almost all teleosts obey Boyle's law perfectly. So if a fish with a gas-filled swimbladder is to remain neutrally buoyant at different depths, it must secrete or absorb gas to keep the swimbladder at constant volume as the ambient pressure changes. To regulate the mass of gas within the swimbladder in this way requires complex mechanisms of great physiological interest.

Fats and oils are also used by fishes as sources of static lift. These are much less efficient and much bulkier for a given amount of lift. However, they have the great advantage that the lift provided varies little with depth, because changes in ambient pressure have relatively little effect on the volume of the fat or oil; if a fish using lipid as a source of static lift is neutrally buoyant at the surface of the sea, it will also be close to neutral buoyancy at the sea bed, even at considerable depths.

In the short term then, lipid provides fewer problems of buoyancy regulation than does gas, but in the longer term, difficulties arise, because the lipids stored to provide lift may have other functions as well. For example, sharks which store oil in the liver and muscles may have to draw on this store as a fuel for continuous swimming, or as food reserve for developing embryos or for the adult. Where lipid is the sole source of static lift, there will certainly be great complexities in the regulation of lipid metabolism and the fish will find it difficult to adjust its density rapidly to cope with the short-term density changes resulting from feeding and parturition.

Fish using gas as a source of static lift can afford to support the heavy components of their bodies without difficulty, but where lipids are used, the safety margin is not so great, and a number of fish using lipid for static lift can only achieve neutral buoyancy by reducing the dense components of their bodies and so reducing the lift they require (Figure 4.1). Their skeletons are reduced and poorly calcified, and they even reduce the amount of protein in the muscles, which are weak and watery. Mesopelagic and bathypelagic fishes without swimbladders have an extremely high water content, e.g. *Melanocetus* (angler fish) 95%, *Eurypharynx* (gulper eel) 94%, *Photostomias* 92%, *Gonostoma elongatum* 90% and *Chauliodus* 88%. Those with swimbladders like the lantern fishes (myctophids) and hatchet fish (sternoptychids) are 70–85% water. The first group have very small hearts, very little red muscle and their low haematocrits (packed red blood cell volumes) average 8%. These fishes are little more than floating traps or ambush predators, attracting their prey with luminous lures and quite unable to pursue potential food. The second group have larger hearts, more red muscle and haematocrits averaging 21%. Even they are poor locomotor performers compared with the mackerel with a haematocrit

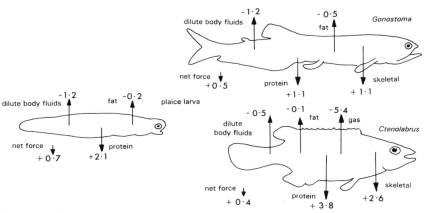

**Figure 4.1** Buoyancy budgets of a mesopelagic fish (*Gonostoma*), a shallow-water acanthopterygian (*Ctenolabrus*) and a fish larva (plaice, *Pleuronectes platessa*). All upward (−) and downward (+) forces are in arbitrary units. After Denton and Marshall (1958) and Blaxter and Ehrlich (1974).

of about 50%! Larval fishes before ossification of the skeleton are watery and have their own characteristic buoyancy balance sheet (Figure 4.1).

## 4.3 Lipid as a source of static lift

In four unrelated fish groups (and perhaps in others) enough lift is provided by stored lipids to achieve neutral buoyancy. Interestingly enough, the lipids stored are not biochemically similar, but they are all of particularly low density and so most efficient in providing lift. Fish lipids vary in density from around 930 g litre$^{-1}$ (cod liver oil and the oil of dogfish livers) to the wax esters of myctophids, gempylids and the coelacanth *Latimeria* (densities around 860 g litre$^{-1}$), and the hydrocarbons of some sharks (860 g litre$^{-1}$). The difference in specific gravity between cod liver oil and the wax esters or shark hydrocarbons may not seem very striking, but 1 g of the less dense oil will provide 0.1675 g of lift in seawater (density 1027.5 g litre$^{-1}$), whereas a gram of the denser oil will only provide about half as much lift (0.0975 g). It is thus well worth while for fishes to store these lighter lipids rather than the more common triglyceride metabolic reserves, as all the fishes that achieve neutral buoyancy using lipids actually do.

### 4.3.1 Squalene

The first fishes found to use lipid to achieve neutral buoyancy were the deep-sea squaloid sharks. Unlike the spur dog of the same family, these

fish live near the bottom in deep water, and the habitat is one where the
family has successfully diversified, for there are many genera known. All
share two striking characteristics: very large livers (making them grossly
corpulent) and very small pectoral fins. In most animals, including
ourselves, the liver is around 4–6% of the total weight, but in these fish, it
may be more than one-quarter of the total weight, because it contains an
enormous amount of pale yellow oil. On this oil the fish literally float—
when their livers are removed, they sink. In all species examined, the liver
oil is of low density (870–880 g litre$^{-1}$) because it is mainly composed of the
hydrocarbon squalene. Squalene, which was the first isolated from such
sharks, is formed by the condensation of isoprene units on the pathway
leading to cholesterol. It has the low density of 860 g litre$^{-1}$, and so is
admirably suited to provide static lift.

We do not yet know how sharks regulate their liver lipids so as to
balance their weight in water, but it appears from experiments on *Squalus*
that the fine adjustments required for neutral buoyancy may depend not on
changing amounts of squalene, but rather on varying the other less-
abundant lipid constituents of the liver oil. By attaching weights to *Squalus*
in the aquarium, it was found that the fish responded by increasing the
amount of low-density alkoxydiglycerides in the liver oil, compared
with the control fishes which had larger amounts of the denser triglycerides.

Deep-sea squaloids bear live young and have very large eggs (about the
size of a billiard ball in a shark a metre long); ingeniously enough, the eggs
contain squalene and are neutrally buoyant themselves, so that pregnancy
does not increase the density of the mother.

It is obvious that these fishes need only have relatively small pectoral
fins; they are used only during manoeuvring. Although the tails in most
genera are markedly heterocercal, they evidently do not generate lift (p.
18). Why do deep-sea squaloids need to be neutrally buoyant if they live
on the sea bed? The answer is that they hover just off the bottom (as we
know from ciné films taken by deep cameras), unlike the dense bottom-
dwelling elasmobranchs of shallow water, which rest on the bottom and are
invariably very dense, like the dogfish.

Deepwater Holocephali evidently live in a similar way to the deep-sea
squaloids, and like them, are close to neutral buoyancy, although they only
manage this by virtue of reduction of dense components, and have poorly
calcified skeletons. Their liver oil consists largely of squalene and this is
also the main source of static lift in some teleosts. Eulachon (*Thaleichthys*),
for instance, contain 20% of lipid by weight. Eulachon do not have a gas-
filled swimbladder, and during their spawning migration, squalene forms a
higher proportion of the total lipid than at other times; it seems that the
fish metabolises reserve triglycerides during these migrations and becomes
denser, so that in this case, lipid metabolism is not sufficiently well

regulated to cope with buoyancy and metabolic demands upon the total lipid pool, and at the same time, maintain neutral buoyancy.

### 4.3.2 Wax esters

A little squalene is also found in the living coelacanth *Latimeria*, but most of the massive amounts of lipid stored are wax esters. These make up 30% of the wet weight of the ventral musculature and over 60% of the wet weight of the swimbladder (which contains only lipid, no gas), and in the pericardial and pericranial tissues, the percentage is even higher. We know very little about the habits of *Latimeria*, but its fins certainly seem unsuited to provide dynamic lift, and the only specimen examined alive floated at the surface, so that it seems safe to assume that it is neutrally buoyant in life (see p. 20).

Wax esters are probably stored to provide lift in many families of mesopelagic teleosts, but the only ones examined so far have been the gempylids and a few of the numerous kinds of myctophids. The gempylid *Ruvettus* (which has the remarkable integument described on p. 75) is loaded with low-density oil (density 870 g litre$^{-1}$) which has purgative properties, hence the common name of castor oil fish. The cranial bones have been modified as oil tanks, and are the least dense tissues of the body. This large (1 m) predatory fish ranges from depths of 15 m to over 500 m, and is very close to neutral buoyancy, feeding on smaller fishes which undertake diurnal vertical migrations.

The most remarkable of these are the myctophids, some undertaking daily a double journey of 500 m up and down in the water column. Although all myctophids have gas-filled swimbladders as larvae, in many species the swimbladder shrinks and becomes invested with more and more lipid as the fish gets older, until no gas remains. As in *Latimeria*, lift from gas is replaced by lift from lipid, which may eventually make up 15% of the wet weight of the fish. As we should expect, the species with a high lipid content are neutrally buoyant, and store low-density wax esters. Why should many myctophids have abandoned gas as a source of static lift? Although the evidence is only circumstantial, it seems a reasonable guess that it is because there are difficulties in regulating buoyancy over a wide range when gas is the source of the lift. Those species which as adults have much low-density lipid and are neutrally buoyant, have a greater depth range than the juveniles, which still have gas in the swimbladder (Figure 4.2), and similarly, species which have gas-filled swimbladders as adults undergo less extensive vertical migrations than those which rely on lipid only.

On the whole, the evidence is that lipid storage for static lift is a secondary phenomenon in myctophids, and that the most 'advanced' species in the family are those which have abandoned gas altogether as

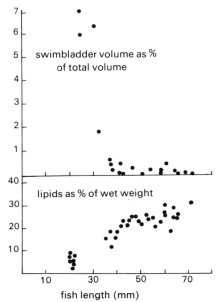

**Figure 4.2** Changes in swimbladder gas and lipid content during growth of the myctophid *Diaphus theta*. After Butler and Pearcey (1972).

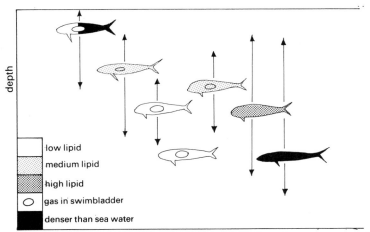

**Figure 4.3** Functional types of myctophids. Vertical arrows indicate known or assumed extent of vertical migration. After Bone (1973).

adults; it is interesting that within the family there are fishes showing all stages in this changeover adapted to different lifestyles (Figure 4.3).

### 4.3.3 Insufficient static lift for neutral buoyancy

So far, we have considered fishes of different groups which store low-density lipids to attain neutral buoyancy. What of fish which use lipid for static lift, but are not neutrally buoyant? Many teleosts are certainly in this category. Mackerel (*Scomber*) vary in density at different times of year because they have mainly lipid stored in the muscles; such a system is evidently a primitive one, for lipid is not stored to maintain a constant density. Rather, the lipid store is at the mercy of metabolic demand, and reduction in density, valuable as it must be, is simply a side effect of the storage of lipid for metabolic purposes.

We know that a wide range of elasmobranchs, including large pelagic sharks, are each of a characteristic density, some being fairly close to neutral buoyancy, others being very dense. Liver lipid is an important factor in determining the density of some species, but in most, it is the density of other tissues (amount of fat in the white muscle, mineralization of the skeleton) which determines the density of the species. Not only do different elasmobranchs store different amounts of lipid in their livers, but the stored oil is least dense in the least dense species. It seems that elasmobranchs have managed to set apart the lipid used for density regulation, whether in the muscles or liver, from metabolic stores, so that the fish can regulate its density to a characteristic figure, whatever metabolic demands are made. How this is done remains to be discovered.

At the beginning of this section, the advantages of neutral buoyancy were extolled, and it seems odd to find fishes using static lift to maintain a fixed density below that of neutral buoyancy. The regulatory mechanisms are there, and at first sight, it seems a harder problem to maintain a constant density (say around 1.03) than to regulate to neutral buoyancy, and what is more, dynamic lift must still be generated, with the penalties that this incurs. However, it is not hard to see why many pelagic sharks use lipid to reduce their density, but not so far as to become neutrally buoyant.

Shark fins vary a good deal in flexibility and in their stiffening by skeletal elements, but the basic design, unlike that of teleosts, makes them impossible to retract; they are not of varying geometry. If we bear this in mind, we can see how considerations of dynamic lift generation lead to different densities in different species, depending on their mode of life. Reduction in density by lipid storage means that less lift need be generated in level swimming, so that the shark can either reduce the size of its pectoral fins, or it can cruise more slowly without stalling. Since sharks use their pectoral fins for manoeuvring (to turn, change in pitch, and even sometimes, as does *Heterodontus*, to creep backwards along the sea bed), there is a limit to reduction in fin size. Reduction in fin area is wholly beneficial (since it reduces drag), but to remain manoeuvrable, the shark must have pectoral fins of a certain size, and below this, no benefit will be gained by reducing density, apart from a slight reduction in vortex drag.

We saw earlier (p. 70) that vortex drag is proportional to $1/V^2$, so that this will be most important at low swimming speeds, and unimportant at high speeds. We should expect from these considerations that sharks which swim fast would be of reduced density, but not neutrally buoyant; those which swim very slowly (for example as part of the feeding pattern) would be close to neutral buoyancy, while those living on the bottom would be very dense. Looking into the matter, this is just what we find. The huge whale-shark and basking shark which sieve plankton while swimming very slowly, are close to neutral buoyancy; the fast pelagic tiger and blue sharks are of reduced density, but are not so close to neutral buoyancy; and the bottom-living dogfish and rays of shallow water have no special arrangements for static lift.

Since teleosts have differently designed fins, which can vary their geometry, the arguments above do not apply, and we should not expect to find any teleosts using lipid to reduce their density to a constant figure not close to neutral buoyancy; so far, none has been found.

## 4.4 Gas as a source of static lift

Fish swimbladders obey Boyle's law, changing volume in proportion to the ambient (hydrostatic) pressure but often retaining a slight positive internal pressure. To be at neutral buoyancy, freshwater fish need a swimbladder occupying about 7% of the body volume while marine fish need a swimbladder occupying 5% of the body volume. Since the ambient pressure increases by 1 atmosphere for every 10 m depth the volume of the swimbladder changes rapidly as fish move up and down, especially near the surface. Sinking from the surface (1 atm) to 10 m (2 atm) causes the swimbladder to halve in volume. To halve the volume for a fish at 200 m (21 atm) would require it to sink to about 400 m (41 atm), so the problems of adjusting the swimbladder to provide neutral buoyancy are much more severe near the surface. Many surface-dwelling pelagic fish have either lost the swimbladder altogether (tunas and some mackerel species) or reduced its importance as in *Scomberomorus*. Similarly, many bottom-dwelling fishes have lost their swimbladders and those that make vertical movements, as we have seen earlier, sometimes make other arrangements. If they retain a gas-filled swimbladder near the surface it will certainly constrain their ability to move vertically, for, if a fish is adapted to a particular depth, it will almost certainly burst its swimbladder if the pressure is reduced substantially (p. 88). The distribution of fishes with

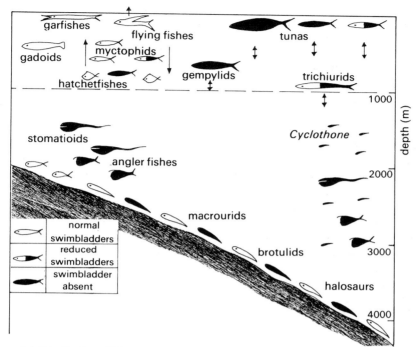

**Figure 4.4** Distribution of fishes with and without swimbladders in the oceans. After Marshall (1960).

gas-filled swimbladders is shown in Figure 4.4. They occur down to the greatest depths. Although the effect of changing depth is much less significant at great depths, the problems of secreting and retaining gas are formidable. Yet the macrourid *Nematonurus armatus* has been caught at depths of 4000–4500 m where the pressure is about 450 atm.

### 4.4.1 Swimbladder structure

This is very variable (Figure 4.5) partly because swimbladders perform other functions in hearing and sound production (p. 229). The swimbladder arises in ontogeny from a diverticulum in the roof of the foregut. In the physostomatous teleosts, the connection with the gut is retained as the pneumatic duct so that the swimbladder lumen is in contact with the environment and gas can be obtained by swallowing air at the surface and passing it down the gut (or ejecting it in the reverse direction). In physoclistous teleosts, the swimbladder may be open for a brief time in ontogeny (allowing the larvae to swallow air to fill the swimbladder for the first time) but in the adult it is closed and is equipped with gas-secretion

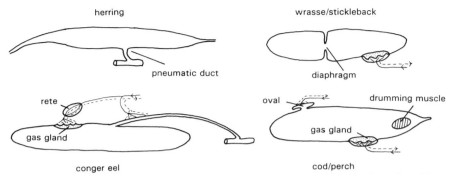

herring

pneumatic duct

wrasse/stickleback

diaphragm

rete

gas gland

conger eel

oval

drumming muscle

gas gland

cod/perch

**Figure 4.5** Swimbladder structure. Left: in two physostomes. Right: in two physoclists. Note that the conger eel is a physostome with a gas gland and rete.

and gas-resorption mechanisms. The physostomatous state is the primitive condition and is found in elopomorphs (eels) and clupeoids.

### 4.4.2 Gas in the swimbladder

The swimbladder is not a static buoyancy chamber. Because the partial pressure of gas in the swimbladder is higher than in the tissues of the fish, gas is continuously lost by diffusion; upward movement needs to be compensated for by gas loss and downward movement by gas gain, if neutral buoyancy is to be maintained. Physostomes have different problems from physoclists. It is difficult for them to remain neutrally buoyant at considerable depths because they cannot swallow sufficient air at the water surface. If they did they would be so bloated with air that they could not swim down anyway! Physoclists must resorb or secrete gas very fast if they are to remain neutrally buoyant following rapid changes of depth, and there is a potential danger of bursting the swimbladder if they move up too fast. Experiments on cod, haddock and saithe show that reducing the pressure to one-third (expanding the swimbladder by three times) causes rupture of the swimbladder. Physoclists thus have a powerful constraint on upward movement under normal conditions. During fishing, however, physoclists caught in deep water are often brought to the surface with burst swimbladders or the viscera protruding through the mouth, blown out by their enormously dilated swimbladders.

Oxygen is the main component of the swimbladder gas of physoclists and, in general, the greater the depth, the higher the percentage of oxygen and the lower the percentage of nitrogen. Clearly, the oxygen has to be secreted into the swimbladder against the concentration gradient; the partial pressure of oxygen may be several tens or even hundreds of atmospheres, compared with 0.2 atm in the tissues and possibly even less in the ambient water.

*Gas secretion.*   Gas enters the swimbladder via blood capillaries that run into a modified area of the inner wall, the gas gland. In actively secreting swimbladders, the surface of the gland is covered with a foamy mucus. Ironically much of the work on gas secretion has been done on the eel, *Anguilla vulgaris*, which has a functional pneumatic duct (and is technically a physostome) but also secretes gas. The gas gland has a glandular cap, which may be extensively folded to increase its surface area, and an associated capillary network, the rete mirabile, working on the well-established countercurrent capillary arrangement found, for example, in the muscles of tuna. The rete consists of thousands of alternately opposed and parallel afferent and efferent capillaries. In the unipolar rete, typical of most physoclists, the capillaries have many hair-pin loops embedded in the glandular cap itself. In the bipolar rete of the eel the afferent and efferent capillaries are apposed in a bundle a short distance from the glandular cap making it possible to cannulate the blood supply entering and leaving the bundle, a very helpful arrangement for physiologists.

How then is oxygen transferred from the blood to the lumen of the swimbladder? This puzzle has exercised the minds of eminent physiologists for many years but the mechanism has only become clearer of late. What is needed is a mechanism that reduces the amount of gas in the venous (efferent) capillaries leaving the rete compared with the arterial (afferent) capillaries entering the rete. This mechanism must be ingenious, for if the partial pressure of the oxygen in the venous capillaries is lower than the arterial capillaries, oxygen might be expected to diffuse from the arterial to the venous capillaries, and the secretory process would never take off. Changes in the properties of the blood, as it passes through the gas gland, should then raise the partial pressure of oxygen but decrease the actual gas content of the blood. Oxygen will then pass into the lumen of the swimbladder and at the same time diffuse from the venous to the arterial capillaries within the rete.

In most fishes we need to seek some change in the blood that will account for a higher partial pressure of oxygen ($Po_2$) in the venous capillaries. Elegant cannulation experiments on shallow-water fish like the eel show that the blood increases in acidity as it passes through the gas gland. Here the cells are rich in glycogen, carbonic anhydrase and lactate dehydrogenase and release lactic acid into the venous capillaries (Figure 4.6). Haemoglobin in the red blood cells unloads its oxygen when acidified by two mechanisms: the Root effect reduces its carrying capacity for oxygen and the Bohr effect its affinity for oxygen. A change of one pH unit can unload 50% of the oxygen even against considerable oxygen partial pressures. Another feature of the secretory mechanism is salting-out in which the solubility of oxygen in the blood plasma is reduced by the increased solute concentration of lactic acid, so releasing further oxygen as gas. Carbonic anhydrase also has a role; it apparently accelerates the

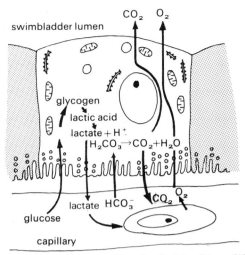

**Figure 4.6** Operation of a gas gland cell. After Fänge (1966).

formation of bicarbonate from excess $CO_2$ which would otherwise buffer the hydrogen ions released by the dissociation of lactic acid.

Although these mechanisms explain how the partial pressure of oxygen is raised to a modest extent in the rete by the changes in blood chemistry, they do not explain how very high partial pressures are achieved. Figure 4.7 shows how a countercurrent multiplication system operates. The parallel and unbranched arrangement of arterial and venous capillaries provides a maximum area for countercurrent diffusion. In the eel, the area for gas exchange is estimated at over $100 \text{ m}^2$. Since the capillaries only contain about $60 \text{ mm}^3$ of blood, the diffusional area is about $1700 \text{ m}^2 \text{ cm}^{-3}$, about 17 times that of the human lung. It is thought that transretial diffusion of lactate is enhanced by the endothelium of the capillaries being very thin. Multiplication of $Po_2$ takes place across the rete as, with a continuous process of lactate secretion, excess oxygen in the efferent capillaries will be added to oxygen already present from previous oxygen enrichment of the afferent capillaries.

It is not surprising that the length and complexity of the retial system increases with the problem of secreting gas against high concentration gradients. The longest retia observed, about 25 mm, were found in the abyssal *Bassozetus taenia* caught at about 5000 m.

Of course other gases are released by salting-out. Nitrogen is secreted, but very slowly. (In freshwater physostomes, with a crude rete, such as *Coregonus*, the main component of the gas is nitrogen.) Carbon dioxide is secreted and mopped up by bicarbonate but also diffuses out of the rete, or indeed through the swimbladder wall, very fast. The transfer of gas from

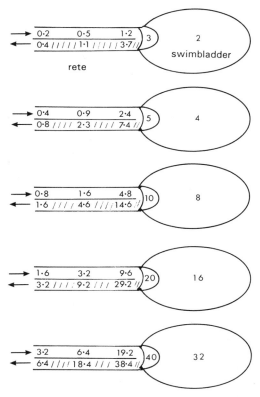

**Figure 4.7** Diagram showing countercurrent multiplication using imaginary values of percentage increase of $Po_2$ above the arterial partial pressure at different positions along the rete, in the glandular gap and in the swimbladder. Hatching shows presence of lactic acid and salting out. Note that the $Po_2$ increases exponentially along the rete and that there is trans-retial diffusion. The $Po_2$ in the swimbladder is below that of the glandular cap so that there will be no reverse diffusion. (Redrawn from Blaxter and Tytler, 1978.)

the blood to the swimbladder lumen via the glandular cap may be enhanced by the presence of $CO_2$ which helps to initiate bubble formation, as do phospholipids which are secreted by the epithelium.

*Gas resorption.* In physostomes, gas can be lost by voiding through the pneumatic duct, and also through an anal duct in some clupeoids (Figure 4.5). In the conger eel, the pneumatic duct is flattened when the fish is in buoyancy equilibrium but if the pressure is reduced the swimbladder gas expands the duct. Since the duct is supplied with a capillary network, gas can then diffuse away via these capillaries; only if the pressure is further reduced does the oesophageal valve open and gas is lost as bubbles through the mouth.

**Table 4.1** Gas secretion and resorption rates (from various authors summarized by Blaxter and Tytler, 1978)

| Species | Secretion rate* | Resorption rate* | Temperature (°C) |
|---|---|---|---|
| *Physoclists* | | | |
| Sunfish | 1.36–1.60 | | 12–32 |
| Saithe | 1.67–2.50 | 7.80 | 9–13 |
| Cod | 1.08–6.42 | 12–36** | 0–15 |
| *Physostomes* | | | |
| Eel | 0.28 | – | 18–20 |
| Goldfish | 0.18–0.48 | – | 29 |

*Expressed in $cm^3$ (STP) $kg^{-1}$ $h^{-1}$.
**Resorption rate pressure-dependent, not temperature-dependent.

In physoclists gas resorption is faster than secretion (Table 4.1) and specialised parts of the swimbladder wall are involved in gas resorption. In perciforms and gadoids, like cod and saithe, there is a vascularised area known as the oval, which can be occluded by a sphincter (Figure 4.5). In the same way, the posterior chamber of the labrid (wrasse) swimbladder is vascularised and can be closed off by a sphincter at the junction with the anterior chamber. When these vascular beds of capillaries are exposed to the swimbladder gases, oxygen passes into the blood because of its high $Po_2$ in the swimbladder. There is some evidence of a shunt mechanism in the gills that allows arterial blood low in oxygen to reach the resorbent capillaries so accelerating the rate of diffusion. Vagal stimulation and injection of adrenaline cause loss of gas by relaxing the sphincter and increasing the capillary circulation; acetylcholine has the reverse effect.

Gas is also being lost continually by diffusion through the swimbladder wall. Experiments on the eel show that carbon dioxide diffuses 40 times faster than nitrogen and oxygen twice as fast as nitrogen. In physostomes, which swallow air, the swimbladder gas will have less than 21% oxygen (its proportion in air) unless the fish has recently visited the surface. The percentage of nitrogen may thus be used to calculate the time of the last visit. Diffusion of gases is reduced by the presence of guanine, the silvery pigment so characteristic of swimbladders, which is distributed in overlapping platelets that increase the path length for gases diffusing through the swimbladder wall. The guanine content of deep-sea fish swimbladders is much higher than surface-living fish. *Synaphobranchus* and *Halosaurus* have 15 times the amount of guanine that eels have and five times that of a herring.

As gas is lost, the swimbladder contracts and the wall thickens so increasing the path length for diffusion. The epithelium of the gas gland is

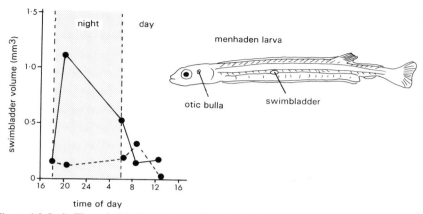

**Figure 4.8** Left: The swimbladder volume of menhaden larvae by night and day with access to water surface (continuous line) and denied access to the surface (dashed line). Right: a 15-mm long menhaden larva showing the swimbladder and otic bulla (p. 226). Redrawn from Hoss and Phonlor (1984).

also impervious to outward diffusion and acts as an additional barrier; it may cover as much as 50% of the internal surface of the swimbladder.

Larval fish have little guanine in the swimbladder and a very high surface:volume ratio, both of which will increase gas loss. Frequent sampling of larval anchovy and menhaden at sea shows that they make daily visits to the surface to replenish their swimbladders (Figure 4.8).

## 4.5 The swimbladder and vertical migration

If physoclistous fishes are placed under increased hydrostatic pressure they secrete gas into their swimbladders and if the pressure is reduced they resorb gas. The same occurs if gas is withdrawn or injected by syringe. Physostomes like herring move rapidly to the surface to swallow air if their swimbladders are artificially emptied or if they have been denied access to the surface for some time. The way in which most fishes monitor the degree of distension in the swimbladder is not known although stretch receptors have been reported in the swimbladder wall of roach (*Leuciscus rutilus*) and rudd (*Scardinius erythrophthalmus*).

Experiments involving cardiac conditioning (p. 297) and observing spontaneous changes in pressure show that fishes can respond to very small changes of pressure even without a swimbladder (Table 4.2). Usually such pressure sensitivity is expressed as the percentage change perceived because this best expresses the degree of change of volume of the

**Table 4.2** Pressure sensitivity thresholds of various species determined experimentally and expressed as the minimum percentage pressure change perceived (from various sources summarized by Blaxter and Tytler, 1978)

| Species | Threshold (%) | Experimental technique |
|---------|---------------|------------------------|
| Minnow | 0.05–0.1 | Operant conditioning |
| Minnow | 0.5 | Spontaneous behaviour |
| Perch | 0.1–0.2 | Spontaneous behaviour |
| Pinfish | 0.02 | Yawning behaviour |
| Cod, saithe | 0.5 | Cardiac conditioning |
| Plaice*, dab* | 1–2 | Cardiac conditioning |

*Flatfish with no swimbladders.

swimbladder. A threshold of 0.1% near the surface means that a fish can perceive a change of depth of 1 cm or at 90 m a depth change of 10 cm.

The swimbladder might be expected to act as a depth gauge were it not for the fact that it is in a dynamic state and is responding to changes of depth by secreting or resorbing gas. It can act as a depth gauge in the short term but in the long term it never act as an absolute depth gauge because the gas secretion/resorption mechanisms act to retain the fish at neutral buoyancy.

Fishes not only move up and down short distances as part of their normal daily routine of feeding and avoiding predators, they also make diel vertical migrations (p. 306), generally moving towards the surface at dusk and towards the bottom at dawn. Do these fishes maintain neutral buoyancy during these migrations? The answer is almost certainly no! Cod and saithe are two gadid species known to make diel vertical migrations. Cod can secrete gas at a sufficient rate to allow them to move down at about 1 m h$^{-1}$ (dependent on the ambient temperature) or can resorb gas to move up at 2.4 m h$^{-1}$ and still remain at neutral buoyancy (Table 4.1). Saithe need 24 h to reach neutral buoyancy if the pressure is doubled and 48 h if it is quadrupled (moving from the surface to 10 m (2 atm) or 30 m (4 atm), respectively). Resorption, as in cod, is much faster for they require only 5 h to reach neutral buoyancy if the pressure is halved (e.g. moving from 10 m to the surface). The perch, a freshwater vertical migrant, requires 24 h to adapt to an increase of pressure from 1 to 2 atm and about 9 h to adapt to a decrease from 2 to 1 atm. In these species the speed of vertical migration far exceeds the possibility of remaining at neutral buoyancy during the ascent, let alone the descent. What almost certainly happens is that the buoyancy of the fishes like cod and saithe always lags behind the optimum. These fish may reach neutral buoyancy after a long spell near the surface during the night (remember that gas resorption is faster than secretion). As they move down at dawn to the daytime depth, they will become negatively buoyant and may not reach neutral buoyancy by the following dusk. This means that they can move up a greater distance without bursting the swimbladder than if they had reached neutral buoyancy at the daytime depth.

## 4.6 The swimbladder as a dynamic organ

Because the swimbladder is changing its volume (and no doubt sometimes its internal gas pressure) its other functions must be affected. The volume influences the resonance frequency and so its role in hearing and sound production (p. 229). The swimbladder is also the main organ returning the echoes from commercial echo-sounders, the larger the swimbladder the greater the echo. Thus the 'target strength' of fish changes as they move vertically and this has all sorts of implications for the estimation of fish biomass by acoustic methods.

## Bibliography

Alexander, R. McN. (1966) Physical aspects of swimbladder function. *Biological Reviews*, **41**, 141–176.

Blaxter, J.H.S. and Ehrlich, K.F. (1974) Changes in behaviour during starvation of herring and plaice larvae. In *The Early Life History of Fish*. (Blaxter, J.H.S. ed) Springer, Berlin pp. 575–588.

Blaxter, J.H.S. and Tytler, P. (1978) Physiology and function of the swimbladder. *Advances in Comparative Physiology and Biochemistry*, **7**, 311–367.

Bone, Q. (1973) A note on the buoyancy of some lantern fishes (Myctophidae). *Journal of the Marine Biological Association of the United Kingdom*, **53**, 619–633.

Butler, J.L. and Pearcey, W.G. (1972) Swimbladder morphology and specific gravity of myctophids off Oregon. *Journal of the Fisheries Research Board of Canada*, **29**, 1145–1150.

Denton, E.J. (1961) The buoyancy of fish and cephalopods. *Progress in Biophysics and Biophysical Chemistry*, **11**, 178–234.

Denton, E.J. and Marshall, N.B. (1958) The buoyancy of bathypelagic fishes without a gas-filled swimbladder. *Journal of the Marine Biological Association of the United Kingdom*, **37**, 753–767.

Fänge, R. (1966) Physiology of the swimbladder. *Physiological Reviews*, **46**, 299–322.

Hoss, D.E. and Phonlor, G. (1984) Field and laboratory observations on diurnal swimbladder inflation – deflation in larvae of gulf menhaden *Brevoortia patronus*. *Fishery Bulletin of the United States*, **82**, 513–517.

Marshall, N.B. (1960) Swimbladder structure of deep-sea fishes in relation to their systematics and biology. *Discovery Reports*, **31**, 1–122.

# 5   Gas exchange and the circulatory system

Fishes show an interesting diversity of approach to the problem of acquiring and transporting oxygen to the tissues. Even though most fishes obtain oxygen from the water, they do so using gills (and sometimes other surfaces) of varied designs, causing water to flow over them in a variety of ways. There are also air-breathing species in no less than 70 freshwater genera; some of these have rather curious methods of aerial gas exchange, involving unexpected structures like the body scales and the hindgut. Many phyletically ancient freshwater fishes breathe air, like *Amia*, *Polypterus* and *Lepisosteus*, as well as lungfish, testifying to its adaptive advantage in some freshwater environments which contain little oxygen or are at risk of drying up.

Compared with air, water contains relatively little oxygen. At the sea surface for example, air-saturated water at 20°C contains only around 3% of the oxygen in the same volume of air. Since oxygen only crosses cell surfaces by diffusion, this means that the gas exchanger to acquire oxygen from the water must have a large area. In the active menhaden (*Brevoortia*), for example, the gill area is over 18 times that of the body surface excluding the fins. Also, because water is dense and viscous, a relatively large expenditure of energy is needed to force water to flow around the gas exchanger. Unlike most terrestrial animals (including ourselves), which breathe air in and out of a lung in tidal fashion, water flow over the fish gill is almost invariably unidirectional, whether it is pumped over the gills or simply flows over the gill via the open mouth as the fish swims forwards. Adult lampreys are the only fishes where flow over the gills is tidal.

We can easily see why unidirectional flow of water is advantageous, for it means that the blood flow in the gill can be arranged in the opposite direction. If blood flow through the gas exchanger was in the same direction as the water flow, the maximum partial pressure ($Po_2$) of oxygen in the blood of the exchanger would be the same as that in the exhalent water stream. On the other hand, if the blood flows in the opposite direction to the water, the maximum $Po_2$ of the blood leaving the exchanger could be very close to that of the incoming (ambient) water, and above that of the exhaled water. Similar counter current systems (between arterial and venous blood vessels) in the muscles of warm fish and in the swimbladder rete are described in Chapters 3 and 4. In gnathostome fishes, 70–80% of the oxygen in the water flowing into the gill chambers can be extracted, a remarkable degree of efficiency. The blood in the gas exchanger is almost always very different in osmolarity from the water, and always different in ion content, but it must be in intimate contact with the water to make the diffusion pathway short. So we might expect to find special arrangements to circumvent (or at least to cope with) water and ion fluxes across the exchanger; some possible mechanisms will be considered in Section 5.4.2.

## 5.1 The origin of respiratory gills

In ascidian tunicates and in amphioxus, the gills are ciliated food-collecting devices, trapping particles on mucous nets produced by the endostyle; blood flowing through them is probably de-oxygenated since the ciliary tracts of the gill bars must use more oxygen than is provided by the water flowing through them. The respiratory gills of larger and more complex chordates are likely to have been derived from a filtering arrangement of the kind seen in amphioxus, the significant step being the change from ciliary to muscular movement of water through the gill. This change, which led to the arrangement seen today in the lamprey ammocoete larva, presumably came about as a consequence of the demand for a higher filtering rate than cilia alone could provide. When this more efficient filtering system allowed increase in body size beyond that where simple diffusion across epithelial surfaces sufficed for gas exchange, respiratory gills became specialised. In lampreys, a significant proportion of the oxygen needed is still gained across the skin, despite the development of respiratory gills. Cutaneous respiration is the only source of oxygen for many larval fishes, and is important for some adult teleosts, for example, the Antarctic fishes lacking haemoglobin (Section 5.6).

## 5.2 Respiration of fish larvae

Elasmobranchs and teleosts with large eggs such as salmon, hatch with functional gills, a well-developed circulatory system and blood cells containing haemoglobin. Most teleosts, however, hatch as much smaller larvae depending on cutaneous respiration across the body surface. Since many of these small transparent larvae live a pelagic existence, where oxygen is plentiful, cutaneous respiration is adequate, so haemoglobin is not needed and might make them conspicuous to predators. Leptocephalus larvae, for example, may be quite large, but are laterally compressed (so diffusion distances are small), and lacking haemoglobin, are exceptionally transparent.

As larvae increase in size, two changes take place that affect respiration

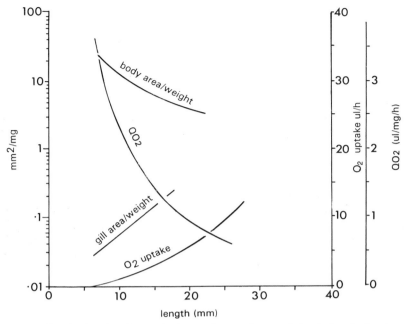

**Figure 5.1** Graphs showing changes in the respiratory characteristics during the development of a plaice larva. As the body area falls in relation to body weight the cutaneous area for respiration also falls but gill area per unit weight increases. Although the total oxygen uptake increases with length, as would be expected, the $Q_{O_2}$ (oxygen uptake per unit weight) falls, a phenomenon found in all animals.

profoundly: the surface-to-volume ratio becomes smaller, so that the surface for cutaneous respiration becomes relatively smaller, and the pathways for the diffusion of gases and metabolites become longer (Figure 5.1). A size is reached when gill respiration (vastly increasing the area for diffusion) becomes essential, especially in very active fishes. The development of haemoglobin about the same time, or later, increases the oxygen-carrying capacity of the blood. The chemical form of haemoglobin then changes with age, as judged by electrophoretic banding patterns.

Larger salmonid alevins at hatching in the spawning redd live in relatively safe conditions of low oxygen, making gills and haemoglobin a *sine qua non* for survival. The same is true of some riverine species from the Amazon whose small larvae have haemoglobin at hatching. The main imperative, however, is size and the need for an increased surface area for respiration as growth proceeds.

A stationary larva in the same mass of water tends to deplete the available oxygen, hence larvae need to swim out of this ambient water to enter a new oxygenated region. Northern anchovy larvae increase the

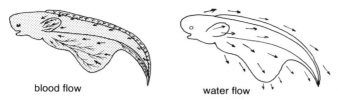

blood flow                          water flow

**Figure 5.2** Countercurrent water and blood flow in the larva of the freshwater symbranchiform teleost *Monopterus*. The larvae live in oxygen-poor water and use the whole body surface to gain oxygen; the current of water flowing over the body surface produced by the movement of the pectoral fins, is countercurrent to skin blood flow. After Liem (1981) *Science*, **211**, 1177.

frequency and duration of swimming bouts in water with oxygen levels below 60% saturation. Larvae living in hypoxic environments may also generate convective water flow along the body, for example in the lungfish *Neoceratodus forsteri* by means of cilia or by movement of the pectoral fins in *Monopterus albus* (Figure 5.2). In Atlantic salmon alevins the pectoral fin movements seem to draw water over the gills.

## 5.3 Respiration in hagfish and lampreys

### 5.3.1 Hagfish

In hagfishes, unidirectional water flow through the serial muscular gill pouches is chiefly brought about by rolling and unrolling of velar folds (Figure 5.3). These lie in a chamber developed from the naso-hypophyseal

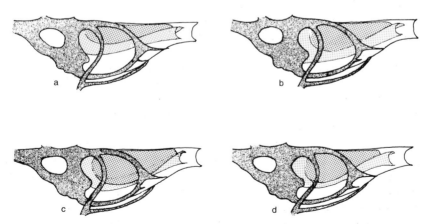

**Figure 5.3** Velar folds of the hagfish *Myxine* showing stages (a–d) in the pumping cycle. Cranial cartilages shaded, velar folds stippled. After Strahan (1963) *The Biology of Myxine*, Universitetsforlaget, Oslo.

tract and are operated by a complex set of muscles inserting onto cartilages of the neurocranium. Peristaltic contractions of the gill pouches and their ducts assist in producing the flow. In *Myxine*, the gill pouches open by a common duct, whilst in *Eptatretus* 5–16 gill pouches (according to species) open directly to the exterior.

Since hagfishes feed half-buried in their prey, and since they survive well after the nostrils have been blocked to interrupt gill ventilation, it is clear that cutaneous respiration is important. Indeed, this is indicated by the vast subcutaneous blood sinuses found in hagfishes; the skin is like a loosely fitting sock and if one holds up a living hagfish, blood flows down to swell the lower end. Blood volume is over twice that of gnathostome fishes, and much greater than that of lampreys. Ingenious measurements of resting oxygen consumption by hagfish on the sea bed at 1230 m (using a respirometer mounted on a remotely-controlled vehicle) gave average values of only 3.1 $\mu$g g$^{-1}$ h$^{-1}$. So cutaneous respiration may well suffice for hagfishes at rest, in *Myxine* the water pumped by the velar folds bypassing the gills.

### 5.3.2 Lampreys

In the ammocoete larva (Figure 5.4), the pharynx is undivided, and unlike the adult, the larva filter-feeds and respires from the same unidirectional inhalent water flow. This is driven partly by the action of the anterior muscular velum, and partly by contractions of the branchial basket brought about by gill muscles. The branchial basket is a continuous cartilaginous meshwork, rather than being jointed, and expands by its elasticity. Valves at the entrance and exit of the gill pouches ensure unidirectional flow during the rhythmic movements of the branchial basket. The possibility for countercurrent flow exists in the ammocoete gill, but this has not yet been demonstrated experimentally, although oxygen extraction rates are about double those of adults where the water flow is tidal.

In the adult, where the mouth and sucker are involved in feeding, the velum is not involved in pumping water (it remains after metamorphosis as a small flap valve), and contraction and expansion of the branchial skeleton pumps water in and out of the gill pouches. The direction of flow is controlled by valves, and is tidal. It obviously has to be tidal when the lamprey is feeding, or moving pebbles with its sucker as it makes its redd to spawn, but it always seems to be tidal even if the lamprey is free-swimming and could inhale through the mouth.

We should expect therefore, that oxygen extraction would be relatively inefficient in adult lampreys, and instead of achieving gnathostome fish values, *Entosphenus* can only manage to extract between 10 and 28% of the oxygen in the inhaled water. Another way of looking at respiratory efficiency is to consider it as the ratio between the amount of oxygen

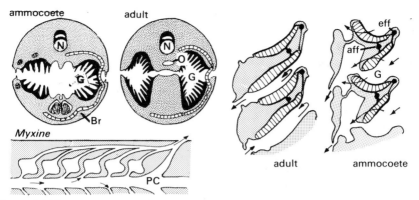

**Figure 5.4** The arrangement of the gills in hagfish and lampreys. Upper left: transverse sections of branchial region of ammocoete and adult lamprey. Right: horizontal sections of left side of branchial region in adult and ammocoete, showing a tidal water flow in the adult gill sac. Br: branchial skeleton; G: gill sac; N: notochord; o: oesophagus. In the ammocoete, the efferents (eff) and afferent (aff) vessels are arranged to permit counter current flow. Bottom left: Horizontal section of right gill sacs in *Myxine* (anterior to left) showing pharyngo-cutaneous duct (PC) and common outflow. After Alcock (1898) *J. Anat. Physiol.*, **33**, 131; Sterba (1966) *Freshwater Fishes of the World*, Studio Vista, London, and Goodrich (1909).

acquired at the gas exchanger available for metabolic purposes, and that used by the respiratory muscles themselves. Lampreys have a relatively large amount of branchial musculature, so here again, respiratory efficiency seems likely to be low.

## 5.4 Gnathostome fishes

### 5.4.1 Gill design

Gill structure is essentially similar in all gnathostome fishes. It is true that elasmobranch gills differ from most teleost gills in the way that the gill filaments remain attached along their length (hence their name, see Chapter 1), but the basic design is the same. There are usually four branchial arches bearing gills in teleosts (the number is reduced in air-breathing fishes like *Anabas* or *Amphipnous*), but in elasmobranchio-morpha and chondrosteans, the hyoid arch bears a posterior respiratory hemibranch, and there are thus usually five gill-bearing arches in sharks, and up to seven in the shark *Heptranchias*. Each arch bears a series of regular comb-like gill filaments, supported by skeletal gill bars, and on each of these there are closely-ranged primary gill lamellae. These in turn bear a number of stacks of smaller secondary lamellae set parallel to the

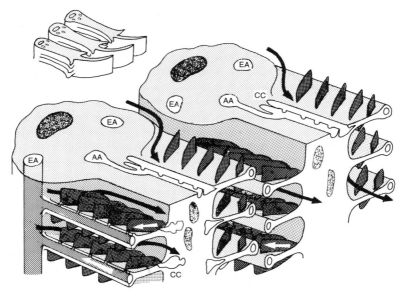

**Figure 5.5** Design of the elasmobranch gill as seen in the dogfish *Scyliorhinus*. Upper left: general view of three gills and their flaps. Main diagram: part of two adjacent gill arches showing alternation of secondary lamellae (dark stipple) on adjacent gill filaments. Afferent blood from the afferent branchial arteries (AA) passes along afferent arterioles to the corpora cavernosa (CC) and thence to the secondary lamellae and to the efferent branchial artery (EA). Note that the direction of water flow (arrows) is counter to the flow of blood in the secondary lamellae. After Wright (1973) *Z. Zellforsch.*, **144**, 489.

water flow through the branchial chamber: these are the sites of gas exchange. Figures 5.5 and 5.6 show the arrangement in elasmobranchs and teleosts, and the way water flows through the gills.

The numbers and dimensions of the secondary lamellae vary between different species according to their activity; typically, in a fish weighing 1 kg there may be up to 18000 cm$^2$ of secondary lamellae, and in very active fishes like tunas, there may be more than 5 million secondary lamellae! This huge area is needed partly because the oxygen content of water is low, and partly because rates of oxygen diffusion are relatively low in animal tissues. In connective tissue for example, the oxygen diffusion rate is only 10$^{-6}$ that in air. It is no surprise then, that the secondary lamellae have very thin walls to minimise the diffusion distance between the water and the blood within them. In fact, they are essentially thin-walled sacs filled with blood flowing around the pillar cell posts that separate the lamellar walls (Figure 5.6). Blood in the secondary lamella therefore flows through an interrupted sinus, rather than through capillaries. In some fishes, like tunas and the bowfin *Amia*, the pillar cells

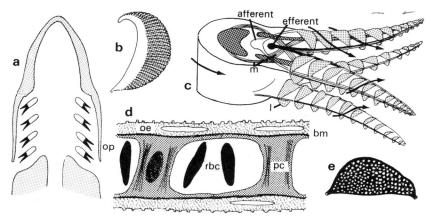

**Figure 5.6** Design of the teleost gill. (a) Horizontal section showing disposition of gill filaments (black) on gill arches; (b) single hemibranch; (c) arrangement of gill filaments with secondary lamellae (light stipple) supplied by efferent and afferent vessels. Water flows (arrows) over lamellae countercurrent to blood flow through them. The gill arch and gill rays (dark stipple) are linked by intrinsic muscles (m) which can change the apposition of the gill filaments (unlike the elasmobranch gill, Figure 5.5). (d) Section across secondary lamella showing pillar cells (pc), red blood cells (rbc) in blood space, and basement membrane (bm) separating pillar cells from outer epithelium (oe) with 'lymph' spaces. (e) Cast of vascular spaces in secondary lamella (the white dots are where the blood space is interrupted by pillar cells). After Munshi and Singh (1968) *J. Zool. (Lond.)*, **154**, 365, and Hughes and Grimstone (1965) *Quart. J. micr. Sci.*, **106**, 343.

are not distributed polygonally, but in discrete rows, so that although the secondary lamella is a sinus, the pillar cell array effectively divides the interior into a series of parallel channels. As we have seen earlier, to enable efficient oxygen extraction, the flow within the secondary lamellae is countercurrent to that of the water passing them, and the tuna arrangement ensures that the flow of blood is exactly parallel to the water flow.

The lining of the sinus is formed by flanges from the pillar cells, so that to reach the blood in the lamella, oxygen in the water has first to pass across an epithelial cell layer, then across the basement membrane of the epithelial cells, and lastly, across the pillar cell flanges lining the lamella. This diffusion barrier differs in thickness in different fishes, and as we might expect, it is thinnest in those fish requiring the greatest rate of oxygen uptake (Table 5.1). Apart from the specialised pillar cells (which are already present in lamprey gills), the arrangement is rather similar to that in lungfish or tetrapod lungs, but there are complications arising from the intimate proximity of the blood to the water flowing by the secondary lamellae.

The large area of the gas exchanger, and the difference in composition

**Table 5.1** Diffusion distances between water and blood in the secondary lamellae of different fishes. Distances in μm. Note minimum distances in the active pelagic skipjack and mackerel, as compared with the other benthic fishes. From Hughes and Morgan (1973)

| Fish | Epithelium | Basement membrane | Pillar cell flange | Total water–blood (mean) |
|------|-----------|-------------------|--------------------|--------------------------|
| Dogfish | | | | |
| (*S. canicula*) | 2.38–18.48 | 0.3–0.95 | 0.37–0.71 | 11.27 |
| *Squalus* | 3.0–22.5 | 0.3–0.6 | 0.12–0.6 | 10.14 |
| *Raja clavata* | 0.5–11.5 | 0.13–0.63 | 0.03–1.13 | 5.99 |
| *Microstomus kitt* | 0.21–16.7 | 0.1–0.69 | 0.1–0.13 | 3.23 |
| Skipjack | | | | |
| (*Katsuwonus*) | 0.013–0.625 | 0.075–1.875 | 0.017–0.375 | 0.598 |
| Mackerel | | | | |
| (*S. scombrus*) | 0.165–1.875 | 0.066–1.0 | 0.033–1.75 | 1.215 |

between the blood and the water mean that the fish possesses a structure that will inevitably not only act as a gas exchanger, but also as an efficient heat, ion and water exchanger. Thermal diffusion is much more rapid than gaseous diffusion, and so fishes can only retain metabolic heat by organising special countercurrent heat exchangers near the organs that are to be kept warm (Chapter 3). What about osmotic and ion exchange? The mechanisms for ion and water exchange in the gills are considered in Chapter 6; here we are concerned with the possibility that ion and water exchange might be minimised when oxygen demand is low, by avoiding so far as possible the intimate blood/water contact in the gills, i.e. by reducing functional gill area.

## 5.4.2 Functional gill area

In principle, functional gill area could be reduced without much difficulty by re-routing blood flow in the gills to non-lamellar pathways, by reducing and re-routing flow within the secondary lamellae themselves, and by altering the water flow past them. Such mechanisms would seem to be sensible, since they would avoid the cost of running the branchial ion and water pumps at maximum levels when oxygen demand is low, i.e. when the fish is at rest or swimming very slowly.

What evidence is there for changes in functional gill area, and how significant might they be? Exercise (i.e. increased oxygen demand) in trout is followed by increased urine production to get rid of the increased entry of water across the gills, and in lampreys, activity is well correlated with urine production. This certainly suggests that functional gill area is related to oxygen demand, but other explanations are possible. In fishes like the eel or dogfish (*Scyliorhinus*) there are anatomical connections between the efferent and afferent arteries in the gill filaments, and other links between the afferent arteries and the central venous space of the gill filaments. Thus, in principle, some proportion of the blood could be shunted via these links to a 'non-respiratory' route, bypassing the 'respiratory' route through the secondary lamellae when oxygen demand is low. But considerations of the dimensions of these links, and calculations of the pressures in the different vascular spaces, has made it unclear whether fishes actually have such a switchable double circulation in the gills. However, direct evidence that in eels at least, this is the case, was provided by ingenious experiments (Figure 5.7) where cardiac output ($Q$) was measured directly by a flowmeter in an external extension fitted to the ventral aorta, and compared to the output calculated by the Fick principle:

$$Q_F = \frac{V_{O_2}}{(C_{aO_2} - C_{vO_2})}$$

i.e. cardiac output ($Q_F$) = oxygen uptake at the gills ($V_{O_2}$) divided by oxygen content difference between arterial and venous mixed blood ($C_{aO_2} - C_{vO_2}$). The oxygen uptake at the gills was calculated from measurements

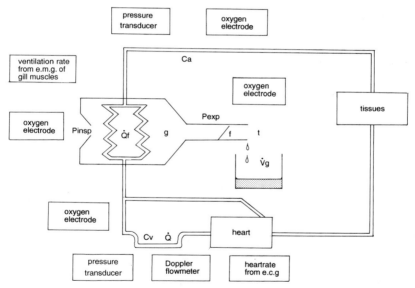

**Figure 5.7** Diagrammatic scheme of the complex experimental arrangement required to study respiratory 'shunting' in the eel *Anguilla*. Water flowing over the gills (g) was collected from tubes (t) in the round opercular openings, which were provided with rubber flaps (f) to mimic the normal opercular valves. The ventral aortic ciculation was extended outside the body to permit blood flow velocity from the heart to be measured accurately by a Doppler flowmeter. Heart rate was monitored by ECG electrodes, and other electrodes near the opercular monitored ventilation frequency. Pressure transducers monitored dorsal and ventral aorta blood pressures, and oxygen electrodes $P_{O_2}$ in these vessels and in the inspired and expired water flowing over the gills. After Hughes *et al.* (1982) *J. exp. Biol.*, **98**, 277.

of ventilatory water flow and the inspired–expired $Po_2$ difference. The result of this experiment was that $Q_F/Q = 0.72$, indicating that in the resting eel about 30% of the mixed venous blood afferent to the gills returns directly to the heart, bypassing the lamellar 'respiratory' route. After injection of adrenaline, $Q_F/Q$ changed to near unity, so it seems that the vascular shunts of the eel gill have adrenergic sphincters, and that circulating catecholamines like adrenaline will fully 'open' the 'respiratory' route as the sphincters are closed. In eels, injection of adrenaline increases arterial $Po_2$, as expected. However, it is not yet clear that this result applies to all fishes. An interesting and rather different possibility for changing the functional surface area of the secondary lamellae is raised by the structure of the pillar cells in the lamellae themselves. These have a ring of connective-tissue supporting columns (Figure 5.8), but also contain arrays of myosin filaments, and the sinuous shapes of the connective-tissue columns in electron micrographs of fixed gills suggest that pillar cells are contractile, shortening to reduce or re-route blood flow through the lamellae. Since they are not innervated, the consensus of opinion now is that they form part of an auto-regulative system in the lamellae, responding to rapid increases of blood pressure by increasing tonus to prevent lamellar swelling and rupture.

In most teleosts, but not in elasmobranchs, intrinsic muscles in the gill filaments can change the angles adopted by the gill filaments on each arch, and so alter the ventilation pattern of the secondary lamellae, but (as might be expected) it is hard to discover whether such changes in gill geometry are significant in changing functional gill area in normal fishes.

We have to conclude that although there is evidence for several kinds of

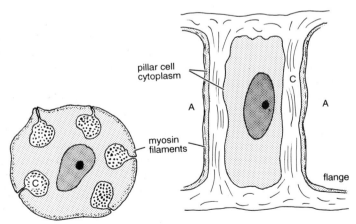

**Figure 5.8** Sagittal and transverse A–A section of pillar cell from secondary lamella. Note *extracellular* connective tissue columns (C) within the cells, and cortical myosin filament zones. After Laurent (1984) in *Fish Physiology* (Hoar and Randall eds), **X**, part A, 73.

mechanism that could change the functional area of fish gills in response to oxygen demand, it is still not known how these are interrelated, nor, indeed, how important they may be in the life of the fish. We have considered them in relation to possible limitation of ionic and osmotic exchange, and a significant part of the energy budget of the fish has to be devoted to the operation of pumps to cope with the exchange problem (Chapter 6). But fish gills offer a significant resistance to blood flow, and the changes in blood and water flow pathways in the gills we have been considering as mechanisms for reducing ion and water exchange, may be equally important in reducing the load on the heart when oxygen demand is low.

### 5.4.3 Branchial pumps

In both teleost and elasmobranchs, the gills are ventilated by water driven across the gill chamber by double pumps: a pressure pump upstream to the gill resistance, and a suction pump downstream. In fishes which swim continually, like scombroids or lamnid sharks, the pumps are found but they are not in use above a certain swimming speed, since forward motion provides sufficient ram gill ventilation. It is, of course, difficult to see how these pumps operate directly (although one can see their result as water is expelled from the branchial chamber), but by a combination of pressure records from strategically-placed cannulae, strain gauge records of the movements of mouth and gill openings, and EMGs (Chapter 3) from different muscles, it is possible to glean a fairly complete picture of the way that the pumps operate. With these techniques, Hughes and his colleagues (1965) have examined various teleosts and elasmobranchs; their work on dogfish is an example of the approach.

Figure 5.9 shows the main muscles and skeletal structures involved, whilst Figure 5.10 shows the relationships of the pressures measured during the three phases of the cycle, with the activity of the different muscles. First, when most of the respiratory muscles are active, the volume of the orobranchial chamber is reduced, and pressure within it rises. After an initial increase (as water flows out of the orobranchial cavity) the parabranchial cavities also decrease in volume, and pressure there also rises. Next, pressure within the orobranchial cavity drops as it passively expands, owing to the elasticity of the skeletal and ligamentous elements compressed during inspiration; no muscles are active in this phase. Finally, there is a short pause, before the cycle begins again, and this may be preceded by a more rapid expansion of the orobranchial cavity at the end of the second phase, during which the hypobranchial musculature may be active.

As we should expect from the anatomical arrangements shown diagram-matically in Figure 5.9, there are many interactions between the different

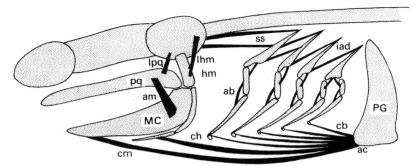

**Figure 5.9** Skeletal structures and muscles involved in the branchial pumps of the dogfish *Scyliorhinus*. ab: adductor branchialis; ac: arcualis communis; am: adductor mandibulae; ch: coraco-hyoideus; cm: coraco-mandibularis; hm: hyomandibula; iad: interarcualis dorsalis; lhm: levator palatoquadrati; MC: Meckel's cartilage; PG: pectoral girdle; pq: palatoquadrate. After Hughes and Ballantijn (1965).

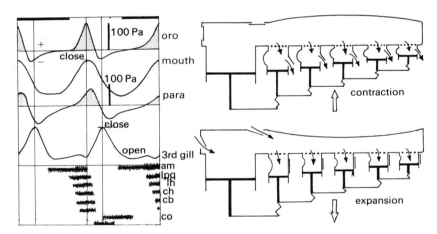

**Figure 5.10** The branchial pumps of *Scyliorhinus*. Left, upper: orobranchial (oro) and parabranchial (para) pressures (above ambient stippled) and the movements of the mouth and third gill slit. Below: the periods of activity (EMG records) of the different muscles driving the pumps during the respiratory cycle (abbreviations as for Figure 5.9). Right: model of the branchial pumps in expiratory (upper) and intake phases. After Hughes and Ballantijn (1965).

parts of the system, and the two pumps are not completely separate. For example, the muscles driving the orobranchial pump also affect the parabranchial pump. Figure 5.10 illustrates diagrammatically a model incorporating some of these interactions. Its most important feature is that by using the interaction between an upstream pressure pump and a downstream suction pump, the dogfish can maintain unidirectional flow over the gills to enable countercurrent water and blood flow in the secondary lamellae.

In teleosts, this dual pump is essentially similar to that in dogfish, but since the skin is stiffer and there is normally a rigid operculum, coupling of the two pumps is closer than in dogfish and both expansion and contraction phases of the pumps are active. The relative contributions of the pressure and suction pumps to gill ventilation differ in different teleosts; in bottom-living teleosts like plaice and sole the opercular suction pump is most important. Increased gill ventilation when ambient $Po_2$ falls or when oxygen demand rises can be brought about by increasing cycle frequency or increasing the stroke volume for each cycle; fishes may do either or both. In trout, for example, cycle frequency alters little, but the volume of water pumped may increase up to five times. An interesting puzzle is provided by the way that the respiratory muscles are controlled, for although it is known physiologically (in trout) that receptors detecting low ambient $Po_2$ are located on the dorsal part of the first gill arch, $Po_2$ receptors are also present in the pseudobranch. Neither have been identified histologically, nor have the length and tension receptors known to be associated with the muscles themselves.

## 5.4.4 Ram ventilation

Many fishes cease respiratory pumping as soon as they are swimming fast enough to ventilate the gills simply by keeping their mouths open, allowing water to flow into the gill chambers. The change takes place when the pressure difference across the gills reaches 2 kPa. Some, like the larger fast-cruising scombroids, can only respire by this ram-jet method, and in skipjack tuna (*Katsuwonus*) the dynamic pressure difference required is around 80 Pa. This is similar to the pressures (50–100 Pa) developed by resting fishes using the branchial pumps to ventilate the gills, but ram-jet ventilation is less energetically costly. For example, a striped bass (*Morone saxatilis*) swimming in a respirometer at 30 cm s$^{-1}$ (1.35 body lengths s$^{-1}$) used 322 mg $O_2$ h$^{-1}$, and used its branchial pumps to respire. At 55 cm s$^{-1}$ (2.47 body lengths s$^{-1}$) when it had switched to ram ventilation, it used 360 mg $O_2$ h$^{-1}$. Although swimming speed had increased over 80%, the $O_2$ used only increased by just under 12%! Estimates for the cost of ram ventilation (where the fish does not have to accelerate and decelerate volumes of dense water) vs. branchial pumping suggest that ram ventilation requires about 9% of the total energy budget, whereas branchial pumping requires nearly double this, about 15%. However, these estimates did not take into account a second advantage of ram ventilation. Not only does it avoid the cost of branchial pumping, but the steady exit of water from the opercula during ram ventilation also helps to

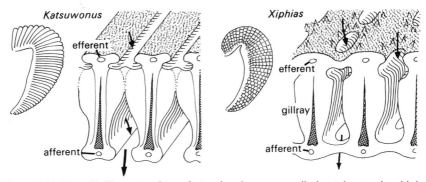

**Figure 5.11** The gill filaments of two fast swimming ram-ventilating teleosts, the skipjack tuna (*Katsuwonus*) (left), and the swordfish (*Xiphias*) (right). Arrows show water flow. Hemibranchs of first gill arches inset. After Muir and Kendall (1968) *Copeia*, **2**, 388.

maintain the boundary layer (Chapter 3), producing a better flow regime with less drag. It is partly this additional advantage that accounts for the unexpectedly small increase in oxygen consumption as the bass were forced to swim more rapidly.

In the special case of remoras (echeneids) which live attached by the dorsal sucker to larger fishes, ram ventilation is free! Obviously, when using ram ventilation, a greater flow rate over the gills can easily be obtained by increasing mouth gape, but this will greatly increase drag. Probably pseudobranch baroreceptors monitor the entry pressure over the gills in order to keep mouth gape as small as possible at different swimming speeds.

Tuna and swordfish gills do not look like those shown in Figure 5.4, because the gill filaments are linked by a series of bridges and the filaments may be fused at their edges (Figure 5.11). A somewhat similar design is seen in the holostean bowfin *Amia*. *Amia* is a lurking predator in freshwaters of North-East America. It would hardly be possible to think of a fish more different in taxonomic position and lifestyle from tunas and swordfish! Tuna and swordfish gills often show damaged and regenerating areas, and it seems clear that the fused design here is to strengthen the gill sieve against damage by floating objects in the rapid inhalent flow. In the sluggish air-breathing *Amia* on the other hand, fusion may have evolved to keep the gill sieve patent in air to assist the swimbladder in gas exchange.

### 5.4.5 Air-breathing fishes

*Amia* is unusual in being an air-breather living in temperate waters, for most of the extremely fascinating variety of air-breathing fishes live in tropical swamps, where a combination of stagnant water, high temperature and abundant microorganisms make the water very acid, with high $P\text{CO}_2$ and low $P\text{O}_2$. These unfavourable conditions for aquatic respiration have led to extraordinary adaptations for acquiring oxygen, also found in fishes which normally live out of the water. In tropical swamps, some fish manage by ventilating the gills with water from just below the surface, where it is oxygenated, but most have to use accessory respiratory organs of various kinds. These are essentially hollow spaces with richly vascularised walls, which can be ventilated periodically. Many air-breathing fishes use the swimbladder as a lung, as well as for buoyancy, and sometimes for sound production (Chapters 4 and 10), and its surface area is much increased by septation (see Figure 5.14 below). *Polypterus* and the African (*Protopterus*) and South American (*Lepidosiren*) lungfishes all have single or double 'lungs', they are obligate air-breathers and drown if denied access to the surface. The Australian lungfish (*Neoceratodus*) only breathes air if stressed, and has a single 'lung' much less septated than those of other lungfishes.

With the exception of the tarpon *Megalops* (whose juveniles live in fresh and brackish water), all teleosts with respiratory swimbladders live in fresh water. Remarkable modifications are found in osteoglossomorphs like the obligate air-breather *Pantodon* where extensions of the swimbladder penetrate the transverse processes of the vertebrae and in *Notopterus* where there are many ventral finger-like extensions. Air is exchanged in such respiratory swimbladders by a variety of methods, the most curious being that in *Polypterus* where the elastic recoil of the scales deformed by the decrease in swimbladder volume caused by intrinsic muscles, provides positive 'recoil aspiration'. Another hollow space commonly used by air-breathing fishes is the gill chamber itself where there are often accessory respiratory organs, as in the climbing perch (*Anabas*) or the walking catfish (*Clarias batrachus*), which ventilates its respiratory trees in the gill chambers during synchronised trips to the surface. The water boils with a mass of fishes for a few moments, and then is undisturbed until the catfish take their next gulps of air, a strategy which apparently confuses predators. Oddly, although at least six of the 31 catfish families breathe air, only the obligate air-breather *Pangasias* uses the swimbladder; other catfishes use the stomach, intestine, or (like *Clarias*) the opercular chambers. In other fishes, as well as catfishes, for example, the loach *Misgurnus* which uses the rectum, or the Alaskan blackfish *Dallia* which uses the oesophagus, other regions of the gut are used to absorb oxygen. Such fish ventilate these parts of the gut by gulping air via the mouth and exhaling by burping through the mouth, or farting via the anus. Mudskippers (periophthalmids), the most terrestrial of all fishes, acquire oxygen via highly vascularised opercular cavities, and can spend extended periods out of water, periodically refilling the enlarged opercular chamber with air (and lying on their sides to dampen the skin).

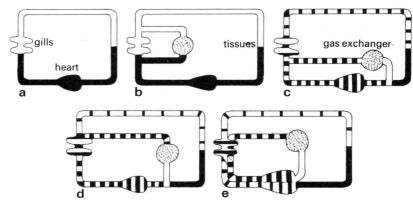

**Figure 5.12** Different circulatory patterns in various air-breathing fishes. (a) Normal fish gaining oxygen from gills in water; the gills are in series with the tissues of the systemic bed. (b) Fishes using the opercular chambers or buccal mucosa as airbreathing organs (*Clarias*, *Saccobranchus*). (c) Fish using the opercular or pharyngeal mucosa as the air-breathing organ (*Electrophorus*, *Anabas*, *Periophthalmus*). (d) Swimbladder used for respiration (holosteans). (e) Lung-like swimbladder, partial division between pulmonary and branchial circulation (lungfishes). Blood oxygen content: black, low; white, high. After Johansen (1970) In *Fish Physiology*, **IV** (Hoar and Randall, eds), 361.

Fish that breathe water perfuse the gills with systemic venous blood, which is oxygenated and sent direct to the systemic arterial system. In air-breathers, the circulatory arrangements are more complicated (Figure 5.12). When the swimbladder is used to acquire $O_2$, afferent vessels from the gill circulation supply it, and oxygenated blood leaves to enter the venous circulation before the heart. In lungfish, oxygenated blood passes from the lung direct to the heart via a pulmonary vein (as it does in ourselves). Some air-breathers have accessory respiratory organs in the buccal and opercular cavities in parallel with the gills, and their blood supply is linked to the gills in such a way that oxygenated blood from the accessory organs joins that from the gills and passes to the systemic arterial circulation. But in other air-breathers, like the electric eel *Electrophorus* the arrangement is less efficient, since the oxygenated blood from the gas exchanger simply enters the systemic venous circulation anterior to the heart. *Electrophorus* is an obligate air-breather, the buccal mucosa which is papillated and highly vascular (Figure 5.13) provides a respiratory surface around 15% of the body surface. It seems extraordinary that such a fish which feeds on living prey can use a delicate respiratory surface in this position, until we recall that it stuns its prey with a powerful electric shock (Chapter 10) and swallows it whole.

Even obligate air-breathing fishes retain the gills (though they may be reduced to avoid loss of oxygen at the gills in waters of low $Po_2$), because together with the skin, they still act as the site of $CO_2$ excretion. Since $CO_2$ diffuses much more rapidly than $O_2$, loss of $O_2$ at the gills can be

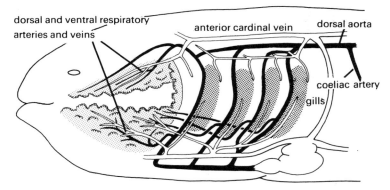

Figure 5.13 The blood supply to the respiratory buccal mucosa of the electric eel *Electrophorus*. Compare with schematic diagram of Figure 5.12. After Johansen *et al.* (1968) *Z. vergl. Physiol.*, **61**, 137.

diminished not only by reducing gill area, but also by increasing the diffusion distance, which will have little effect on $CO_2$ excretion. Thus in the climbing perch (*Anabas*), the diffusion distance is 15 µm, compared to the 1–3 µm found in the gills of fish respiring in water (Table 5.1).

### 5.4.6 Lungfishes

Lungfishes are unlike most air-breathing fishes, because they can respire with their lungs and gills simultaneously, just as in many amphibians there is both lung and cutaneous respiration. To do this, they can adjust the circulation to favour gas exchange by one or other route, according to external conditions, as Johansen and his colleagues showed. In the African lungfish, *Protopterus*, and in the normally water-breathing Australian *Neoceratodus*, they measured blood $Po_2$ at different points (Figure 5.14) and converted the values obtained to $O_2$ content (to account for the $O_2$ combined with haemoglobin). From the $O_2$ content at these different points, they were able to estimate the degree to which blood was selectively passed through different circulatory routes, and the relative

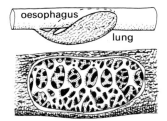

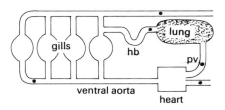

Figure 5.14 Left: lung of *Protopterus* showing septated structure. Right: sampling sites (spots) in the experiments of Johansen *et al.* (see Table 5.2). ab: anterior branchial; hb: hemibranch; pv: pulmonary vein. After Spencer (1898) *Denkschr. Med. Nat. Ges. Jena*, **4**, and Johansen *et al.* (1968) *Z. vergl. Physiol.*, **59**, 157.

importance of the gills and lungs in $O_2$ uptake. Table 5.2 shows the results they obtained.

In well-oxygenated water, *Neoceratodus* does not ventilate its single lung, which has no respiratory function. Nevertheless, rather surprisingly, the fish sends about the same amount of blood to the heart from the pulmonary vein as from the vena cava. But when ambient water $Po_2$ is lowered to 5.3–10.6 kPa, the fish begins to breathe air, and as Table 5.2 shows, by far the most important site of $O_2$ uptake is the lung. Blood in the anterior branchial arteries (which supply the systemic circulation) is now made up of about 5 parts of pulmonary vein blood to 1 part of blood from the vena cava. So *Neoceratodus* is able partially to separate the blood leaving the heart into streams flowing to the anterior and posterior branchial arches. This it can do because there is a rudimentary spiral valve in the sinus and conus of the heart (Figure 5.15), foreshadowing that of amphibians. *Protopterus* (an obligate air-breather) has more efficient separation of the two streams of blood, and blood in the anterior branchial artery contains about 10 parts of pulmonary vein blood to 1 part of vena cava blood.

Lungfishes breathe through the mouth, gulping air into the expanded buccal cavity (previously emptied of water). They then deflate the lung with intrinsic muscles and force air into it by closing the mouth and opercula and raising the floor of the buccal cavity. In water, *Protopterus* breathes every 5–7 minutes, but if kept out of water (which they do not seem to mind greatly), they breathe every 1–3 minutes.

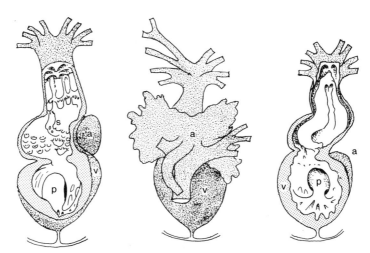

**Figure 5.15** Lungfish hearts. (a) *Neoceratodus* opened, ventral view; (b) *Protopterus* dorsal view; (c) *Protopterus* opened, ventral view. Note that the conus is partially separated in *Neoceratodus* by a row of special large valves (s), whilst in *Protopterus* separation is more complete. a: atrium; v: ventricle; p: plug in atrio-ventricular junction. After Goodrich (1930).

**Table 5.2** Blood $O_2$ content in two lungfishes under different conditions. From Johansen *et al.* (1968)

| Species | Condition | $O_2$ content (vol%) | | | | | Pulmonary venous blood/ vena cava blood | |
|---|---|---|---|---|---|---|---|---|
| | | Pulmonary artery | Pulmonary vein | Anterior branchial | Vena cava | | Anterior branchial | Pulmonary artery |
| *Neoceratodus* | In aerated water | 7.3 | 7.25 | 5.0 | 3.4 | | 5/4 | |
| | In hypoxic water | 6.0 | 7.9 | 6.75 | 0.8 | | 5/1 | 3/1 |
| *Protopterus* | In aerated water | 4.3 | 6.05 | 5.5 | 0.15 | | 10/1 | 7/3 |

### 5.4.7 Aestivation

*Neoceratodus* lives in deep pools of permanent rivers where ambient water $Po_2$ is high, but both the other lungfishes can not only live in waters of low $Po_2$, and possibly flounder across from a drying pool to another where water remains, but they can also survive a prolonged dry season by aestivating. As the water dries up and becomes more and more muddy, the fish burrow into the mud and become torpid, reducing their metabolic rate and oxygen demand until re-awakened by the first rains. *Protopterus* makes a bottle-shaped burrow lined by mucus secreted by the skin to form a cocoon; the nares are plugged with mucus, and the fish breathes air through its mouth once an hour or so, via the tube leading to the surface. In nature, aestivation lasts 4–6 months, but aestivating fish taken into the laboratory have survived in their cocoons for several years. In water, *Protopterus* excretes nitrogen as ammonia across the gills, but when aestivating this is no longer possible, and nitrogen (from the muscle proteins metabolised) is converted in the liver to non-toxic urea which reaches high levels in the blood. When water enters the tube to the surface and reaches the mouth, the fish makes breathing movements and after a series of convulsive jerks, swims out of its burrow. Aestivation is not peculiar to lungfish. The enigmatic little Western Australian *Lepidogalaxias salamandroides* (it is not a galaxeid, but does look much like a salamander), which was discovered in 1961, aestivates in the mud of dried stream beds curled up in a pear-shaped burrow connected to the surface by a thin tube. More spectacular are the New Zealand mudminnows (*Neochanna*) which aestivate up to 2 m below the ground for 1–2 months in summer and autumn. This habit gave rise to the early comment that 'the colonists obtained a bounteous harvest of potatoes and fish at one digging'!

## 5.5 The circulatory system

The oxygen acquired (from water or air), and the carbon dioxide excreted at the gills, have to be transported around the body by the circulation of the blood. In fishes using the gills as a gas exchanger, the circulation is single, blood leaves the heart to pass first through the gill capillary bed, thence to the systemic capillaries, and back to the heart. Usually the gills account for approximately 30% of the total resistance to blood flow, the remainder being in the visceral and somatic vasculature, but in tunas, which have high oxygen requirements and large gill areas, the gills account for up to 57% of the total resistance. In most fishes, blood pressures in the circulation are relatively low (very low in hagfishes), though once again tunas are the exception, and pressures in the ventral aorta in resting tunas reach 87 mmHg. In the venous system, pressures are very low, and may be

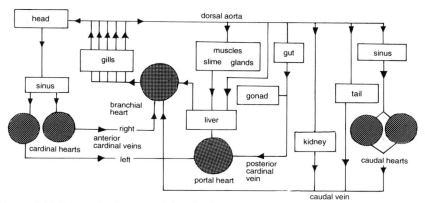

**Figure 5.16** Schematic diagram of the circulatory system of a hagfish, showing accessory hearts. After Satchell (1992) in *Fish Physiology*, **12A** (Hoar, Randall, and Farrell eds), 141.

sub-ambient; venous return is assisted by an unusual variety of pumps including accessory hearts. Hagfishes, for example, have no fewer than five 'hearts' in addition to the usual one (Figure 5.16).

In general, the teleost circulatory system is more efficient than that of elasmobranchs, blood volume is lower, as is its cardiac output ($Q$), and narrower veins occur instead of venous sinuses. There are interesting exceptions to some of these generalisations, which will be considered later, for example, in the Antarctic icefish (*Chaenocephalus aceratus*), which lacks haemoglobin, $Q$ is exceptionally high.

### 5.5.1 The heart

Fish hearts are S-shaped and four-chambered with, from behind forwards, sinus venosus, atrium, ventricle and either a bulbus or conus leading to the ventral aorta (Figure 5.17). We have already seen that some air-breathing fishes have a double circulation (the Japanese mudfish, *Channa argus*, even having a double ventral aorta), but only lungfishes have a morphologically partially divided heart. The teleost bulbus is an elastic reservoir passively enlarged by blood driven forwards out of the ventricle, but the equivalent conus (in elasmobranchs, *Amia*, *Lepisosteus* and *Polypterus*) is contractile, contracting in sequence with the rest of the heart. Valves at the junctions between the different regions assure unidirectional flow, and pocket valves are also found along the conus (Figure 5.17), sometimes in large numbers; *Lepisosteus* has no less than 72 valves in eight rows. Heart mass in most fishes scales as body mass (as in other vertebrates), but the size of the ventricle differs a good deal, being largest in tunas and icefishes where $Q$ is greatest.

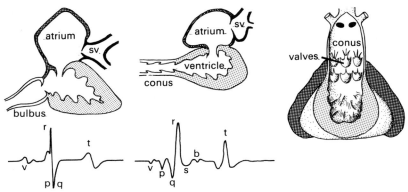

**Figure 5.17** Fish hearts. Above, left: teleost; mid: shark; right: valves in the conus of the mako shark (*Isurus*). sv: sinus venosus. Below: electrocardiograms of trout (left) and Port Jackson shark (*Heterodontus*). v, p, q, r, s, t and b: depolarisations associated with different regions of the heart (see text). After Daniel (1992); Bennion (1968) (in Randall (1970) *Fish Physiology*, **IV**, 133); and Holst (1969) (in Satchell (1972) Circulation in fishes. *Cambridge Monogr. Exp. Biol.*, **18**, Cambridge University Press).

The cardiac cycle consists of systole, when the ventricle is emptied, and diastole when it is refilled; it is accompanied by a progression of electrical events along the heart resulting from the depolarization and repolarization of the cardiac muscle cells. The sum of these cardiac action potentials, which are relatively easy to record *in situ* with electrodes which need not be in or on the heart itself, are electrocardiograms (ECGs) (Figure 5.17). Fish ECGs are essentially the same as those of mammals: atrial contraction produces the P wave, ventricular contraction the QRS complex, and ventricular relaxation the T wave. But fish heart ECGs can be more complex than in mammals. Contraction of the conus in elasmobranchs adds a small B wave, and contraction of the sinus venosus adds a V wave, prominent in the hagfish (*Eptatretus*) and in the eel. Contractions of the atrium and ventricle have to be coordinated in such a way that there is a delay between atrial and ventricular contraction, and a rapid synchronous contraction of the ventricle. It is peculiar that the specially modified conduction pathways (Purkinje fibres) found in higher vertebrates seem lacking in fish hearts, and it would certainly be worth a special search for them in fishes like tuna whose hearts have a near-mammalian output.

Ventricular contractions naturally lead to cyclic variations in pressure and flow in the ventral aorta from which blood flows through the serial capillary resistances of the gill and systemic capillary beds. Such oscillations are damped in teleosts by the elastic bulbus (constant flow is what is optimal (Figure 5.18) for the gill gas exchanger), but it does not seem that the elasmobranch conus can act in this way, although it has been claimed to

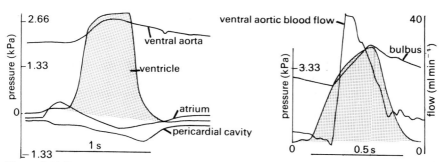

**Figure 5.18** Pressures recorded in different regions of the elasmobranch and teleost heart. Left: the small shark *Mustelus* (note sub-ambient pressures in pericardial cavity); right: in the lingcod (*Ophidion*). After Sudak (1965) *Comp. Biochem. Physiol.*, **15**, 199, and Randall (1970) *Fish Physiology*, **IV** (Hoar and Randall eds), 133.

do so; conus contraction is too slow to do other than make the pocket valves close together to prevent backflow, and pressures oscillate much more in the ventral aorta than in teleosts.

### 5.5.2 Cardiac output and its control in fishes

Obviously enough, the amount of blood entering the ventral aorta is determined (1) by the volume ejected from the ventricle at each stroke, and (2) by heart rate. In mammals, increase in heart rate is the most significant response to a demand for increased cardiac output, but in many fishes, stroke volume changes are more important (as in sharks, for example). The ventricle is filled by the contraction of the atrium (which is itself filled both by expanding as the ventricle contracts and pericardial pressure decreases, known as force from in front, *vis a fronte*, and by force from behind, *vis a tergo*, from the pressure in the venous return to the sinus venosus). The relative importance of these two mechanisms is presently unclear. In benthic teleosts, only the *vis a tergo* mechanism seems to be used, but in active teleosts like tunas, the *vis a fronte* mechanism seems important, at least at high stroke volumes. In order for *vis a fronte* to operate, the pericardium clearly has to be rigid (as it is in tunas), for if not, as the ventricle contracted, the pericardium would simply follow ventricular contraction and blood would not be sucked into the atrium by sub-ambient pericardial pressures.

It is always interesting when new work upsets long-held dogmas, and this seems to be the case for the operation of the shark heart. In sharks, the pericardium is thick and rigid, and in almost all texts, it is stated that pericardial pressures are sub-ambient, and that the atrium is filled by the *vis a fronte* mechanism. Recent work on leopard sharks (*Triakis semifasciatus*) set up to swim in a respirometer and appropriately cannulated to measure blood and pericardial pressures, has shown that when swimming, as stroke volume increases, pericardial pressures rise and the pericardio-peritoneal canal opens to reduce pericardial volume. In accord with this, puncturing the pericardium does not reduce stroke volume. So at least when they swim, sharks seem to fill their hearts by the *vis a tergo* mechanism, rather than mainly by the *vis a fronte* mechanism.

The more blood enters the atrium and ventricle, the more are their muscle fibres stretched, and the more powerfully they contract in accord with the Frank–Starling mechanism of the heart which states that 'the energy of contraction is a function of the length of the muscle fibre'. So stroke volume changes depend on the atrial–ventricular blood flow, and are controlled by circulating catecholamines, and by vagal and autonomic nerve supply as are heart rate. The American physiologist Greene discovered in 1902 that hagfishes are unique in completely lacking any heart innervation, when he set up a class experiment to demonstrate vagal control of the heart using *Bdellostoma*! Fortunately he had been conscientious enough to try the experiment himself before giving it to his students.

Intrinsic or resting heart rate varies from around 15 beats $min^{-1}$ (bpm) in hagfishes to 30–50 bpm in most elasmobranchs and teleosts, but in skipjack tuna (*Katsuwonas pelamis*), intrinsic heart rate is around 120 bpm, whilst in swimming skipjack, rates up to 240 bpm have been recorded. During long-sustained aerobic swimming, cardiac output naturally rises to meet increased tissue oxygen demand (the so-called scope for activity, Chapter 3). In trout it triples, and in tuna doubles from a basal level of 132 $ml^{-1}$ $min^{-1}$ $kg^{-1}$ (at 26°C) which is about half that of a mammal at 37°C. Contrary to what might be expected perhaps, heart rate and cardiac output decline during (anaerobically driven) burst exercise.


### 5.5.3 Accessory pumps

A remarkable variety of accessory pumps occur in the venous circulation of different fishes. These range from the portal heart of hagfishes behind the liver (Figure 5.16) which has cardiac-type muscle and resembles the atrium of the main heart, with an ECG with P and T waves, to the haemal arch and fin pumps and the caudal hearts of elasmobranchs and teleosts. Apart from the hagfish portal heart, all of these interesting devices are driven by skeletal muscles. For example, in the shark haemal arch pump (Figure 5.19) venous blood from the myotomes is driven into the caudal vein past ostial valves as the myotomes contract and compress the vascular bed, and so when the fish swims, there are cyclical pressure pulses in the caudal vein, and venous blood flow increases (just when cardiac output increases and a higher venous return is required).

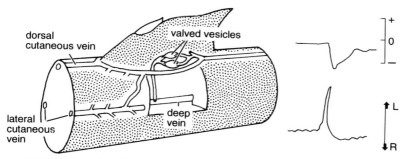

**Figure 5.19** The haemal arch pump of the shark *Heterodontus*. After Satchell (1992) in *Fish Physiology*, **12A** (Hoar, Randall, and Farrell eds), 141.

## 5.6 Fish blood and gas transport

Different fishes have very different lifestyles, so it is not surprising that the properties of their blood vary according to metabolic demands, and the way that the fish acquires $O_2$ and excretes $CO_2$. For example, blood in active fishes like scombroids must have a much higher $O_2$ capacity than in sluggish fishes like angler fish; in obligate air-breathers it must be less sensitive to $CO_2$ content that it is in water-breathing fishes.

Table 5.3 gives $O_2$-capacity values of whole blood in different fishes. Note that the $O_2$ capacity of whole blood comprises $O_2$ in solution in the blood plus $O_2$ combined with haemoglobin; this red cell respiratory pigment raises the $O_2$-capacity up to 40 times.

**Table 5.3** Blood $O_2$ capacity and gill areas in fish of different habit. After Steen (1971) *Comparative Physiological Respiratory Mechanisms*, Academic Press, London and New York. (See also Figure 5.1).

| Species | $O_2$ capacity (vol%) | Gill area (mm² g body wt⁻¹) | Habit |
|---|---|---|---|
| Bonito (*Sarda*) | 18.0 | 595 | |
| Mackerel (*Scomber*) | 19.6 | 1158 | Very active |
| Menhaden (*Brevoortia*) | 16.2 | 1773 | |
| Butterfish (*Pholis*) | 10.7 | 598 | |
| Sea-robin (*Prionotus*) | 9.3 | 360 | Active |
| Eel (*Anguilla*) | 8.0 | 302 | |
| Goosefish (*Lophius*) | 5.7 | 196 | |
| Toadfish (*Opsanus*) | 5.3 | 200 | Sluggish |
| Sand-dab (*Hippoglossoides*) | 4.6 | 188 | |

Remarkably, in several Antarctic icefishes in the family Chaenichthyidae, blood haemoglobin is much reduced or totally lacking, and there are no red blood cells. All the $O_2$ reaching the tissues must do so in solution in the blood, which has the same $O_2$-carrying capacity as seawater, viz. around 0.7 vol%, compared with around 8 vol% in many normal fishes with haemoglobin. To overcome the low $O_2$-capacity of the blood, icefishes have relatively large gills, well-vascularised skin for cutaneous respiration, large hearts with large cardiac output and large-diameter blood vessels; to reduce $O_2$ demand they have reduced their red aerobic myotomal musculature. The resting $O_2$ uptake of icefishes is from one-half to two-thirds that of fishes in the same habitat which possess haemoglobin, and they survive by a combination of low metabolic rate, high cardiac output and living at low temperatures (when blood $O_2$ capacity is high). It remains a mystery as to why they should have lost haemoglobin. The early larvae of many teleosts also lack haemoglobin (though of course it appears later in development), and so do the long-lived leptocephali larvae of eels, where it is presumably lacking to complete their glassy transparency. But even in fish like trout, pike (*Esox*) and goldfish, $O_2$ dissolved in the blood suffices for resting metabolism. When these fishes are poisoned with CO (so that the haemoglobin cannot combine with $O_2$), they survive well until exercise increases $O_2$ demand, when they perish.

### 5.6.1 Fish haemoglobins and oxygen transport

Apart from the special case of the icefishes, most oxygen in the blood is carried by red cell haemoglobin, oxygenated when $Po_2$ is high and de-oxygenated when it falls, according to the oxygen dissociation curve (blood $Po_2$ plotted against the amount of $O_2$ bound to the haemoglobin). Oxygen dissociation curves are always non-linear, and in active fishes, usually sigmoid (Figure 5.20), this shape representing a compromise between high $O_2$ affinity needed for loading at the gills and lower affinity for unloading at the tissues. The normal working range in the fish usually lies on the steep part of the curve so that much $O_2$ can be unloaded for small changes in $Po_2$. The slope of the dissociation curve can be changed by changes in pH; with increase in acidity, the curve is usually shifted markedly to the right. This is the Bohr shift (defined as the shift or change in log 50% $O_2$ saturation divided by the pH change causing it; see Figure 5.20). It results from pH-dependent configurational changes in the haemoglobin molecules which inhibit $O_2$ binding; what it means in practice is that $O_2$ is unloaded at sites where $Pco_2$ is high, just where it is needed by the fish.

In many fishes, increase in $Pco_2$ not only shifts the dissociation curve to the right, but it also prevents complete oxygenation of the haemoglobin, thus depressing the curve. The lowering of blood $O_2$-carrying capacity (rather than $O_2$ affinity) in this way is the Root shift (we saw in Chapter 4

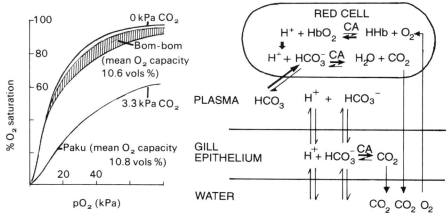

**Figure 5.20** Left: Oxygen dissociation curves of blood from fishes living in oxygenated and oxygen-depleted water. Note difference between the blood of the paku (*Pterodoras granulosus*) which lives in well-oxygenated water and shows a large Root effect, and the bom-bom (*Myleus setiger*) which lives in oxygen-depleted water and shows only a small Root effect (stippled). After Willmer (1934) *J. exp. Biol.*, **11**, 283. Right: Schematic diagram to show $CO_2$ and $H^-$ movements between plasma, red cells, the gill epithelium, and the ambient water. CA: carbonic anhydrase. After Randall and Daxboeck (1984) in *Fish Physiology*, **XA** (Hoar and Randall eds), 296.

that the Root shift is used by the fish to drive $O_2$ into the swimbladder from the rete). The Root shift is really an extreme case of the Bohr shift, and it is found in the blood of fishes with swimbladders or other $O_2$-concentrating retia (like those of the choroid plexus of the eye) but is absent from elasmobranchs that do not possess such retia.

Air-breathing fishes like lungfish or the electric eel (*Electrophorus*) have haemoglobins that are rather insensitive to $P_{CO_2}$, and they need to have this reduced Bohr shift, because $P_{CO_2}$ at the gas exchanger and in the blood will be higher than in water-breathing fishes. Hagfishes and lampreys have monomeric haemoglobins, but in all other fishes the haemoglobins are tetrameric (as they are in mammals), and polymorphic. Several different haemoglobins may occur in one fish, perhaps to adapt the gas-transport system to changing conditions, as in the American eel (*Anguilla rostrata*), where one type has a high $O_2$ affinity in seawater, the other in freshwater.

## 5.6.2 $CO_2$ transport

Only a small proportion of the $CO_2$ diffusing into the blood at the tissues remains dissolved in the plasma; most is hydrated to the bicarbonate ion (about 95% of $CO_2$ in the venous blood is plasma $HCO_3^-$), and so $CO_2$ from the tissues is transported in the blood mainly as $HCO_3^-$. Rehydration of $CO_2$ to $HCO_3^-$ is slow in the veins, taking place after the venous blood has left the respiring tissue, but is rapidly catalysed by carbonic anhydrase in the red cells, where $O_2$ is driven off the haemoglobin in the respiring tissues as it binds the resulting protons. Bicarbonate entry in the red cells is accompanied by water entry (to rehydrate the $HCO_3^-$ and by $Cl^-$ to maintain electroneutrality). This chloride shift increases the osmolarity of the red cells, which therefore swell slightly so that their volume becomes 2–3% greater in venous than in arterial blood. The upper part of Figure 5.20 shows the situation schematically.

At the gills, total blood $CO_2$ is reduced by 10–20%, mainly because $HCO_3^-$ falls by 20% in the plasma. One scheme by which this could occur is shown in the lower half of Figure 5.20, where carbonic anhydrase-catalysed $CO_2$ produced in the red cell from $HCO_3^-$ diffuses away across the plasma and gill epithelium to the water flowing over the gills. Carbonic anhydrase is present in the gill epithelium, but does not appear to play a role in $CO_2$ excretion.

Oxygen and carbon dioxide transport are complementary, and combine to make a system efficient enough to satisfy the gas transport demands of such active fishes as tunas. Probably it is generally true that fishes use most oxygen in the red muscle driving cruising swimming (Chapter 3), and in the fast-swimming scombroids this tissue is extremely well vascularised. Capillary fibre ratios up to 7:1, and external diffusion distances of just over 10 µm compare very favourably with those of mammalian muscle.

## Bibliography

Davenport, J. (1993) Ventilation of the gills by the pectoral fins in the fangtooth *Anoplogaster cornutum*: how to breathe with a full mouth. *Journal of Fish Biology*, **42**, 967–970.

El-Fiky, N. and Wieser, W. (1988) Life styles and patterns of development of gills and muscles in larval cyprinids (Cyprinidae: Teleostei). *Journal of Fish Biology*, **33**, 135–145.

Goodrich, E.S. (1909) Vertebrata craniata. Fasc. 1. Cyclostomes and fishes. In *A Treatise on Zoology* (Lankester, E.R. ed), A. &. C. Black, London.

Goodrich, E.S. (1930) *Studies on the Structure and Development of Vertebrates*, Macmillan, London.

Greene, C.W. (1902) Contributions to the physiology of the Californian hagfish *Polistotrema stouti*. II. The absence of regulative nerves for the systemic heart. *Amer. J. Physiol.*, **6**, 318–324.

Houde, E.D. (1989) Comparative growth, mortality and energetics of marine fish larvae: temperature and implied latitudinal effects. *Fishery Bulletin of the United States*, **87**, 471–495.

Hughes, G.M. and Ballantijn, C.M. (1965) The muscular basis of the respiratory pumps in the dogfish (*Scyliorhinus canicula*). *Journal of Experimental Biology*, **43**, 363–383.

Hughes, G.M. and Morgan, M. (1973) The structure of fish gills in relation to their respiratory function. *Biol. Revs.*, **48**, 419–475.

Johansen, K., Lenfant, C., Schmidt-Nielsen, K. and Petersen, J. (1968) Gas exchange and control of breathing in the electric eel, *Electrophorus electricus*. *Z. Vergl. Physiol.*, **61**, 137–163.

Kamler, E. (1992) *Early Life History of Fish: An Energetics Approach*. Chapman & Hall, London.

Lai, N.C., Graham, J.B., Bhargava, V., Lowell, W.R. and Shabetai, R. (1989) Branchial blood flow distribution in the blue shark (*Prionace glauca*) and the leopard shark (*Triakis semifasciata*). *Experimental Biology*, **48**, 273–278.

Liem, K.F. (1981) Larvae of air-breathing fishes as countercurrent flow devices in hypoxic environments. *Science*, **211**, 1177–1179.

Perry, S.F., Kinkead, R. and Fritsche, R. (1992) Are circulating catecholamines involved in the control of breathing by fishes. *Reviews in Fish Biology and Fisheries* (Pitcher T.J. ed), **2**, 65–83.

Rombough, P.J. (1988) Respiratory gas exchange, aerobic metabolism and effects of hypoxia during early life. In *Fish Physiology*, **XIA** (Hoar, W.S. and Randall, D.J. eds), Academic Press, San Diego, pp. 59–161.

Sayer, M.J.D. and Davenport, J. (1991) Amphibious fish: Why do they leave the water? *Reviews in Fish Biology and Fisheries* (Pitcher T.J. ed), **1**, 159–181.

Tsuneki, K. and Koshida, Y. (1993) Structural organization of the blood-sinus systems in lampreys and hagfish: functional and evolutionary interpretations. *Acta Zoologica (Stockholm)*, **74**, 227–238.

Val, A.L., Affonso, E.G., and de Almeida-Val, V.M.F. (1992) Adaptive features of Amazon fishes: blood characteristics of curimata (*Pronchilodus* cf *nigricans*, Osteichthyes). *Physiological Zoology*, **65**, 832–843.

Wright, D.E. (1973) The structure of the gills of the Elasmobranch, *Scyliorhinus canicula* (L). *Zeitschrift für Zellforschung*, **144**, 489–509.

# 6 Osmoregulation and ion balance

## 6.1 The osmotic problem

Different fishes live in waters ranging from nearly pure (virtually distilled-water quality) to hypersaline ponds where salinity is so high that they have difficulty in swimming below the surface. What is more, there are a good many euryhaline species able to move between seawater and freshwater. Fish skin is as a rule rather impermeable to water, even in fishes where there is significant cutaneous $O_2$ uptake (which remains rather puzzling), but all fishes have large areas of permeable epithelia in contact with the water. Apart from the vast area of the gill lamellae, there are the oral and narial mucosae, and in addition, some water will also inevitably be swallowed into the stomach during feeding. Very few fishes are isosmotic with the external medium, so there will be osmotic gradients across these permeable surfaces.

All freshwater fishes have body fluids more concentrated than the medium, and so will tend to gain water and lose ions, whilst in the sea, the body fluids of most fishes are more dilute than the seawater, and so they will tend to lose water, and gain ions (Figure 6.1). It is indeed remarkable that the osmoregulatory mechanisms for coping with these opposite problems in freshwater and seawater can be switched by euryhaline fishes as they migrate between the two. Changes in body fluids do in fact occur as these fishes are confronted by salinity changes; for example in eels, flounders (*Platichthys flesus*) and salmon, increase in plasma sodium mainly accounts for the 20% increase in plasma osmolarity when they are

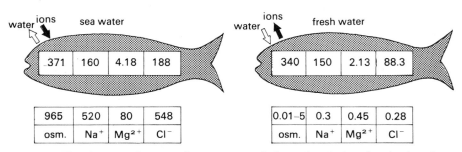

**Figure 6.1** The plasma and surrounding water osmolarity and molarity of major ions in representative marine and freshwater teleost fishes, showing the opposite tendencies of water loss and ion gain in the former, and water gain and ion loss in the latter. After Pang *et al.* (1977).

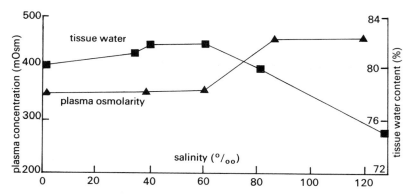

**Figure 6.2** Tissue water content and plasma osmolarity in killifishes (*Fundulus*) adapted to different salinities. After Feldmeth and Waggoner (1972) *Copeia*, 592.

adapted to seawater from freshwater. The little killifish, *Fundulus*, can surmount the greatest osmotic challenge, being able to live in freshwater and (in Southern California) in hypersaline pools (128‰ salinity). Between 0 and 60‰ NaCl, *Fundulus* can regulate body water and plasma salt concentration; higher salinities are tolerated by accepting tissue water loss (5%), and increase in blood osmolarity (around 30%), as shown in Figure 6.2.

The large areas of permeable epithelium in contact with the water in fishes, in particular the gills, mean that extrarenal routes of excretion and ion exchange can be exploited, as well as the renal route of terrestrial animals, but before considering the mechanisms involved, we can first examine the situation in hagfishes, which are unique in being not only isosmotic with the seawater in which they live, but in having blood sodium and chloride levels similar to seawater.

## 6.2 Hagfishes, lampreys and the origin of the glomerular kidney

In nature, hagfishes live in an environment of unchanging salinity, yet *Eptatretus* can tolerate gradual experimental changes in salinity within certain limits, swelling and shrinking as more or less perfect osmometers since they are freely permeable to water. Hagfish have a large blood volume (see p. 100), and the blood is isosmotic or very nearly so to seawater (Table 6.1). Sodium and chloride concentrations in the blood are similar to those in seawater, but higher than in the tissues; internal osmotic balance is maintained by high levels of intracellular amino acids, which can be regulated to some degree when hagfishes are exposed to osmotic stress.

Because the blood is isosmotic (or very nearly so) to seawater, there is little or no osmotic exchange of water, but because they are freely

**Table 6.1** Composition of plasma and urine (mM) in some marine fishes. Values for seawater representative. From Pang et al. (1977) and Griffith and Pang (1979)

|  |  | Agnatha | | Chimaera | Elasmobranchiomorpha | | | Coelacanth (Latimeria) | | Teleostei | | | |
|  |  | | | | | Squalus | | | | blood | | urine | |
| Ion | Sea water | Myxine blood | Eptatretus urine | Chimaera blood | Hydrolagus urine | Squalus blood | Squalus urine | Latimeria blood | Latimeria urine | Fundulus blood | Muraena blood | Paralichthys urine | Lophius urine |
|---|---|---|---|---|---|---|---|---|---|---|---|---|---|
| $Na^+$ | 470 | 487 | 553 | 338 | 162 | 296 | 240 | 197 | 184 | 183 | 212 | 59 | 11 |
| $K^+$ | 10 | 8.4 | 11 | 11.7 | 7.8 | 7.2 | 2.0 | 5.8 | 9 | 4.8 | 2.0 | 3.4 | 2.0 |
| $Ca^{2+}$ | 10 | 4.8 | 4 | 4.3 | 17 | 3.0 | 3.0 | 4.8 | 2.0 | 2.3 | 3.9 | 11 | 7 |
| $Mg^{2+}$ | 54 | 9.3 | 15 | 6.1 | 69 | 3.5 | 40 | 5.3 | 30 | 2.1 | 2.4 | 78 | 137 |
| $Cl^-$ | 548 | 500 | 548 | 353 | 268 | 276 | 240 | 187 | 15 | 146 | 188 | 124 | 132 |
| $HCO_3^-$ | – | 7.2 | – | 2.6 | – | – | – | 9.6 | – | 13.3 | – | – | – |
| $PO_4^{3-}$ | – | 0.4 | 9 | – | 25 | 2.4 | 33 | 5.1 | 38 | 5.3 | – | 11 | 2.0 |
| $SO_4^{2-}$ | 28 | 3.7 | 7 | 5.2 | 26 | 3.1 | 70 | 4.8 | 104 | – | 5.7 | 28 | 42 |
| Urea | – | 2.8 | 9 | 332 | 52 | 308 | 100 | 377 | 384 | 4 | 9.1 | – | 0.6 |
| TMAO | – | – | – | 0.0 | – | 72 | 10 | 122 | 94 | – | – | – | 13 |
| Amino acids | – | – | – | – | – | 11.6 | – | 16 | – | 8.5 | – | – | – |
| Osmolarity (mosmol) | 1011 | 969 | – | 1046 | 820 | 998 | 800 | 932 | 962 | 363 | – | 295 | 406 |

–, no data.

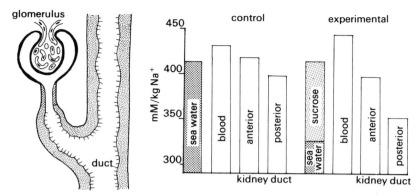

**Figure 6.3** The hagfish kidney. Left: single kidney tubule of *Myxine*. Right: Sodium levels in control and experimental *Eptatretus* (see text). After McInerney (1974) *Comp. Biochem. Physiol.*, **49A**, 273.

permeable to water, very high water exchange rates (2287 ml kg$^{-1}$ h$^{-1}$) have been observed using tritiated water as a tracer. As expected, the kidney produces little urine. Most of the kidney is essentially of the usual vertebrate mesonephric type though the tubules are little more than short junctions with the longitudinal archinephric duct (Figure 6.3); there is also a persistent pronephros but the pronephric tubules do not retain their original connection with the kidney duct, and play no role in urine production. Specialised chloride cells (see p. 133) have recently been reported in the gills, but extrarenal routes of ion exchange seem unimportant.

Since Macallum first suggested (in 1910 and 1926) the attractive but possibly incorrect idea that the blood ionic composition and osmotic level of vertebrates reflected their past history and that of the environment their ancestors lived in, it has seemed reasonable to suppose from their blood composition that hagfishes (all marine today) were from their origin a marine group. Yet careful experiments on *Eptatretus*, where blood and kidney filtrates were sampled after placing the hagfish in seawater diluted with sucrose (to retain osmolarity but diminish sodium content) have shown that they responded to low external sodium by increasing renal sodium uptake (Figure 6.3). Re-absorption of sodium from the glomerular filtrate is a necessary trick for any fish living in freshwater, and it is hard to think how this 'surplus' capacity of the sodium recovery mechanism could be useful in seawater.

If it were a leftover from an original freshwater ancestry, this would fit with Homer Smith's famous hypothesis (Marshall and Smith, 1930) that the glomerular kidney first arose in freshwater as a device for excreting water. However, it now seems more likely that the glomerular kidney first arose in seawater as a device to regulate ions by producing a filtrate which could be selectively altered (by tubular secretion and absorption). Divalent cations like $Mg^{2+}$ and $Ca^{2+}$ are at much lower levels in hagfish plasma and urine than they are in seawater (Table 6.1), but $Mg^{2+}$ is apparently mainly excreted by the liver via the gall bladder. The low systemic blood pressure in hagfishes (5–7 mmHg) and counteracting plasma osmotic pressure due to organic solutes mean that glomerular ultrafiltration is hardly possible when the hagfish is quiescent. So it seems that the hagfish kidney is an on/off kidney, filtering only when blood pressure rises during activity! That is, it is primarily a volume and ion-regulating system, and not a water-regulating device. On this view of the origin of the glomerular kidney, the common ancestor of hagfishes and all other fishes were 'pre-adapted' to entry into freshwater.

How do lampreys fit into this scheme? In freshwater, like all other freshwater fishes, both adult and ammocoete larva tend to gain water and lose salts; they excrete large amounts of dilute urine, at an osmolarity around 20–30 mosmol. The mesonephric kidney (see Figure 6.12 below) has much longer tubules that those of hagfishes, and sodium, chloride and (less efficiently) potassium are absorbed from the glomerular filtrate as far as possible. Tubular absorption, however, cannot prevent significant loss of these and other ions, and so unlike hagfishes, lampreys have special ion-uptake cells in the gill epithelia, as in teleosts (see p. 133) to make good the loss in the urine. This extrarenal ion-uptake route is remarkably efficient, as was shown in isotope studies of ammocoete larvae, which were able to maintain stable blood composition even in solutions containing only 30 $\mu$mol $Na^+$ litre$^{-1}$. In the sea, lampreys face the reverse problem of losing water. Unfortunately adult lampreys are rarely caught at sea, and although the large *Petromyzon marinus* can usually be seen attached to any basking shark encountered, they drop off if the shark is caught or stranded. Similarly, the marine phase of the smaller *Lampetra fluviatilis* is only found as it enters rivers to spawn, when the marine osmoregulatory mechanism has already changed so that they can no longer survive in seawater. It is only on such partially freshwater-adapted lampreys that experiments have been done on lamprey marine osmoregulation. They swallow seawater which is absorbed in the anterior intestine by the active uptake of $Na^+$ and $Cl^-$ ions followed passively by water along the osmotic gradient across the gut wall. These monovalent ions are excreted across the gills by chloride cells (again best known in teleosts, p. 134); divalent ions are excreted by the kidneys in the small amounts of urine produced. Lampreys are today invariably freshwater in the ammocoete larval stage, and those which pass downriver to the sea upon metamorphosis are assumed to have done so secondarily to seek more abundant host fishes. The essential similarity between the kidneys of hagfish, lampreys and higher vertebrates strongly suggests that the glomerular kidney arose on the line before hagfish separated as the sister group to lampreys and other vertebrates, and that this origin was in the sea.

## 6.3 Teleosts

Teleosts show a wide spectrum of morphological and physiological adaptations to waters of different osmolarity and ionic composition, and what is more, some are known to have become adapted to one environment, and then secondarily and relatively recently entered another (like the fishes with aglomerular kidneys that are known from freshwater).

### 6.3.1 Marine teleosts

Marine teleosts have much lower ion concentrations in the body fluids than seawater (Figure 6.1), and to overcome osmotic water loss, they drink seawater, like marine lampreys. Drinking rates vary in marine teleosts, as we should expect, since relative gill areas differ according to activity levels; in *Serranus* for example, 12% of the body weight is drunk each day. About 75% of the water drunk is absorbed in the gut, and since urine flow is small, this can maintain water balance. In silver eels in seawater, chronic oesophageal perfusion experiments (Figure 6.4) have shown that the oesophagus is impermeable to water, but permeable to $Na^+$ and $Cl^-$, which therefore diffuse into the blood down their concentration gradients. This means that the water entering the intestine is less concentrated than

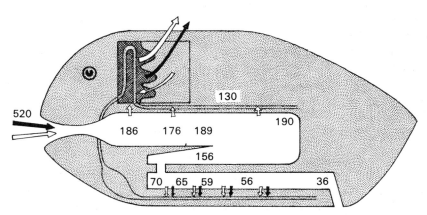

**Figure 6.4** Schematic diagram showing gut osmoregulatory activity in the silver eel *Anguilla* (see text). The eel has been shortened to fit onto the page. Open arrows: water movement; solid arrows: salt movement. The numbers are milliequivalents of $Cl^-$ per litre: note that salts are absorbed in the oesophagus which is impermeable to water. After Kirsch *et al.* (1981) *Bull Soc. Zool. France*, **106**, 31.

seawater, and nearly isotonic with the blood. The intestine is permeable to water, and so water is taken up there. But because water uptake is coupled to salt intake (as we have seen in marine lampreys), this process replaces the osmotic problem the fish faces with an ionic problem! The story of the gradual unravelling of the way that marine fishes solve this ionic problem is an interesting one, depending on advances in technique and to some degree, on fashions in other fields.

Early experiments by Homer Smith using rubber partitions to separate the head of the fish from the rest of the body (i.e. separating urine outflow from the water that had flowed over the gills), were the first to show that extrarenal routes of salt excretion were important, and this result was soon confirmed by gill-perfusion experiments (using isolated heart–gill preparations), which showed that salt levels decreased in the perfusate, and increased in the external seawater. Special columnar cells were found in the gills, which looked like secretory cells, and these were called chloride cells, since they were suggested to secrete chloride actively from the blood into the seawater.

It took nearly 20 years before any direct evidence was provided for chloride secretion by these cells, but at about this time, radioactive tracers like $^{24}$Na became available to monitor salt fluxes across fish, and largely due to a long series of tracer experiments by the distinguished French physiologist Jean Maetz and his colleagues, the subject was revolutionised. It was soon found that a completely unsuspected massive salt influx took place across the gills (Figure 6.5). This was found to be 5–10 times greater than salt entry from drinking seawater, and it was obvious that salt excretion from the gills was much greater than previously supposed (Table 6.1). Perhaps because in other epithelia like frog skin, active chloride transport was absent, attention focused on sodium rather than chloride, and sodium transport seemed to be the main driving force for salt secretion across the gills. It was only in the late 1970s (after Maetz's untimely death in a road accident) by using the much less complex opercular membrane

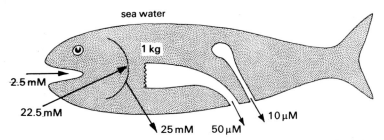

**Figure 6.5** Salt balance in a marine teleost as revealed by radioactive tracer experiments. After Potts (1976) *Perspectives in Experimental Biology*, **1**, 65.

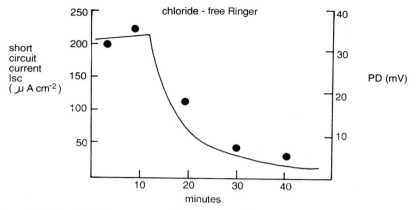

**Figure 6.6** Potential difference across isolated opercular membrane and its rapid decrease in $Cl^{2-}$-free Ringer. After Degnan *et al.* (1977) *J. Physiol. Lond.*, **271**, 155.

preparation instead of the gill, it was shown that active chloride transport by the chloride cells was the driving force for salt secretion by the gills, as had originally been suggested almost 50 years before!

The gills and opercular membrane are around 20 mV positive with respect to the external seawater, and this potential difference rapidly drops when the isolated opercular membrane is placed in chloride-free Ringer solution (Figure 6.6). From such experiments it has become clear that the transport of chloride from blood to seawater is an active electrogenic system, and that sodium simply passively follows the gradient created by chloride movement. Excellent agreement between the chloride flux and numbers of chloride cells present, makes it clear that it is the chloride cells which are involved.

### 6.3.2 Chloride cells

Chloride cells have a very particular structure (Figure 6.7). The basal (blood) and lateral sides are very extensively infolded, to make a complicated system of smooth branching tubules extending almost to the apex of the cell which is exposed to the seawater flowing over the gills. As seen in Figure 6.7, the smooth tubular system (STS) represents extracellular space; the cytoplasm of the cell is filled with mitochondria closely packed in amongst the STS, leaving a small apical region exposed via a crypt partially roofed by overlying epithelial cells, to which the chloride cell is linked by deep tight junctions. Chloride cells are often found in association with developing chloride cells, the junction between the two being leaky. The STS membranes are lined with closely packed regular particle arrays, which appear to be almost solid masses of the enzyme $Na^+/K^+$-activated ATPase.

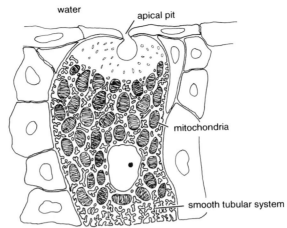

**Figure 6.7** Ultrastructure of chloride cell. After Degnan *et al.* (1977) *J. Physiol. Lond.*, **271**, 155.

Direct measurements over the surface of the epithelium have shown that negative current peaks are correlated with chloride cell apices (visualised with a fluorescent ionophore), so the chloride cells are certainly secreting chloride at their apices. The rate of movement of salt across the gills of marine teleosts is very high; across the apices of the chloride cells it is continuously about the same as that across the squid giant axon membrane at the height of the action potential! How is this massive flux brought about?

Unfortunately, since intracellular ionic activities are yet to be measured in this remarkable cell, models of how it works are still only tentative. One possible scheme is shown in Figure 6.8, based on work on the opercular epithelium of *Fundulus*, and on the better known rather similar chloride-secreting cells of the shark rectal gland (see Section 6.4.2).

The primary driving force for $Cl^-$ secretion is the STS $Na^+/K^+$-activated ATPase sodium pump which keeps cytoplasmic $Na^+$ low. In the shark rectal gland cells, cytoplasmic $Na^+$ is 20 mM, compared to 280 mM in the body fluid. This large $Na^+$ gradient provides the energy for driving carrier-mediated electroneutral entry of $Cl^-$ and $Na^+$. In the shark rectal gland, $2Cl^-$ are co-transported with each $Na^+$ and $K^+$ ion (as in the mammalian kidney), but it is not yet clear if this may be the case in the teleost chloride cell. Chloride ions then diffuse to the apex of the chloride cell, where they pass out of the cell via $Cl^-$ ion channels. Since the membrane potential of the chloride cell is negative (with respect to the seawater), any increase in cytoplasmic $Cl^-$ will lead to corresponding $Cl^-$ exit by electrical forces. Sodium ions on the other hand, leave passively via the cation-selective paracellular pathway ending in the leaky tight junction between chloride

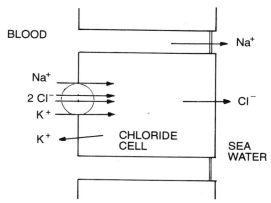

**Figure 6.8** A possible scheme for chloride secretion by the chloride cell in a fish in seawater, based on experiments with *Fundulus* operculum and the shark rectal gland. After Greger and Kunzelmann (1990) *Basic principles in transport* Comparative Physiology, **3** (Kinne ed), Karger.

cells. Sodium ions thus follow the transepithelial potential gradient established by the movement of Cl⁻.

However, chloride-secretion in marine teleosts is not the only function of the chloride cell. We might suppose that chloride cells would be less abundant or even absent in freshwater fishes, where this role is exactly the opposite of what is needed, and indeed, they are less conspicuous in freshwater fishes, and multiply in the gills of anadromous fish like salmon as the smolts prepare to pass downriver into the sea. In the freshwater *Tilapia*, however, recent evidence has shown that chloride cells take up $Ca^{2+}$ from solutions as dilute as 0.2 M $Ca^{2+}$; perhaps in freshwater they should rather be termed calcium cells.

### 6.3.3 Freshwater teleosts

In freshwater, water enters across all permeable surfaces, and there is a large concentration gradient favouring diffusion of salts across these surfaces. So drinking rates are low, and water influx is met by the excretion of large amounts of dilute urine (between 0.1 and 1.4 ml 100 g body weight⁻¹ h⁻¹, about ten times the values for marine teleosts). The urine is much more dilute than the plasma (see Table 6.2) because, whilst little water is absorbed from the glomerular filtrate as it passes along the tubule, salts are very efficiently resorbed. For example, measurements on some North American freshwater fishes have shown that over 99.9% of the $Na^+$ and Cl⁻ ions passing into the glomerular filtrate are resorbed, the filtrate osmolarity falling from 220–320 mosmol so low as 20–80 mosmol. Water balance can be maintained by the excretion of copious urine, but to

**Table 6.2** Composition of plasma and urine (mM) in lampreys and in some freshwater fishes. From Robertson (1974), Holmes and Donaldson (1969) in *Fish Physiology* (Hoar and Randall ed), Vol. 1, 1–89; and Hickman and Trump (1969) in *Chemical Zoology* (Florkin and Scheer ed), Vol. 5, 149–193.

| | | Agnatha (landlocked sea lamprey) | | | | Chondrostei | Holostei | Teleostei | | | |
| | | Blood | | | Urine | | | Charr (Salvelinus) | | Catfish (Ameiurus) | |
| Ion | Lake Huron water | ammo-coete | parasitic adult | spawning adult | (adult in FW) | (sturgeon in FW) blood | (Amia) blood | blood | urine | blood | urine |
|---|---|---|---|---|---|---|---|---|---|---|---|
| Na$^+$ | 0.02 | 103.0 | 137 | 136 | 4.8 | 155.8 | 132.5 | 161 | 17.4 | 122 | 12.2 |
| K$^+$ | 0.05 | 3.4 | 3.3 | 5.1 | 0.99 | 4.3 | 2.0 | 2.8 | 2.5 | 2.7 | 1.61 |
| Ca$^{2+}$ | 0.9 | 2.4 | 2.2 | 1.8 | – | 2.3 | 5.3 | 2.05 | 0.95 | – | – |
| Mg$^{2+}$ | 0.25 | 1.6 | 2.0 | 2.7 | – | 1.47 | 0.4 | 0.75 | 0.55 | – | – |
| Cl$^-$ | 0.05 | 91.0 | 122.0 | 112.0 | 4.7 | 119.7 | 119.5 | 140.6 | 8.1 | 110.0 | 18.0 |
| SO$_4^{2-}$ | 2.3 | 0.1 | 0.1 | 0.7 | – | 0.7 | 2.2 | – | – | – | – |
| HCO$_3^-$ | 1.75 | 6.0 | 5.0 | 5.2 | – | – | – | – | – | 3.4 | 0.4 |
| Osmolarity (mosmol) | – | – | – | 241.0 | 36.0 | 318.0 | – | 328.0 | 36.2 | – | – |

–, no data. FW = freshwater.

maintain salt balance, freshwater teleosts (like freshwater lampreys) have to have a high-affinity salt-uptake mechanism at the gills. The efficiency of this mechanism is shown by the low loss rate of salts in freshwater teleosts as compared with marine teleosts (Table 6.1), and by the accumulation of $Na^+$ from very dilute solutions ($\geqslant 10^{-4}$ M).

Various lines of evidence indicate that $Cl^-$ and $Na^+$ uptake are independent; selective blocking of one does not affect the uptake of the other. The Danish physiologist August Krogh suggested in 1939 that salt uptake in freshwater fishes was linked to acid–base metabolism, $Na^+$ being exchanged for $NH_4^+$, and $Cl^-$ for $HCO_3^-$. Certainly, when fish are in a steady state with the water they are living in, the rate of $NH_4^+$ loss is similar to that of $Na^+$ uptake.

But this cannot be the whole story, for in seawater, marine teleosts excrete $NH_4^+$ and at the same time excrete $Na^+$ rather than absorbing it. Injection of acetazolamide (which blocks $Na^+$ uptake) has no effect on $NH_4^+$ excretion. Furthermore, in brown trout (*Salmo trutta*) in acid freshwater (ca. pH 4), sodium influx is reduced almost to zero, whilst ammonia excretion increases. Under these conditions, ammonia is excreted by non-ionic diffusion rather than being coupled with $Na^+$ entry.

This puzzle was resolved by the discovery that $Na^+$ can be also exchanged for $H^+$. Figure 6.9 compares two goldfish which use these two different mechanisms. $Na^+$ uptake in each is well correlated with the sum of the two processes. $H^+/Na^+$ exchange in acid water is likely blocked by

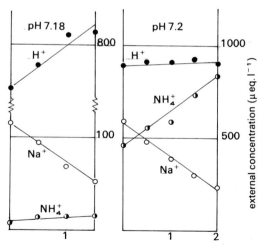

**Figure 6.9** Sodium uptake in two goldfish. Measurements were made of external $Na^+$, titrable acidity, and total ammonium ($NH_4^+$). Note that the measurements for the fish on the left show $Na^+$–$H^+$ exchange, whilst those for the fish on the right show $Na^+$–$NH_4^+$ exchange. After Maetz (1974) *Biochem. Biophys. Perspect. Mar. Biol.*, **1**, 1.

the high $H^+$ concentration ratio between plasma and water, which explains why in acid waters $Na^+$ uptake fell markedly in brown trout.

The hypothesis that $Cl^-$ is exchanged for $HCO_3^-$ has withstood experimental attack. As we have seen (Chapter 5), the gills are the major route of $CO_2$ excretion, and the gill epithelium contains much carbonic anhydrase, the enzyme concerned with catalysis of conversion of carbon dioxide into bicarbonate. For obvious reasons, experiments on $Cl^-$ uptake are usually made in $Na^+$-free solutions, but when both ions are present in the external medium, $Cl^-$ uptake is facilitated.

It is important to remember that these processes take place across the outer membrane of the respiratory epithelial cells of the secondary gill lamellae, in contrast to the active $Cl^-$ and passive $Na^+$ loss in seawater, which takes place in the chloride cells.

### 6.3.4 The kidney and salt balance

Although we have looked briefly at the hagfish and lamprey kidney, so far we have been considering almost entirely extrarenal routes of excretion and ion balance, and the role of the kidney has hardly been mentioned. The elongate fish kidneys are typical mesonephroi (like those of adult lampreys) retaining a segmental structure. This is most evident in their blood supply, which is essentially venous (Figure 6.10), with a renal portal system and operating at low systemic blood pressures not above 20 mmHg. Fish kidneys are particularly interesting because the nephrons in different fishes are remarkably diverse, so it is possible to infer the function of the different segments by comparing the nephrons of fishes living in different habitats (Figure 6.10).

The most striking modification of the nephron is found in some marine teleosts (for example, the angler fish, *Lophius*) where the glomerulus is reduced or even completely lost. So far, 30 species of marine teleosts in six families have been found to have aglomerular nephrons, and there are even some aglomerular Siamese syngnathids which have secondarily returned to freshwater! As so often in physiology, the study of special cases has proved particularly fruitful, and the aglomerular kidney has been no exception in its contribution to renal physiology, for it was here that tubular secretion was first demonstrated in 1928 (by the excretion of phenol red in *Lophius*) and insulin found to be the tracer of choice for determining the glomerular filtration rate (GFR). Another special case (yet to be investigated physiologically) is provided by the kidneys of cottids, where there are some enormous 'giant' nephrons ideally suited for micropuncture and analysis of fluid in different regions of the tubule.

Wide simultaneous variations in GFR and urine production are found in normal fishes without change in urine osmolarity, and it seems that these involve recruitment of glomeruli under varying conditions. Since systemic

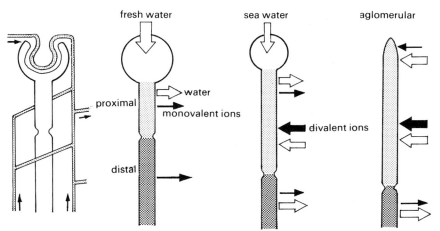

**Figure 6.10** The teleost nephron. Left: blood supply (venous except to glomerulus); right: three functional types of nephron in freshwater and marine teleosts. Note development of proximal region in marine teleosts. In the aglomerular nephron, the distal region includes also the collecting ducts and bladder. After Lahlou (1981) *Bull. Soc. Zool. France*, **106**, 21.

blood pressure is always low, even small variations in pressure can 'shut down' glomeruli, or bring them into use, and coupled with linked changes in tubular absorption, can greatly change urine production without change in urine osmolarity. GFR measurements in single nephrons of the trout, for example, have shown that, in freshwater, 45% of the nephrons are filtering, whereas in seawater only 5% filter. Such a flexible system is well suited to the mainly automatic regulation of urine production as euryhaline fishes suffer changing osmotic loads when they move between waters of varying salinity. The changes in GFR leading to changes in urine production are under the control of pituitary hormones such as prolactin, isotocin and arginine vasotocin (Figure 6.11), hormonal release itself being controlled by a central nervous system osmoreceptor.

### 6.3.5 Tubular structure and function

Tubular structure and arrangement varies greatly in different fishes, as shown in Figure 6.12. Apart from the hagfish, much the simplest is the teleost kidney (hardly surprising remembering the importance of extrarenal excretory routes). In freshwater fishes, the proximal segment absorbs some water and monovalent ions, and the distal segment absorbs monovalent ions. In marine fishes, where the glomeruli are often much reduced, urine flow is significantly greater than GFR; the proximal tubule secretes $Cl^-$ and $Na^+$ (driven by water excretion), but it also secretes divalent ions, in particular $Mg^{2+}$ and $SO_4^{2-}$, which enter the plasma via the seawater the fish

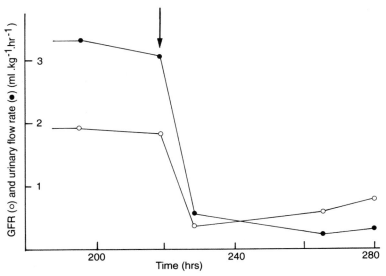

**Figure 6.11** Pituitary control of glomerular filtration rate (GFR) in the freshwater eel. The effect of arginine vasopressin arrowed on GFR (open circles) and urinary flow rate (solid circles). After Babiker and Rankin (1978).

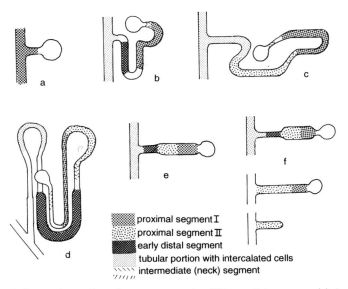

**Figure 6.12** Comparison of nephron structure in different fish groups. (a) hagfish; (b) lamprey; (c) lungfish; (d) elasmobranch; (e) sturgeon; (f) teleost (note differences between tubule structure of glomerular and aglomerular teleost nephrons). Light stipple: collecting duct. After Hentschel and Elger (1989) In structure and function of the kidney, *Comparative Physiology*, **1**, p85, (Kinne ed), Karger.

drinks. Both ions are at much higher levels in the sea (seawater contains around 50 mM $Mg^{2+}$ and 25 mM $SO_4^{2-}$) than in the plasma. These ions are not excreted by extrarenal routes, and although it has been remarked that it would scarcely do the marine teleost kidney justice to consider it purely a magnesium sulphate pump, this is certainly one of its main roles. The distal segment (lacking in many acanthopterygians) mainly absorbs water, and although some monovalent ions are also excreted, extrarenal routes are more important for these. After the distal region of the tubule, collection ducts of various sizes in different species pass the urine to a urinary bladder, and in some species at least, this may play an important role in modifying the ionic content of the urine. For example, in the angler fish (*Lophius*) in seawater, water is resorbed in the bladder, and the bladder of trout in freshwater resorbs and conserves ions.

### 6.3.6  Teleosts in alkaline saline lakes

Interesting very special cases are provided by the teleosts which live in alkaline lakes, like the tilapia *Oreochromis alcalicus grahami* of Lake Magadi in Kenya (pH 10) and *Chalcalburnus tarichi*, a small cyprinoid which lives in Lake Van in eastern Turkey (pH 9.8). Lake Van is a soda lake which has a salinity of 22‰, and is so soapy that the locals can wash their laundry in the lake without need of soap! The tissue fluids of both of these teleosts contain urea at 20–32 mM (probably acting as an osmolyte) and their plasma is nearly isosmotic with the lake water. In most teleosts, $NH_4^+$ is excreted via the gills, but in these fishes, living in a remarkably hostile environment, it is probably excreted at least in part, via the kidneys.

### 6.4  Osmoregulation in elasmobranchiomorphs

Blood ionic composition in marine Holocephali and elasmobranchs is somewhat higher than that of teleosts, but its osmolarity is very much higher, being close to that of seawater (Table 6.1). This is because the blood contains, in addition to the usual ions, large quantities of low-molecular-weight nitrogenous solutes, in particular urea. First found in elasmobranchs in 1858, urea is usually present at around 0.4 M. Various methylamine substances such as trimethylamine oxide (TMO), betaine and sarcosine, and some free amino acids like taurine and $\beta$-alanine are also in fairly high concentration, in total around 0.2 M. Over half of the blood osmolarity is thus due to these nitrogenous solutes, and with the inorganic ions, elasmobranch blood is osmotically close to seawater. Probably marine elasmobranchs are always slightly *hyperosmotic*, as for example at Plymouth, where dogfish (*Scyliorhinus*) in the aquarium circulation seawater (1154 mosmol) were found to have the serum at 1243 mosmol. The gills are permeable to water and so under normal conditions there will always be a slight influx of water excreted as urine that is more dilute than the serum or seawater. The difficulty that elasmobranchs face is that urea is a small molecule (MW = 60) and in other animals rapidly diffuses across cell membranes.

So how do elasmobranchs avoid urea loss across the gills and in the urine? Almost all urea loss is across the gills, around 20–70 $\mu$mol g$^{-1}$ h$^{-1}$ in those species examined, and it is not yet clear how the rate of loss is kept to this level. It is, however, understood how the kidney conserves urea. The most striking feature of the elasmobranch kidney, is that 95% of the urea in the glomerular filtrate is resorbed in the distal tubule, by a countercurrent system. The initial thinner part of the distal tubule is applied to the proximal tubule, making up lateral bundles in the kidney (Figure 6.13). The end result of this folding of the tubule back on itself is that a countercurrent system of five parallel tubular segments is formed (Figure 6.13), effectively resorbing urea and other organic nitrogenous solutes in the GFR as it passes down the tubule.

### 6.4.1  Urea and proteins

Urea retention at near half molar concentration would be a very unexpected adaptation if we were not so familiar with adding urea to dogfish Ringer solutions, for at such a concentration urea disrupts proteins in most animals. Mammalian collagen, haemoglobin and many enzymes are denatured by 0.5 M urea. Either elasmobranchs have in some way modified their proteins to resist the effects of urea (as the Na$^+$ channels of

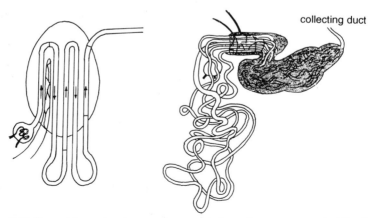

**Figure 6.13** Ray nephron showing countercurrent absorption arrangement. After Lacy *et al.* (1985).

puffer fishes resist the effects of their own tetrodotoxin although this rapidly poisons $Na^+$ channels in other animals), or they protect their proteins otherwise. Some elasmobranch proteins, like their haemoglobins and certain enzymes, are resistant to urea denaturation, and indeed, the eye lens protein and the M4 lactate dehydrogenase (LDH) actually require urea to function properly. Yancey and Somero (1980) working at the Scripps Marine Laboratory in California have shown that there is also another scheme involved. Several muscle enzymes (such as LDH, creatine kinase and pyruvate kinase) are actually protected against the destabilising effects of urea by the other nitrogenous solutes, especially TMO. Maximum protection was found to be given by urea:TMO ratios of 2:1, i.e. the ratio actually found in the blood. So elasmobranchs osmoregulate in the sea in an economical way (see p. 141) by using waste nitrogen products – a strategy similar to the much more wasteful one used by several invertebrates which increases serum osmolarity with high concentrations of free amino acids.

### 6.4.2 Extrarenal salt excretion and the rectal gland

Blood and urine values for $Na^+$ and $Cl^-$ are similar (Table 6.1) and lower than in seawater, so there must be extrarenal routes for NaCl excretion. Elasmobranchs have relatively far fewer chloride cells on the gills than teleosts, and branchial $Na^+/K^+$-ATPase activity is relatively low, so the gills seem likely to be a site of net salt uptake rather than extrusion. This rather puzzling situation was resolved by the unexpected discovery that a rectal gland in the spur dogfish (*Squalus acanthias*) secreted a fluid which was almost pure NaCl at a concentration (ca. 550 mM litre$^{-1}$) about twice that in the body fluids, whilst urea was only at 10–20 mM litre$^{-1}$. The rectal gland (Figure 6.14) open by a short duct from just behind the spiral valve, and consists of a mass of tubules composed of cells very similar in ultrastructure to the chloride cells of fish gills (and with the same high activity of $Na^+/K^+$-ATPase). In Holocephali the rectal gland is represented by nodules in the rectal wall, whilst in *Latimeria* it is similar to sharks. Because the chloride-secreting cells of the rectal gland are a single population and their tubules are accessible for experiment, much of what is known of chloride-cell operation in fishes has come, as we saw in Section 6.3.2, from studies on the elasmobranch rectal gland.

### 6.4.3 Freshwater elasmobranchs

A small number of elasmobranchs are euryhaline, including sawfishes (*Pristis*), sting rays (*Dasyatis*), and blacktip (*Carcharinus melanopterus*) and bull sharks (*C. leucas*). They enter estuaries, and are able to live for long periods in brackish or even freshwater. Bull sharks, for example,

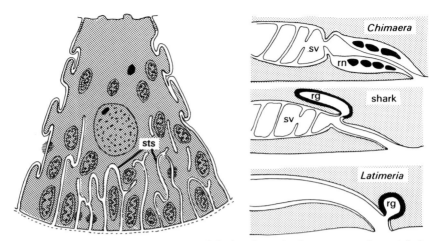

**Figure 6.14** The salt-secreting rectal gland. Left: schematic ultrastructure of a rectal gland cell in *Latimeria* (cf. Figure 6.7). sts: smooth tubular system. Right: rectal glands (rg) in elasmobranchiomorphs and *Latimeria*. In Holocephali, the rectal gland tissue forms nodules (rn) in the rectal wall; sv, spiral valve. After Lagios (1979) The biology and physiology of the living Coelacanth. In *Occ. Papers Calif. Acad. Sci.*, **134**, 25–44.

swim up the Central American San Juan River into Lake Nicaragua, 350 km from the sea. In freshwater, they are much more hyperosmotic to the water than freshwater teleosts, and survive by reducing urea and ion levels in the blood (Figure 6.15), and by secreting a copious flow of dilute urine.

Euryhaline elasmobranchs like these were first studied by Homer Smith in the Malaysian Perak River, but the most interesting freshwater elasmobranchs are the stingrays found in the Orinoco and Amazon drainages. They live up to 4500 km from the sea, and since fossils of the same family (Potamotrygonidae) have been found in Tertiary deposits of Rio Parana Basin, it seems probable that the family has lived in freshwater for millions of years. Blood ion and urea levels are entirely different to those in euryhaline elasmobranchs in freshwater habitats for *Potamotrygon* and its relatives have done what we should have advised them to do – reduced urea to around 1 mM litre$^{-1}$ (compared to around 400 mM litre$^{-1}$ in marine forms), and reduced Na$^+$ and Cl$^-$ to levels similar to those of freshwater telosts (Table 6.2). The rectal gland is superfluous and is vestigial. The GFR is very high, suggesting high rates of production of dilute urine, as we should expect, but this, and the mechanisms which presumably exist for ion uptake at the gills, await investigation. Interestingly, *Potamotrygon* nephrons lack the anatomical countercurrent loops which conserve urea; in consequence even when transferred to half seawater (they die in more concentrated solution) they cannot increase blood urea significantly.

## 6.5 Latimeria

*Latimeria* blood is similar in composition to marine elasmobranch blood (Table 6.1). Unlike elasmobranchs, however, total osmolarity seems to be slightly less than seawater, so if this is correct, *Latimeria* resembles marine teleosts in facing a slight water loss and gain in ions across the gills and other permeable surfaces. As in sharks, $Na^+$ is at the same level in urine as blood (although $Cl^-$ is lower), but what seems strange is that urea levels in blood and urine are the same. These measurements were made at the ureter, and it may be that urea is resorbed up in the bladder, as it is in the urea-containing crab-eating frog, *Rana cancrivora*, which lives in estuaries. Chloride cells are not abundant in the gill epithelium, but there is a rectal gland with cells of similar ultrastructure to those of the shark rectal gland. The existence of urea retention in *R. cancrivora* has helped to dismiss ideas that the similarities between *Latimeria* and elasmobranchs (apart from urea retention, and a rectal gland, there are similarities in the histology of the pituitary and pancreatic islets), imply that they are related, since it is impossible to imagine that the frog should be related to either!

## 6.6 The 'efficiency' of urea retention

It seems clear that urea retention is an adaptation to the marine environment, and we might wonder whether it is a less or more costly way of coping with seawater than that used by marine teleosts. Although calculations are only 'order of magnitude', they suggest that in terms of mM ATP $kg^{-1}$ $h^{-1}$ used, $Cl^-$ excretion by the teleost is about 20 times more costly than urea synthesis and limited $Cl^-$ excretion in the elasmobranch. Yet the urea retention solution requires several features not typical of most fishes. For example, the ornithine–urea enzyme cycle is incomplete or inactive in lampreys and teleosts, but this is probably secondary, and it was likely present in ancestral gnathostomes. Secondly, the urea produced must be retained, and it is usually stated that this will be easier for fishes which are large and have internal fertilisation, for the surface:volume ratio will be greatest in the embryonic stages when it will be less easy to produce sufficient urea to counter its loss across permeable surfaces. However, the egg cases of ovo-viviparous elasmobranchs like rajids and the dogfish *Scyliorhinus* are completely permeable to urea, and the tiny developing embryo can already osmoregulate and retain urea, so the force of the usual argument about size and urea retention is not as great as was supposed.

## 6.7  Plasma ion content and the evolutionary history of different groups of fishes

Different fish groups have body fluids of characteristic ionic composition, for example, lungfishes and *Polypterus* are the most dilute, and hagfishes the most concentrated. We saw in Section 6.2 that the isosmolarity of hagfishes with seawater might be taken to indicate their marine origin, and zoologists of a speculative turn of mind have naturally wondered whether the plasma ionic composition of the other fish groups can be explained in terms of their evolutionary history. Figure 6.15 shows an attempt to work out the ways in which the body-fluid composition (taken as the sum of the major osmotic constituents $Na^+$ and $Cl^-$) suggest that the different groups have moved between seawater and freshwater. A good case could be made that the lowest values represented the longest history of evolution in freshwater, and that where a secondary marine phase appeared, the difference between marine and freshwater plasma concentrations was least in the groups which had most recently re-entered the sea, like the sturgeons. Lampreys do not fit too well into this scheme since seawater values (which are insecurely based on lampreys that had already entered estuaries) should perhaps be even higher, yet in freshwater, blood osmolarity is nearly as low as in lungfishes. Since it seems that lamprey migrations into the sea are of relatively recent origin, perhaps the freshwater value is a more reliable guide to lamprey evolutionary history.

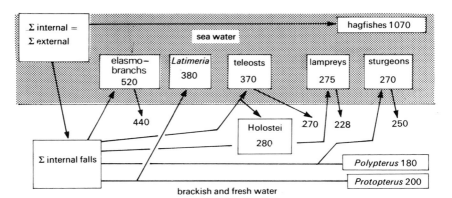

**Figure 6.15** Evolutionary sequence of major fish groups, showing values of sodium and chloride in plasma (approximately equivalent to total ion content). See text. After Lutz (1975) *Copeia*, 369.

## 6.8 Osmoregulation in eggs and larvae

Finally, we consider osmoregulation in fish eggs and larvae. One might expect that being smaller than adult fish that they would have more severe osmotic problems than adults because their surface:volume ratio is more unfavourable. Larval fishes, like adults, are almost all found in freshwater in salinities near zero, or in seawater approaching 35‰ (3.5% salt). Since the developing eggs and larvae of cod, herring, plaice and many other species, like adults, osmoregulate to hold their body-fluid ion concentrations equivalent to between 11 and 14‰ (350–440 mosmol), intermediate salinity conditions in the environment would be osmotically optimal, but they are found only in enclosed or inland seas like the Baltic Sea, Black Sea and Aral Sea.

During oogenesis, the developing oocytes are very permeable and are protected osmotically by the regulatory system of the mother. After shedding, the eggs may be osmotically vulnerable before their own regulatory mechanisms start to operate. Before fertilisation, the chorion (egg case or capsule) is closely apposed to the plasma (or vitelline) membrane that surrounds the yolk, cytoplasm and nucleus of the future embryo. Following fertilisation a perivitelline space usually appears between the chorion and plasma membrane, as a result of imbibition of water. The perivitelline fluid, together with the chorion, which tends to harden, cushion the embryo from shocks caused, for example, by waves breaking or movements of the substratum.

The plasma membrane develops as the site of osmoregulation; at 12 h after fertilisation, plaice eggs can osmoregulate successfully in external salinities between 5 and 50‰, a range far in excess of that found in natural conditions, the yolk osmolarity being regulated between 10 and 20‰. After hatching, plaice larvae survive for 24 h in salinities between 5 and 65‰ and for 1 week between 15 and 60‰. After metamorphosis they can survive for 1 week at salinities between 2.5 and 45‰—altogether a remarkable euryhalinity for a marine fish. Herring sometimes spawn at much lower salinities than plaice. The yolk is regulated to an internal osmolarity of 12–15‰ whatever the external salinity. Following hatching, herring larvae can survive in salinities of 1.4–60.1‰ for 24 h and 2.5–52.5‰ for 1 week, their body fluid concentrations ranging from 8.7 to 27.5‰. After metamorphosis, the salinity tolerance lies between 6 and 45‰.

The mechanisms allowing such impressive osmoregulatory performance are being clarified in current research which shows that embryonic and larval tissues can survive a wide range of ionic concentrations. It is also clear that the embryos, and especially the larvae, have much of the osmoregulatory equipment of the adults. Chloride cells are found in the larvae, and often in the embryos before hatching, in many freshwater or estuarine species such as the carp, rainbow trout, killifish, molly and ayu (*Plecoglossus*) and in the marine plaice, turbot, flounder, anchovy, herring, sardine, puffer and Red Sea bream. After hatching, chloride cells are concentrated in the pectoral region and later can be seen in the branchial region close to where the pseudobranch later develops (Figure 6.16). New techniques involve the microscopic identification of chloride cells by fluorescent markers attached to ouabain which is a specific inhibitor of their $Na^+/K^+$-ATPase system (the basis of the sodium pump) or by the use of fluorochromes specific for mitochondria (which are present in large numbers in chloride cells). Monoclonal antibodies are also becoming available which will locate the sodium pumps that are the underlying mechanism of the chloride cell. After this mechanism is blocked by incubation *in vitro* with 1 mM ouabain, the late-stage embryos of the rockfish *Sebastes* showed a high mortality.

The use of inert substances such as dextran marked radioactively or by fluorescence shows that larvae of herring, plaice, cod and halibut drink seawater just as the adults do and there is evidence that water absorption in the gut is very efficient. In the first three species, the drinking rate is only half in 16‰ that in 32‰ seawater (Table 6.3).

**Table 6.3** Drinking rates of larvae in different salinities related to wet weight (from Tytler and Blaxter, 1988)

| Species | Salinity (‰) | Drinking rate ($\mu l\ g^{-1}\ h^{-1}$) |
|---------|--------------|------------------------------------------|
| Cod | 16 | 42.9 |
|  | 32 | 111.6 |
| Herring | 16 | 10.8 |
|  | 32 | 25.9 |
| Plaice | 16 | 12.1 |
|  | 32 | 17.5 |

A drinking rate of 26 $\mu l\ g^{-1}\ h^{-1}$ for a herring larva can be compared with 3–11 $\mu l\ g^{-1}\ h^{-1}$ in adult fish. The drinking rates, when corrected for differences in body area, are lower in larvae than adults. Experiments using radioactive sodium, chloride and water show that eggs of cod and plaice are least permeable, larvae of cod and herring have intermediate permeability and adult eels and *Serranus* have the highest permeability (Table 6.4). It seems likely that eggs and larvae reduce their osmotic problems by having a relatively impermeable integument. The vitelline membrane probably endows the egg with its low permeability, while the high permeability of adults is presumably due to a large area of gill and

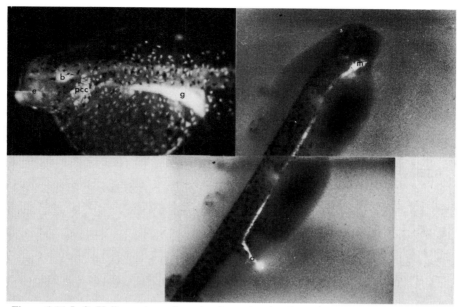

**Figure 6.16** Left: Yolk sac stage of a turbot larva showing the cutaneous chloride cells rich in mitochondria stained with DASPMI which fluoresces in UV light under confocal microscope. Right: Yolk sac turbot larva which has been drinking seawater containing rhodamine-labelled dextran. The contents of the gut fluoresce brilliantly under an epifluorescence microscope. a: anus; b: brain; e: eye; g: gut; m: mouth; pcc: prebranchial chloride cells; ys: yolk sac. Photograph kindly provided by Dr Peter Tytler, Stirling University.

**Table 6.4** Diffusion permeability coefficients ($P_{diff}$ in cm s$^{-1}$ $\times$ 10$^6$) for $^3$H$_2$O for eggs, larvae and adult fish (from Tytler and Bell, 1989)

| Species | Eggs | Larvae | Adults |
|---|---|---|---|
| Cod | 0.2–1.2 | 2.4 | – |
| Herring | – | 1.0 | – |
| Plaice | 0.2 | – | – |
| Eel | – | – | 9.0 |
| *Serranus* | – | – | 9.0 |

–, no data.

'leaky' junctions of branchial chloride cells. Nevertheless larval skin, considering how thin it is and lacking in scales, has a low permeability which is clearly an adaptation to reduce osmotic exchange. The distance for diffusion through the skin has been estimated as 2–5 μm. In herring larvae there are plates of fibrous material and in turbot larvae a high density of cells producing mucus either of which might act as barriers to diffusion.

Larval fish have a functional pronephric kidney, although little is known about the early development of kidney function in fishes. A single nephron was observed in the pronephros of newly hatched brown trout (*Salmo trutta*) and of embryo chum salmon (*Oncorhynchus keta*) 33 days after fertilisation and in the mesonephros at hatching. Flagella are seen beating in the archinephric duct of herring larvae after hatching. In newly hatched turbot, the archinephric ducts join and pass to the gut near the anus but older larvae develop a urinary bladder with a separate opening.

## Bilbiography

Alderdice, D.F. (1988) Osmotic and ionic regulation in teleost eggs and larvae, in *Fish Physiology*, **XIA** (Hoar, W.S. and Randall, D.J. eds) Academic Press, New York, pp. 163–251.

Babiker, M.M. and Rankin, J.C. (1978). Neurohypophysial hormonal control of kidney function in the European eel (*Anguilla anguilla*) adapted to sea-water or fresh water. *Journal of Endocrinology*, **76**, 347–358.

Conte, F.P., Takano, K., Takemura, A. and Boehlert, G.W. (1991) Ontogeny of the sodium pump in embryos of rockfish of the genus *Sebastes*. *Environmental Biology of Fishes*, **30**, 127–133.

Feldmeth, C.R. and Waggoner, J.P. (1972) Field measurements of tolerance to extreme hypersalinity in the California killifish, *Fundulus parvipinnis*. *Copeia*, **1972**, 592–594.

Griffith, R.W. (1985) Habitat, phylogeny and the evolution of osmoregulatory strategies in primitive fishes, in *Evolutionary Biology of Primitive Fishes* (Foreman, R.E., Gorbman, A., Dodd, J.M. and Olsson, R. eds), NATO ASI series A, **103**, 69–80.

Griffith, R.W. and Pang, P.K.T. (1979) Mechanisms of osmoregulation in the coelacanth: evolutionary implications, in *The Biology and Physiology of the Living Coelacanth* (McCosker, J.E. and Lagios, M.D. eds), Occasional Papers of the California Academy of Science, **134**, 79–93.

Hickman, C.P. and Trump, B.F. (1969) The kidney, in *Fish Physiology* (Hoar, W.S. and Randall, D.J. eds), Academic Press, New York, pp. 91–240.

Hughes, G.M., Peyraud, C., Peyraud-Waitzenegger, M. and Soulier, P. (1982) Physiological evidence for the occurrence of pathways shunting blood away from the secondary lamellae of eel gills. *Journal of Experimental Biology*, **98**, 277–288.

Hwang, P-P. (1989) Distribution of chloride cells in teleost larvae. *Journal of Morphology*, **200**, 1–8.

Kirsch, R., Meens, R. and Meister, M.F. (1981) Osmoregulation chez les teleostéens marins: rôle des branchies et du tube digestif. *Bulletin de la Société Zoologique de France*, **106**, 31–36.

Krogh, A. (1939) Osmotic regulation in aquatic animals, in *Cambridge Comparative Physiology* (Borcroft, J. and Saunders, J.T. eds) Cambridge University Press, pp. 242.

Lacy, E.R., Reale, E., Schlusselburg, D.S., Smith, W.K. and Woodward, D.J. (1985) A renal countercurrent system in marine elasmobranch fish: a computer assisted reconstruction. *Science*, **227**, 1351–1354.

Lacy, E.R., Schmidt-Nielsen, B., Galaske, R.G. and Stolte, H. (1975) Configuration of the skate (*Raja erinacea*) nephron and ultrastructure of two segments of the proximal tubule. *Bulletin Mount Desert Island Biological Laboratory*, **15**, 54–56.

Lahlou, B. (ed) (1980) *Epithelial transport in the lower vertebrates*, Cambridge University Press, Cambridge.

Lutz, P.L. (1975) Adaptive and evolutionary aspects of the ionic content of fishes, *Copeia*, **1975**, 369–373.

McDowall, R.M. (1993) A recent marine ancestry for diadromous fishes? Sometimes yes, but mostly no! *Environmental Biology of Fishes*, **37**, 329–335.

Macallum, A.B. (1910) The inorganic composition of the blood in vertebrates and invertebrates and its origin, *Proc. Royal Soc. Lon. B*, **82**, 602–624.

Macallum, A.B. (1926) The palaeochemistry of the body fluids and tissues. *Physiol. Rev.*, **6**, 316–355.

Maetz, J. (1974) Aspects of adaptation to hypo-osmotic and hyper-osmotic environments, in *Biochemical and Biophysical Perspectives in Marine Biology* (Malins, D.C. and Sargent, J.R. eds) Academic Press, London.

Marshall, E.K. and Smith, H.W. (1930) The glomerular development of the vertebrate kidney in relation to habitat. *Biological Bulletin Woods Hole*, **59**, 135–153.

Pang, P.K.T., Griffith, R.W. and Atz, J.W. (1977) Osmoregulation in elasmobranchs. *American Zoologist*, **17**, 365–377.

Potts, W.T.W. (1976) Ion transport and osmoregulation in marine fish, in *Perspectives in experimental biology* (Davies P.S. ed) Pergamon Press, Oxford, pp. 65–75.

Rankin, J.C., Henderson, I.W. and Brown, J.A. (1983) Osmoregulation and the control of kidney function, in *Control Processes in Fish Physiology*, (ed Rankin, J.C., Pitcher, T.J. and Duggan, R.) Croom Helm, London, pp. 66–88.

Smith, H.W. (1961) *From Fish to Philosopher*, Doubleday, New York.

Thorson, T.B., Cowan, C.M. and Watson, D.E. (1967) *Potamotrygon* spp.: elasmobranchs with low urea content. *Science*, **158**, 377–375.

Tytler, P. and Blaxter, J.H.S. (1988) The effects of external salinity on the drinking rates of larval herring (*Clupea harengus*), plaice (*Pleuronectes platessa*) and cod (*Gadus morhua*). *Journal of Experimental Biology*, **138**, 1–15.

Tytler, P. and Bell, M.V. (1989) A study of diffusional permeability of water, sodium and chloride in yolk-sac larvae of cod (*Gadus morhua* L.). *Journal of Experimental Biology*, **147**, 125–132.

Tytler, P., Bell, M.V. and Robinson, J. (1993) The ontogeny of osmoregulation in marine fish, in *Physiological and Biochemical Aspects of Fish Development* (Walter, B.T. and Fyhn, H.J.), University of Bergen Press, pp. 249–258.

Wood, C.M. (1988) Fish gill and kidney ion exchange after exercise. *Journal of Experimental Biology*, **136**, 461–481.

Yancy, P.H. and Somero, G.N. (1980) Methylamine osmoregulatory solutes of elasmobranch fishes counteract urea inhibition of enzymes. *J. Exp. Zool.*, **212**, 205–213.

# 7   Food and feeding

Adult fishes feed in a wide variety of ways, ranging from sieving phytoplankton or grazing algae, to suction feeding on benthic invertebrates, and to devouring other fishes whole or in portions. Some, like the phytoplankton filtering menhaden (*Brevoortia*) feed continuously, whilst the great white shark (*Carcharodon*) is said to sate itself on average with one large meal every 6 weeks, spending the intervening period hungrily looking for the next victim. As we might guess, some with peculiar diets have devised very specialised methods of obtaining them. For example, insectivores like the archer fish (*Toxotes*) knock insects off overhanging vegetation by squirting a jet of water at them, whilst the much larger *Osteoglossum* leaps up to 2 m out of the water to snatch them off branches. Other fishes have a much more general diet – the blue shark (*Prionace*) feeds on dead whales, fish, cephalopods, adult ascidians, gastropods and crabs! Rather surprisingly perhaps, although some fishes seem to have specialisations that suggest a particular diet and mode of feeding, this does not prevent them from turning these adaptations to other uses. Thus the cichlid *Petrotilapia* of Lake Malawi is seemingly specialised for scraping algae off rocks with its trifid teeth, but Liem (1980) found that it could also feed in seven other ways, each involving a different pattern of jaw muscle activity. For instance, *Petrotilapia* bites scales and fins from other fishes, collects floating food, sucks invertebrates from bottom mud, and catches small fishes in midwater (Figure 7.1). This kind of versatility in using what seems to be a morphology adapted for a single purpose may be uncommon, but it is a salutory warning to anyone attempting to interpret function from morphology alone! In a fascinating study of the cichlid fishes of the African Great Lakes, Fryer and Iles discuss other ways of cichlid feeding, including such extraordinary specialisations as shamming dead on the bottom to decoy other fishes, and as in *Haplochromis compressiceps*, sucking the eyes out of other fishes.

## 7.1  Food of larval and young fishes

In the Lower Mesozoic seas, copepods began to diversify and become dominant forms, and today they are the largest source of animal protein in the world. In freshwaters, cladocerans and rotifers dominate the zooplankton. This immense source of conveniently small food particles is ideal

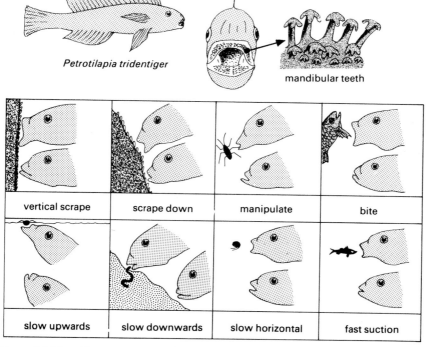

*Petrotilapia tridentiger*

mandibular teeth

| vertical scrape | scrape down | manipulate | bite |

| slow upwards | slow downwards | slow horizontal | fast suction |

**Figure 7.1** The trifid mandibular teeth and versatile feeding techniques of the Lake Malawi cichlid, *Petrotilapia*. The three slow methods (bottom left) involve slow controlled suction. After Liem (1980) *Amer. Zool.*, **20**, 295.

for fish larvae, and the larvae of most bony fishes feed on planktonic crustaceans, whatever they eat as adults. There has been a huge amount of research on larval fishes. A recent bibliography by Hoyt listed 13 717 papers up to 1988 and there have been many hundreds published since. One reason for this interest is explained on page 191 – it is because fishery biologists, trying to unravel the problems of survival, have worked extensively on the early life-history stages. Starvation is one of the causes of the heavy mortality of larval fish, so there are numerous studies of larval feeding and energetics which have sought for mechanisms controlling survival.

Most fish larvae at first feeding are very small (<5-mm long) and show a regular rhythmic swimming pattern, alternating a few tail beats with periods of rest. Their small size and low swimming speed mean that they operate at low Reynolds numbers (p. 55), and so the viscosity of the water is the dominant factor in their locomotion. This short swim-and-rest pattern of searching the surrounding water for food uses the minimum amount of energy. When the larvae grow and swim faster, the Reynolds number increases and so they change to beat-and-glide swimming which is now energetically optimal.

Larvae of species like anchovy, herring, plaice, sole, turbot and cod sight their prey at a distance of only a few millimetres (usually less than one body length), and bend their bodies into an S-shape before darting forward to seize it. The ability to catch the prey depends on its size compared with the larva's jaw gape, and on the experience of the larva, catching efficiency increasing with age as well as larval size. The volume of water searched for food obviously depends on the sighting distance, the field of view of the larva and its swimming speed (Figure 7.2). Since the volume searched depends on the size of the visual field, adaptations to increase the size of the visual field should be at a premium. Some mesopelagic fish larvae, like *Idiacanthus* have extraordinary stalked eyes supported by cartilaginous rods that probably increase the visual field by 80

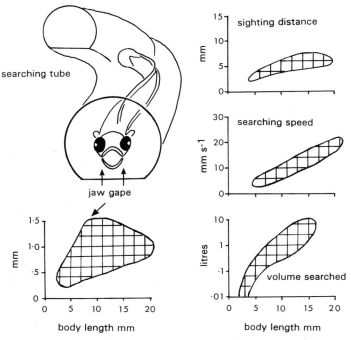

**Figure 7.2** Top left: the feeding 'tube' searched by a fish larva. Bottom left: graph showing how the gape of the jaw changes as larvae grow. The shaded area in this and the other graphs is an envelope containing a number of regression lines showing the relationship between jaw gape and body length in several species. Right: three graphs, the relationship between prey sighting distance, searching speed and volume of water searched as the larvae of several species grow. Redrawn from data in Rosenthal and Hempel (1970), Hunter (1981), and Blaxter (1985).

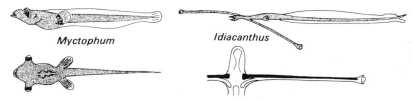

**Figure 7.3** Mesopelagic fish larvae with extended eyes which increase search area. After Weihs and Moser (1981) *Bull. Mar. Sci.*, **31**, 31.

times, but even a much smaller peduncle, as in *Myctophum*, provides a significant improvement in the field (Figure 7.3). Remarkably enough, stalked eyes have been evolved entirely independently not only in the hammerhead sharks (*Sphyrna*), but also in the Australian platystomatid flies like *Achias*!

How much food do fish larvae need? Some earlier experiments on the growth of larvae in tanks suggested that a food density of at least 1000 food organisms per litre was needed. Later results suggested 100–1000 organisms per litre might suffice (depending, of course, on their size and nutritional value). Such high densities of microzooplankton are rather rare in the natural world and the best survival of larvae is likely to be in food patches at various types of frontal systems, thermoclines or other interfaces where food organisms aggregate.

The diet of larval fishes at first-feeding largely consists of copepod nauplii, the eggs and larvae of polychaetes, and small invertebrates such as the pelagic larvacean tunicate *Oikopleura*. Plaice larvae in the North Sea feed almost exclusively on *Oikopleura*, and although these fragile animals are soon digested in the larval gut, their faecal pellets are not. Since the size of the faecal pellets is related to the size of the tunicate, it was possible to show that as the plaice larvae grew, they ate larger and larger *Oikopleura* (Figure 7.4), in the same way that copepod feeders gradually take larger and larger copepods. As they grow, the larvae gradually change their feeding methods, although even quite large larvae and juveniles can continue to snap up small particles. In the pomacentrid *Amphiprion*, the youngest larvae feed by approaching the prey, and opening their mouths whilst moving forwards, to engulf the prey, the so-called ram feeding method. At 8 days, following remodelling and change in shape of the skull and buccal cavity, the larvae switch to suction feeding, where the prey is approached and the jaws protruded concomitantly with increase in size of the buccal cavity, so sucking the prey in, a more efficient technique.

When the later life-history stages move away from the nursery grounds, they adopt different diets and tend to feed on organisms that give the maximum nutritional return for the minimum expenditure of energy – a sensible optimal foraging strategy, one of most return for least effort.

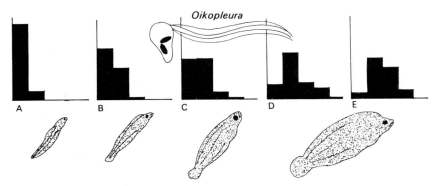

**Figure 7.4** *Oikopleura* eaten by plaice larvae in the North Sea. The histograms show the numbers of *Oikopleura* of different overall lengths taken by different stages of plaice larvae. After Shelbourne (1957, 1962) *J. Mar. Biol. Ass. UK*, **36**, 539, and **42**, 243.

Food passes along the larval gut (as in adults) by peristalsis, but the guts of plecoglossid, osmerid, clupeid and salangid larvae also move food along the gut by ciliary action. These groups of salmoniform and clupeiform fishes are so-called 'primitive' teleosts (see Figure 1.2, p. 4) suggesting that ciliary transport of food in the gut is the primitive condition (it is also seen in the adult sea lamprey, *Petromyzon marinus*). However, anomalously, the larvae of Atlantic halibut (*Hippoglossus hippoglossus*), a more 'advanced' teleost, also have cilia in the gut. The rate that food passes through the gut depends on the ambient temperature and the previous feeding history of the larva. The youngest larvae do not have a stomach and food passes quickly along the gut to be digested near the anus. The guts of most larvae are short, but those of the midwater predatory stomiatoids like astronesthids and melanostomiatids (see Figure 8.12d) are extraordinarily long, trailing far out from the body to end in a specialised terminal region. The function of these bizarre 'extracorporeal' guts, which are muscular and capable of writhing movements, is unknown, but presumably indicates that such larvae feed on food particles requiring long digestion.

With the usual short larval gut, easily digested food is at a premium, the more so since not all the adult digestive enzymes are necessarily present at first feeding. Trypsin and amylase are usually present but pepsin appears later, except in the salmonids, whose larvae hatch at a more advanced stage. There is a distinct possibility that the digestive enzymes of the food organisms themselves (so-called exogenous enzymes) also play a part in digestion. Food intake is often cyclical with bursts of feeding at dawn and dusk. This may be a feature of the availability of food but larvae feeding at dusk must digest their food in the dark and so may be less vulnerable to predators (food in the gut of transparent larvae making them more conspicuous to predators); larvae feeding at dawn after a night of fasting may just be hungry.

Cyclical feeding is less evident in tank experiments where larval fishes show few signs of satiation. In fact, herring larvae continue to feed so speedily in high concentrations of brine shrimp nauplii that live nauplii may be seen to exit from the anus! Apart from the special case of the large lamprey ammocoete larva, we know of no fish larvae that filter-feed on phytoplankton. This is not to say that fish larvae do not utilise smaller planktonic organisms than copepod nauplii as food; rather that they are taken by snapping or possibly during osmoregulatory drinking (p. 148). Many organisms, such as dinoflagellates, have a cellulose cell wall, requiring the larvae to secrete a digestive enzyme, cellulase, to break down the wall before digesting the cell contents. Although some dinoflagellates have an armoured exterior, the unarmoured species *Gymnodinium splendens* is utilized as a food by larval northern anchovy (*Engraulis mordax*) both in tanks and in the sea. It has the great advantage of being small – about 50 μm – and so is ideal for first-feeding larvae. In the same way, small rotifers such as *Brachionus* make good larval food and are widely used in mariculture.

Several attempts have been made to develop microparticles to feed young larvae in rearing tanks, but so far with limited success, owing to various practical problems. For example, quite apart from being nutritionally adequate, they must be neutrally buoyant so that they remain floating in midwater. Although larvae will eat non-motile invertebrate eggs, they prefer moving food.

There has been much debate about whether marine organisms in general, and fish larvae such as eel leptocephali in particular, can utilise dissolved organic matter. Pütter's theory, as this idea was known in the older literature, is difficult to test rigorously since high levels of organic matter encourage bacterial growth, and the larvae may simply be utilising ingested bacteria. The leptocephalus larva has seemed a good candidate for taking up dissolved organic matter. No food is ever seen in their guts, in many species the teeth project forwards in a seemingly inefficient way (but like that of many pterosaurs) for capturing food, and until recently leptocephali caught in good condition did not feed. However, recent work on Japanese leptocephali has shown that they will feed on squid paste and in nature they probably feed on microplankton.

Other recent work using seawater spiked with a radioactive isotope of the amino acid alanine, has shown that turbot larvae take this into the gut, presumably by drinking. It would then be available for uptake across the wall of the gut as in more normal digestion.

## 7.2 Adult fishes

### 7.2.1 Plankton feeders

The protochordate ancestors of fishes were certainly, as adults, suspension feeders, sieving phytoplankton and detrital particles from the water drawn into the pharynx by ciliary action, much as amphioxus does today. A muscular pump to provide the feeding flow, like that of the lamprey ammocoete larva came later, presumably permitting increase in size. However, if conodonts are really allied to chordates, their tooth apparatus seems to show that feeding on larger prey was an early adaptation. The advantages of filter-feeding are twofold: very small prey can be taken by much larger predators and feeding can continue day and night since vision plays no role, as it so often does in particle feeding. The filter-feeders *par excellence* are the clupeoids (herrings, sardines and anchovies) which suck in water, and pass it over the gills, sieving out the particles on the gill rakers. But almost all such clupeoids are 'facultative' filter-feeders, capable also of snapping up particles singly, and switching between the two modes of feeding depending on prey size, prey density and light intensity, almost certainly optimising their intake of food for minimum effort (optimal foraging). Scombrids like the Pacific mackerel (*Scomber japonicus*) and fishes from other groups are also facultative filter-feeders. The only reported obligate adult filter-feeders are the menhaden (*Brevoortia*), and the African freshwater cichlid *Sarotherodon galilaeum*, although the Peruvian anchoveta (*Engraulis ringens*) seems to feed only on phytoplankton, and almost certainly is an obligate filter-feeder.

While the menhaden swims with its mouth permanently open, most filter-feeders open the mouth only for short feeding bouts lasting 0.5–3.0 s. The minimum size of the particles retained depends on the spacing of the elements of the branchial sieve (Figure 7.5); menhaden and anchoveta can retain food like the diatom *Skeletonema* as small as 17 μm. The gill-raker spacing depends on the size of the fish and the extent to which the mouth is open, but many clupeoids have spiny denticles on the rakers that reduce their mesh size further. Generally, clupeoid fishes particle-feed on large prey and filter-feed on small prey. In northern anchovy, the change in feeding mode occurs at a prey size of about 1 mm. Although the efficiency of filter-feeding depends on particle size, it ceases when the particle density drops to a threshold level – for example 5–18 nauplii per litre in the northern anchovy, 18 nauplii per litre in menhaden and just below 50 nauplii per litre in the herring. Filtering continues at much lower densities when the prey are larger. Herring, thread-fin shad (*Dorosoma petenense*) and alewives (*Alosa pseudoharengus*) continue to filter-feed in the dark and it is likely that this also occurs in other clupeoid species. Many fishes abandon school formation to particle-feed, often in a competitive feeding frenzy, but filter-feeding is achieved at a slower pace. Schooling is often retained and the fish do not need to change direction to feed, so reducing vulnerability to predators.

The way in which sieved particles are collected from the gill rakers and transferred to the oesophagus is something of a mystery. In menhaden there are no mucous cells associated with the denticles (spinules) suggesting that sieving is purely a mechanical process. Thereafter food can probably be passed along one side of the gill raker, which has mucus cells, either over the mucus layer or complexed with it. The other side of the raker is grooved but lacks mucus cells and the particles may move down the raker in a vortex formed in the groove. In the thread-fin shad and the anchovy (*Cetengraulis*), both clupeoids, in the osteoglossid (*Heterotis*) and in the milkfish (*Chanos*), paired epibranchial organs are present. These are muscular diverticula containing mucus cells at the back of the pharynx. They probably squeeze concentrated food particles into a bolus before swallowing. These fishes feed on phytoplankton and mud particles, triturated by the same kind of gizzard as in mullets (Mugilidae). Although mullets typically browse on the algal film over mud (leaving a characteristic series of depressions where they have gulped in surface mud), they can also feed planktonically on algae at the surface–air interface using the gill raker sieve (Figure 7.5).

Some filter-feeders, especially those in which phytoplankton is an important source of food, have exceedingly long guts. The gut in the menhaden is ten times the length of that in the herring and has 400 pyloric caecae compared with 20 in the herring, reflecting the greater difficulty of digesting plant food. Filter-feeding appears during ontogeny; it is not found in young larvae, which are usually microzooplankton particle-feeders. In herring, menhaden, in the paddle fish *Polyodon* and in certain

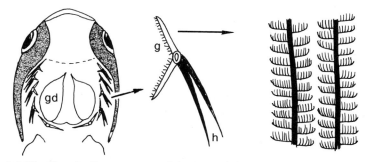

**Figure 7.5** The filter-feeding apparatus of the mullet (*Liza*), a facultative filter feeder. Left: Schematic horizontal section of head showing position of gill rakers on branchial arches. Right: The filtering gill rakers. (g) gill rakers; (gd) gizzard; h: gill hemibranchs. After Guinea and Fernandez (1992) *J. Fish. Biol.*, **41**, 381.

cichlid species filtering appears as the gill rakers develop. Most of the clupeoids have reached a length of 80–100 mm before starting to filter.

Filter-feeding is likely to be adopted most often to exploit phytoplankton which is in sufficient density to make filtering worth while. Densities of 10–50 zooplankton organisms per litre are not particularly common and it is essential for many fishes to have the ability to switch their feeding mode to particle-feeding. In African lakes, tilapias sometimes filter phytoplankton but can switch to sedimented material of planktonic origin. The cichlid *Sarotherodon galilaeum* is, however, an obligate filter-feeder when adult. Particle-feeding is highly efficient but often also highly competitive, leading to feeding frenzies. A lake whitefish (*Coregonus clupeaformis*) has been seen to catch more than 600 *Daphnia* in 15 minutes while herring can take about 60 prey per minute.

Particle-feeding on zooplankton is seen also in some rather unexpected fishes. The deep-sea tripod fishes (see Figure 2.6 on p. 30) pick copepods from the benthopelagic plankton, as they sit perched on their fins facing into the current. Curiously, the spur dog (*Squalus*) may also feed in this way on the planktonic ctenophore *Pleurobrachia*, which it sometimes takes individually in such amounts as to fill its gut, and it has recently been suggested (from their parasite communities) that eels of all sizes in rivers and lakes in England feed regularly on planktonic copepods.

### 7.2.2 Large zooplankton filter-feeders

There are a few large fishes which are able to filter sufficient water that they can filter-feed on zooplankton when this is abundant. In freshwater, the paddle fish (*Polyodon*) of the Mississippi Basin filters cladocerans, cruising around with its lower jaw dropped almost to 90° (Figure 2.14 on p. 39). In the oceans, the huge whale sharks (*Rhincodon*), basking sharks (*Cetorhinus*) and manta rays (*Mobula*) all filter-feed. In whale sharks and manta rays, the filter is spongy tissue formed from modified denticles, whilst in basking sharks, there are long gill rakers (again modified denticles; Figure 1.14 on p. 16) on each gill arch. Whale sharks sometimes filter in the vertical position, pushing their heads slowly out of the water, allowing the pharynx to drain, and then subsiding slowly again below the surface. Manta rays and whale sharks have larger-mesh sieving mechanisms than basking sharks and so retain larger particles, even small fish! Basking sharks feed on copepods and live in temperate waters, where zooplankton varies widely in abundance during the year. Calculations based on zooplankton calorific value and abundance on the one hand, and estimates of the energy expenditure of the shark when swimming with wide open mouth whilst filtering on the other, lead to a curious conclusion. They suggest that these enormous fish are actually balanced on the knife edge of using more energy capturing their food than they can obtain from it. The drag incurred (i.e. energy expenditure) when filtering is much increased over swimming with the mouth closed, so it may be that they only filter-feed when plankton is abundant, perhaps deciding when to filter by sampling the copepod content of the water, using their ampullae of Lorenzini (p. 230) to detect copepod-muscle electrical activity. Surprisingly, basking sharks shed these filtering rakers each autumn and regrow them again in the spring, so they cannot feed in the winter when copepods are scarce. It has even been suggested that they hibernate during the winter, resting quietly on the bottom, but this seems rather unlikely, as rakerless sharks (with empty stomachs) have been caught several times off Plymouth in the winter when they blundered into nets. We still know very little of their biology, but basking sharks evidently offer an interesting challenge to optimality modellers.

## 7.2.3 Herbivorous fishes

Overall, herbivorous fishes are very much in the minority, and are least common in the sea and in temperate freshwaters. Only some 6% of 600 species whose diet was analysed by Love are herbivores, compared with 85% of carnivores. Perhaps this is because more energy has to be expended by herbivores foraging for their food than carnivores expend on feeding. Certainly, compared with carnivores herbivores have to spend a good deal more of their lives feeding. Some phytoplankton feeders filter continuously, whilst a parrot fish spent $8 \text{ h day}^{-1}$, and *Haplochromis* and *Tilapia* $14 \text{ h day}^{-1}$, this in comparison with $1–3 \text{ h day}^{-1}$ for salmonids. Whatever the diet, all fish have a high protein requirement (around 50% of dry weight of the diet). How do herbivores satisfy this requirement? As detritivores and herbivores they ingest large amounts of attached protein-rich microorganisms, even if they do not eat protein-rich algae.

Most marine herbivores are found around the coral reefs of the tropics, where surgeon fishes (Acanthuridae) crop algae growing over corals and on sandy patches near the reef, whilst parrot fishes (Scaridae), rabbit fishes (Siganidae), and damselfishes (Pomacentridae) browse algae scraped off coral surfaces. Some temperate shore fishes, such as blennies and gobies feed largely on seaweeds, and the much prized temperate freshwater ayu (*Plecoglossus*) of Japanese rivers, feeds solely on moss scraped from its territory of mossy stones. Ayu are an expensive delicacy since they are caught in a curious and somewhat inefficient way; fishermen place a small fish on their line, below which is a series of unbaited hooks, and the ayu is foul-hooked as it comes out to defend its feeding territory.

Herbivory is, however, much more common in tropical freshwaters, where seasonal flooding inundates forests and plains. Here fishes like the grass carp (*Ctenopharyngodon*) can feed on grasses, and decaying vegetation, whilst fruits, flowers and seeds that fall into the water are seized by fishes such as the cyprinid *Puntius*. As in phytoplankton, filter-feeders and mammalian herbivores, herbivorous fish have longer guts with greater surface area for absorption than carnivores or omnivores. As a rough rule, the ratio of gut length to body length is greater than 3 in herbivorous fishes, from 1 to 3 in omnivores and less than 1 in carnivores.

## 7.3 Jaw mechanisms

### 7.3.1 *Protrusible jaws of elasmobranchs*

All elasmobranchs are carnivorous, and have relatively simple jaw mechanisms involving the two jaw cartilages (Meckel's cartilage and the palatoquadrate), and (in living forms) the hyomandibula. Early Palaeozoic sharks had more or less terminal jaws, with grasping 'cladodont' teeth, and the upper jaw (the palatoquadrate) articulated with the neurocranium via otic and orbital processes (Figure 7.6). This ancestral *amphistylic* jaw suspension became modified as the jaw shortened and the mouth became inferior rather than terminal, to the modern more mobile *hyostylic* jaw suspension on the neurocranium. Although modern elasmobranchs feed in a variety of ways, it has been suggested that the original importance of this change in jaw suspension may have been to allow suction feeding from the benthos, in the way still seen in rays. In most modern elasmobranchs, the upper jaw has considerable mobility in the vertical plane, and the entire jaw system is braced against the otic region of the neurocranium by the hyomandibular (Figure 7.6). The angel fish (*Squatina*), an ambush predator which is perfectly camouflaged on a sandy bottom, engulfs small fish by rapidly lowering the lower jaw and so vacuuming them into its mouth as it rears upwards. To avoid lateral movement of the lower jaw during this violent movement, Meckel's cartilages have curved upward projections which pass through the orbit in connective-tissue tubes, achieving (by a quite different means) the same limiting of lateral movement as in the jaws of the badger *Meles*.

Feeding mechanisms are best known in carchariniform and lamniform sharks, where there are prominent pre-orbital processes on the upper jaw, which is only loosely connected to the front of the braincase by short ethmo-palatine ligaments. As the shark opens its jaws to feed, the hyomandibulae swing outward to brace the jaws firmly and laterally against the skin behind the mouth. This rotates the upper jaw outwards and downwards, and with large prey, leads to a deep gouging bite, by the cutting teeth of the upper jaw. These cutting teeth are often serrated like a breadknife, and are remarkably sharp, as can be seen by examining the skulls of great white (*Carcharodon*) or tiger sharks (*Galeocerdo*). The combination of a relatively long jaw and the efficient cutting teeth

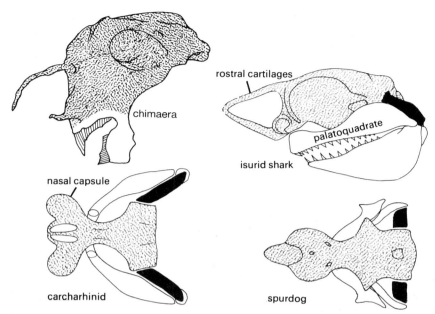

**Figure 7.6** Cranium and jaws of elasmobranchiomorphs. Lateral views above, dorsal below. Hyomandibula, black; neurocranium and capsules, shaded. Note bizarre shape of holocephalan skull with palatoquadrate fused to braincase and different angles of hyomandibula in the carcharinid and spur dog (*Acanthias*). After Daniel (1922) *The Elasmobranch Fishes*, U. Cal. Press, Berkeley; Devillers (1958) in *Traité de Zoologie* (Grassé ed) **13**, fasc. 1, 551 Masson et Cie, Paris; Young (1981) *The Life of Vertebrates*, 3rd edn, Clarendon, Oxford; and Moss (1977).

produces a feeding system unusual in predators – large chunks can be sheared out of prey (like marine mammals) that are too large to be swallowed entire. Such sharks also feed on small fishes and squid, which are swallowed whole.

In squaloid sharks like the spur dog (*Squalus*), the Greenland shark (*Somniosus*), and such deep-sea genera as *Dalatias*, *Etmopterus*, and *Centroscymnus*, the hyomandibulae extend outwards at right angles from the braincase to the hinge of the relatively short jaws (Figure 7.6). The upper jaw is protractile, but the biting mechanism does not have the forward rotation given by the long hyomandibulae of the lamnids and carcharinids. Squaloids feed largely on small fish and invertebrates and seem able to suck in benthic prey. The Greenland shark (*Somniosus*) is sluggish and inactive, and it is rather puzzling how it can capture fishes, though it has been suggested to use the parasitic luminous copepods attached to its cornea as lures to attract fish within range of the jaws! Most rays, and some galeomorph sharks like heterodontids (Port Jackson sharks) and orectolobids (carpet sharks, nurse sharks) have a grasping and/ or crushing jaw mechanism bearing setiform holding teeth or molar-like crushing teeth (see Figure 1.14 on p. 16). The hyomandibulae are vertical and extend to short inferior jaws which may bear powerful muscles. In nurse sharks (*Ginglymostoma*) cranial muscles raise the hyomandibulae and the palatoquadrates, whilst antagonistic muscles raise or depress the jaws. We have already seen how this system in *Squatina* is designed to suck in small fishes; in nurse sharks and rays, small invertebrates are sucked in and, if necessary, crushed before being swallowed. Dasyatid and myliobatid stingrays forage by jetting water out of the gill openings to blow large pits in the bottom, and these have been recognised in fossil sediments.

### 7.3.2 Jaw mechanisms of teleost fishes

Most acanthopterygian fishes have protrusible jaws, and it seems that the evolution and retention of protrusible jaws has been of great importance in teleost evolution. Indeed, the major levels of adaptation in actinopterygian evolution can be defined in terms of jaw morphology, and the radiation of teleosts has been accompanied by increasing complexity of the jaw apparatus, much more complex than that of elasmobranchs. Most planktivorous fishes like clupeids or coregonids have non-protrusible jaws, as do freshwater herbivorous catfishes and characins, and the acanthopterygian marine herbivores such as acanthurids, siganids and scarids, so it seems that the evolution of protrusible jaws was linked to other diets.

As the ray-finned bony fishes evolved, their jaws became more mobile through successive changes to three main adaptive levels. The first innovation took place during the changes from the chondrostean to the holostean forms, as the maxillae were freed from their junctions with the cheek bones, and the mandible acquired larger and more complex adductor muscles, coronoid processes and increased torque around the jaw articulation. The holostean–teleost transition did not involve any major changes to jaw mobility, but the early teleosts soon evolved a ball-and-socket joint between the maxillary and palatine, which in holosteans are simply linked by connective tissue.

In the early teleosts and the least advanced recent forms (like clupeoids, characoids, salmoniforms and osteoglossiforms) both maxillary and pre-maxillary bones make the biting edge of the upper jaw. In more advanced teleosts (scopeliforms to acanthopterygians) the pre-maxillae extend backwards just below the maxillae, so excluding the maxilla from the gape. This was the step which permitted the evolution of protrusible jaws as seen most notably in the acanthopterygian teleosts. The main changes involved were the freeing of the pre-maxillae from the rostral region, coupled with the acquisition of backwardly directed and closely adjacent pedicels sliding

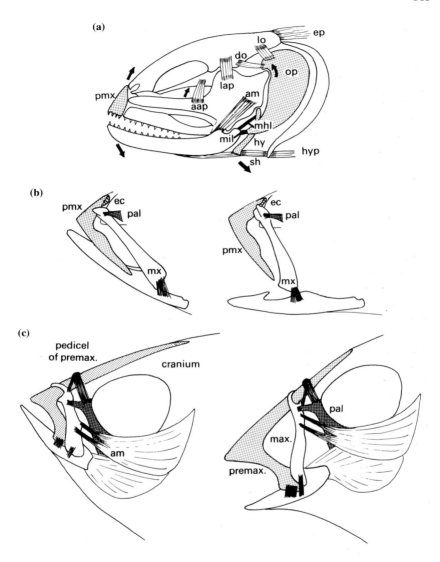

**Figure 7.7** Protrusible and nonprotrusible jaws in teleosts. (a) The jaw muscles, ligaments and jaw movements during feeding in the Arctic char (*Salvelinus*). (b) Open and closed jaws of unspecialised gadoid or percoid, showing how the pedicel of the premaxilla rocks on the rostral part of the cranium via the ethmoid cartilage as the lower jar is depressed. (c) Advanced protrusible jaw mechanism of the mojarra (*Gerres*). The premaxillary pedicel slides under the palatine ligament to protrude the mouth, and as it protrudes the maxilla rotates. aap: adductor arcus palatini; am: adductor mandibularis; do: dilator operularis; ec: ethmoid cartilage; ep: epaxial muscles; hy: hyoid; hyp: hypaxial muscles; lap: levator arcus palatini; lo: levator operculi; mhl: mandibulohyoid ligament; mil: mandibulo-interopercular ligament; mx: maxilla; op: operculum; pal: palatine; pmx: premaxilla; sh: sterno-hyoideus.
After Alexander (1970), Lauder and Liem (1980), and Schaeffer and Rosen (1951).

over the rostrum, the development of upper-jaw ligaments limiting the forward extension of the maxillae, and lastly, a forward shift in lower-jaw suspension. These changes resulted in a mouth in which the protractile upper jaw moves forwards as the mandible is lowered (Figure 7.7). In some advanced acanthopterygians like the John Dory (*Zeus faber*), the extent of the forward protrusion is remarkable, and makes the fish a most efficient predator since it slowly sidles up to small fish (presumably seeming to them to be out of range) and then they are suddenly engulfed as the telescopic jaw system shoots forward and sucks them in. A protrusible jaw system is not, however, a prerequisite for suction feeding; in a fish like the charr (*Salvelinus*) which does not have protrusible jaws, the depression of the buccal floor and lifting of the head, as it opens its mouth, together with the lateral expansion of the head, enlarges the buccal chamber (Figure 7.7) and pressure is suddenly reduced within it so that water and prey are vacuumed in. The more advanced forms with protrusible jaws increase the efficiency of suction feeding by enlarging the volume of the buccal cavity as the upper jaw moves forwards, and the suction can be more accurately directed. Negative buccal cavity pressures in excess of 50 kPa measured in the pumpkinseed (*Lepomis*), led to the inertial suction model of water flow through the orobranchial chamber proposed by Lauder (Figure 7.8) in 1980.

Such inertial suction is the dominant teleost feeding strategy in teleosts, and is used to inhale all sizes and kinds of prey, but not all fish eaters inhale their prey like the John Dory. Long-jawed fish like *Lepisosteus* lie in ambush amongst weeds to catch fish with a sideways strike of the head, and others like *Luciocephalus* open their mouths and then strike forwards to take their prey without suction (Figure 7.9). Tiger musky or pike (*Esox*) strike in the same way at the centre of mass of the prey fish (Figure 7.9).

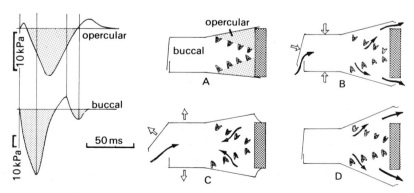

**Figure 7.8** Suction feeding. Left: buccal and opercular pressure records from pumpkinseed (*Lepomis*) – note large negative pressures giving high inflow velocity. Right: model showing successive stages in feeding, incorporating backflow from opercular chamber (lightly stippled in A). After Lauder (1980) *J. Exp. Biol.*, **88**, 49.

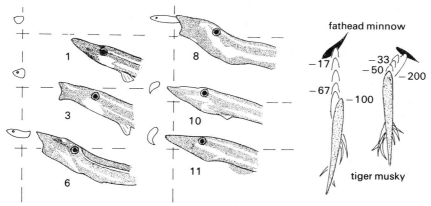

**Figure 7.9** Predator strike patterns. Left: frames (numbers) taken at intervals of 5 m from film of *Luciocephalus* unsuccessfully attacking a guppy (*Poecilia*). Note cranial flexure and widely-opening mouth as fish attacks; no suction is involved. Right: two attack patterns of musky – numbers represent milliseconds before prey contact. Note that it aims for centre of mass of prey. After Lauder and Liem (1981) and Webb and Skadsen (1980).

Distinguished electromyographic and high-speed ciné studies by Liem and his colleagues (1980) at Harvard on primitive actinopterygians and on cichlids have thrown much light on the evolutionary changes as jaw protrusion evolved, and on the functional changes in the musculature operating the jaws; the reader is recommended to consult these studies as excellent examples of functional morphology.

## 7.4 Feeding and digestion strategies

The results of experiments on fishes usually accord well with the predictions from optimality models; when fishes feed, they maximise their energy intake whilst using the least energy to do so, as optimal foraging theory (OFT) suggests. This is considered briefly under fish behaviour (see p. 302). What about how fish extract energy from the food taken in? The ability of the fish to digest and absorb the nutrients in the diet (the absorption efficiency) is of considerable practical importance in aquaculture, where the aim is to produce the largest yield of marketable flesh in the minimum time at minimum cost. Since the components of commercial diets are finely ground up and carefully balanced, high absorption efficiency is expected. Absorption efficiencies of between 90 and 95% are found in eel farming.

Absorption efficiencies of carnivorous and herbivorous teleosts differ significantly. In carnivorous species the energy absorption efficiency is around 80% (up to 97% for adult sea lampreys feeding on blood), and for herbivores much less, around 40–50%. The only elasmobranchs examined have been lemon sharks (*Negaprion brevirostris*) which have been found to have absorption efficiencies up to 83%. As has been pointed out, it is interesting that lemon sharks and carnivorous teleosts should have similar energy absorption efficiency values, for the shark digestive strategy is very different to that of most teleosts. Thus compared to teleosts, these sharks have a low rate of food intake (1% of body weight per day in the sandbar shark, *Galeaspis*), extended retention time, greatly increased surface area with the spiral valve at the hinder end of the gut and grow slowly.

## Bibliography

Alexander, R. McN. (1970) Mechanics of the feeding action of various teleost fishes. *Journal of Zoology, London*, **162**, 145–156.

Almeida, P.R., Moreira, F., Costa, J.L., Assis, C.A. and Costa, M.J. (1993) The feeding strategies of *Liza ramada* (Risso, 1826) in fresh and brackish water in the River Tagus, Portugal. *Journal of Fish Biology*, **42**, 95–107.

Blaxter, J.H.S. (1985) Development of sense organs and behaviour of teleost larvae with special reference to feeding and predator avoidance. *Transactions of the American Fisheries Society*, **115**, 98–114.

Blaxter, J.H.S. and Hunter, J.R. (1982) The biology of clupeoid fishes. *Advances in Marine Biology*, **20**, 1–223.

Crowder, L.B. (1985) Optimal foraging and feeding mode shifts in fishes. *Environmental Biology of Fishes*, **12**, 57–62.

Drenner, R.W., Vinyard, G.L., Gophen, and McComas, S.R. (1982) Feeding behaviour of the cichlid *Sarotherodon galilaeum*: selective predation of Lake Kinneret zooplankton. *Hydrobiologia*, **87**, 17–20.

Gibson, R.N. and Ezzi, I. (1992) The relative profitability of particulate and filter-feeding in the herring *Clupea harengus* L. *Journal of Fish Biology*, **40**, 577–590.

Govoni, J.J., Boehlert, G.W. and Watanabe, Y. (1986) The physiology of digestion in fish larvae. *Environmental Biology of Fishes*, **16**, 59–77.

Hart, P.J.B. (1993) Teleost foraging: facts and theories, in *Behaviour of Teleost Fishes* (Pitcher, T.J. ed), Chapman & Hall, London. pp. 253–284.

Houston, A.I., McNamara, I.M. and Hutchinson, M.C. (1993) General results concerning the trade-off between gaining energy and avoiding predation. *Philosophical Transactions of the Royal Society of London*, B, **341**, 375–397.

Hunter, J.R. (1981) Feeding ecology and predation of marine fish larvae, in *Marine Fish Larvae; Morphology, Ecology and Relation to Fisheries* (R. Lasker ed), University of Washington Press, Seattle. pp. 32–77.

Kennedy, C.R., Nie, P., Kaspers, J. and Paulisse, J. (1992) Are eels (*Anguilla anguilla*) planktonic feeders? Evidence from parasite communities. *Journal of Fish Biology*, **41**, 567–580.

Lauder, G.V. and Liem, K.F. (1980) The feeding mechanism and cephalic myology of *Salvelinus fontinalis*: form, function and evolutionary significance, in *Charrs: salmonid fishes of the genus Salvelinus* (Balon, E.K. ed), W. Junk, The Hague. pp. 365–390.

Liem, K.F. (1980) Adaptive significance of intra- and interspecific differences in the feeding repertoires of cichlid fishes. *American Zoologist*, **20**, 295–314.

Liem, K.F. (1991) A functional approach to the development of the head of teleosts: implications on constructional morphology and constraints, in *Constructional Morphology and Evolution* (Schmidt-Kittler, N. and Vogel, K. eds), Springer Verlag, Berlin. pp 231–249.

Milinski, M. (1993) Predation risk and feeding behaviour, in *Behaviour of Teleost Fishes* (Pitcher, T.J. ed), Chapman & Hall, London, pp. 285–305.

Moss, S.A. (1977) Feeding mechanisms in sharks. *American Zoologist*, **17**, 355–364.

Nelson, G.J. (1967) Epibranchial organs in lower teleost fishes. *Journal of Zoology, London*, **153**, 71–89.

Parker, H.W. and Boeseman, M. (1954) The basking shark *Cetorhinus maximus* in winter. *Proceedings of the Zoological Society of London*, **124**, 185–194.

Rosenthal, H. and Hempel, G. (1970) Experimental studies in feeding and food requirements of herring larvae (*Clupea harengus* L.), in *Marine Food Chains* (J.H. Steele ed), Oliver and Boyd, Edinburgh. pp. 344–364.

Schaeffer, B. and Rosen, D.E. (1961) Major adaptive levels in the evolution of the actinopterygian feeding mechanism. *American Zoologist*, **1**, 187–204.

Webb, P.W. and Skadsen, J.M. (1980) Strike tactics of *Esox*. *Canadian Journal of Zoology*, **58**, 1462–1469.

Wetherbee, B.M. and Gruber, S.H. (1993) Absorption efficiency of the lemon shark *Negaprion brevirostris* at varying rates of energy intake. *Copeia*, **1993**, 416–424.

Wootton, R.J. (1991) *Ecology of Teleost Fishes*. Chapman & Hall, London.

# 8 Reproduction and life histories

## 8.1 Types of life history

In most fishes the males and females are separate individuals. Fertilisation is external and the large number of eggs produced by each female (on an annual basis, its 'fecundity') are left to develop, hatch and grow without parental care. The newly-hatched young are a few millimetres long, usually in the form of larvae quite unlike the adult, with a yolk sac and relatively undeveloped body form (Figures 8.1 and 8.2). Once the yolk is resorbed, the larvae must find food on its own. After a period of growth the larvae metamorphose into the juvenile, immature adult, form.

This typical reproductive or life-history strategy is open to many modifications. Some species are hermaphrodite or change sex during their lifetime; others have internal fertilisation, usually leading to live-bearing; others guard their eggs or young for varying periods of time. Balon (1985) describes 'guilds' of fishes grouped into a classification of reproductive styles such as non-guarding, guarding and bearing which emphasise the very wide range of strategies evolved by fishes for the care of their eggs. There is also a very significant relationship between fecundity and egg size. Since the size of the ovary is limited by the size of the female, high-fecundity females must necessarily have small eggs and vice versa. A further strategy is determined by the spawning season. In the tropics, spawning may be year-round but in more temperate regions, spawning is usually seasonal most often in the spring. Some species have one-off spawning, the eggs all being produced in a single batch; others are batch spawners, producing several batches of eggs over a more prolonged spawning season.

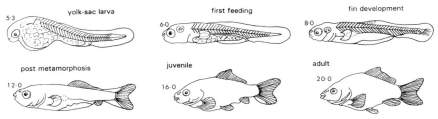

**Figure 8.1** Stages in development of carp. Lengths in millimetres given beside each figure. Redrawn from Auer (1982) *Great Lakes Fish. Comm., Special Publns*, **82–3**, 1 and Kamler (1992).

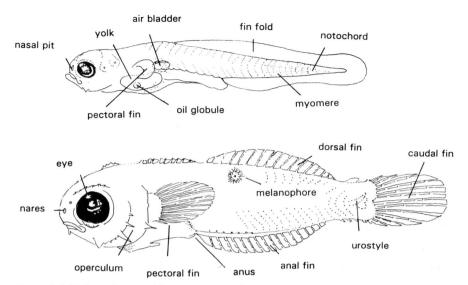

**Figure 8.2** Yolk-sac larva and late stage larva of a blenny. Redrawn from Fahay (1983) *J. N. Atl. Fish. Sci.*, **4**, 1.

Because fish are poikilotherms, the rates of development and growth are very dependent on temperature (Figure 8.3). Generally speaking, tropical fish have rapid growth, small final size and a short generation time although there are obvious exceptions like whale sharks, marlin, barracuda and jacks. Fish in the deep sea or at high latitudes often grow slowly to a considerable size and can live to a greater age. Determination of age depends on counting rings on the scales or otoliths (p. 193) and it is often far from certain whether each ring corresponds to a year of age or not. A lifespan of 10–25 years is common for many of the commercial species with which we are familiar, but some species (e.g. the redfish *Sebastes*) live for 50 years or so, and others possibly very much longer.

Age determination from scales or otolith annuli has been supplemented for some species by radiometric analysis of otoliths. This involves determining the ratio of the naturally occurring radionuclide $^{226}$Ra and its decay product $^{210}$Pb in the otolith. The ratio of the two (which have different half-lives) depends on their decay rates since the time that $^{226}$Ra was incorporated into the otolith. Remarkably enough, such determinations have shown that the orange roughy (*Hoplostethus atlanticus*) from South-East Australian waters matures at around 32 years, when it is 32 cm long, and is very slow growing, fish 38–40 cm long turning out to be no less than 77–149 years old.

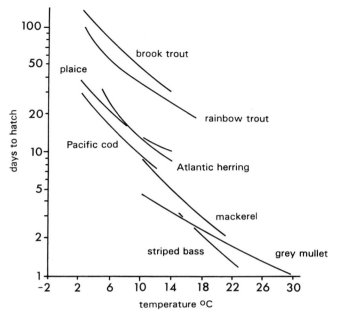

**Figure 8.3** Time from fertilisation to hatching in various species depending on temperature. Redrawn from Blaxter (1988).

## 8.2 Fecundity and egg size

It is not at all clear why some species have evolved a strategy of producing many small eggs, and others the opposite strategy of fewer, larger ones. The relationships between these characters and the implication for the species can be summarized as follows:

- Fecundity and egg size are inversely related when comparisons are made between species (Figure 8.4). In batch spawners, egg size tends to decrease as the spawning season progresses.
- Fecundity increases with age and size of the female within a species (Figure 8.4).
- Large eggs take longer to develop than small eggs, when interspecific comparisons are made.
- Large eggs produce larger larvae at hatching with a longer period of feeding on yolk reserves.
- Fecundity tends to be high in marine fish that release their eggs into open water; it is lower in freshwater species and those species that provide parental care.

These strategies are applicable to many phyla and have been discussed in terms of K- and r-selection by MacArthur and Wilson and other theoretical

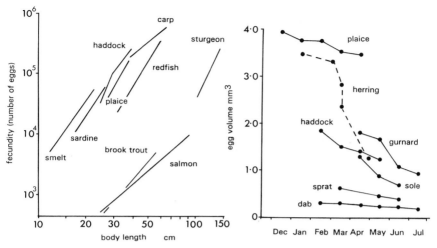

**Figure 8.4** Left: the fecundity of various species depending on their length. Redrawn from Blaxter (1969). Right: the size (volume) of the eggs of various batch spawners as the spawning season progresses. Data from Bagenal (1971) *J. Fish. Biol.*, **3**, 207.

ecologists. A K-strategy is appropriate in stable crowded environments where a low fecundity (large egg) and long developmental period may be favoured. In less-stable, uncrowded environments, where there are chances for maximal population growth, the r-strategy is optimal with high fecundity (small egg) and a short developmental period to exploit any opportunities for expansion.

## 8.3 Maturation

The gonads are mesodermal in origin and develop in close association with the nephric system. In elasmobranchs a lateral cortex gives rise to the ovary and a more medial area becomes the testis. One of these usually atrophies early in development when the sex is determined. In cyclostomes and teleosts the gonads derive from a single region equivalent to the cortex. While most fishes have paired gonads, fusion of two primordia in the cortex of lampreys leads to a single gonad whereas in the hagfishes, one gonad fails to develop, and there is only a single ovary in some sharks.

The gametes first appear as primordial germ cells. The ovary is usually hollow although the wall may become folded as its internal surface area increases. The oogonia become surrounded by a single layer of follicular cells in cyclostomes and teleosts but a multilayer in elasmobranchs. In the next, oocyte, stage the developing eggs are supplied with yolk by the follicular cells during vitellogenesis. At the completion of maturation the oocyte becomes free of the follicle during ovulation and water is taken up, causing the egg to swell. In batch spawners, populations of different-sized oocytes can be seen, each corresponding to a future batch of eggs for release. In some species the mature eggs are released into the body cavity of the fish and pass into the funnel-shaped opening of the oviduct and so to the exterior; in other species the ovarian lumen and oviduct are continuous.

The testis tends to be more divided than the ovary – into compartments called ampullae or acini. Here the spermatogonia pass through a spermatocyte and spermatid stage before becoming spermatozoa. During spawning these pass directly to the exterior via the seminiferous tubules and vas deferens. En route the spermatozoa are diluted with secretions of seminal fluid. Some elasmobranchs and live-bearing teleosts produce spermatophores or packets of spermatozoa, as does the basking shark.

## 8.4 Intersexes

The sexuality of teleost fishes is extraordinarily complex. Although hermaphroditism is almost unknown in the ostariophysans, the dominant freshwater group, it is common in marine actinopterygians, especially in families represented in coral reef fauna such as labrids (wrasses), sparids (sea-breams), serranids (sea-basses), scarids (parrotfishes), pomacentrids (damselfishes) and polynemids (threadfins). In the deep sea, intersexuality is found in a number of groups, for example *Gonostoma* and *Cyclothone* in the bathypelagic stomiatoids and in the benthic chloropthalmid tripod fish *Bathypterois*. Hermaphroditism may be successive (the individual starting as a male and becoming a female as in *Gonostoma gracile* and *Cyclothone microdon*) or synchronous where both parts of the ovitestis mature together as in *Bathypterois*. One of the potential advantages of being synchronously hermaphrodite, viz. to allow self-fertilisation in a sparsely distributed population, does not seem to occur in the tripod fish but, perhaps, the meeting of two individuals allows the fertilisation of two batches of eggs. In the successive hermaphroditism of the stomiatoids the males are obviously smaller than the females. Size is much more at premium for the females so that they can accommodate a large number of eggs. The small male, having fertilised the eggs of its older and larger conspecifics, can then be saved to become a female itself. Successive hermaphroditism is found in many coral reef fish; in some parrotfishes the hermaphrodites are protogynous (where the individuals change from females to males). In this instance the larger males may be less susceptible to predation when searching for females or better at defending their territories.

## 8.5 Fertilisation to hatching (incubation)

The released egg is protected by a fairly tough chorion or egg case. Within this the cytoplasm and yolk are contained by a vitelline membrane. Often one or more oil globules are present. Fertilisation occurs by a spermatozoon passing through a funnel-shaped micropyle (Figure 8.5) leading to a fusion of the pronuclei of the sperm and the egg. This leads to activation when the oocyte, arrested before fertilisation, then resumes development. The vitelline membrane separates from the chorion creating a perivitelline space and the micropyle is plugged, preventing further spermatozoa entering. Some elasmobranchs have polyspermy in which a number of spermatozoa may enter the micropyle but only one fuses with the nucleus of the egg, the rest being resorbed and perhaps used as an additional nutrient. In salmonids, water activation takes place as the egg absorbs water regardless of whether fertilisation takes place or not. In river water, the vitelline membrane becomes opaque and its permeability changes. If spermatozoa are not present, the fertilisability of the egg is soon lost. After fertilisation the chorion hardens, so protecting the egg. This is especially valuable in waves or surf or if the eggs are buried in gravel. The chorion

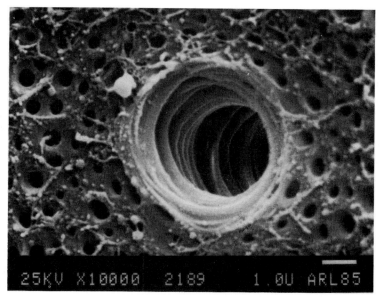

**Figure 8.5** Scanning electron micrograph of the surface of a herring egg showing the micropyle. Photograph by kind permission of Prof. H. Rosenthal, Kiel.

remains permeable to water and small molecules but the site of osmoregulation (p. 147) is the vitelline membrane. Fish eggs are usually round although the hagfish (*Myxine*), anchovy (*Engraulis*) and bitterling (*Rhodeus*) have ovoid eggs and in some gobies they are pear-shaped (Figure 8.6). Most species are telolecithal with yolk concentrated at the vegetative pole and the cytoplasm at the animal pole giving a 'polarity' to the egg. The egg goes through a process of cleavage and morphogenesis as the cells divide, form layers and then organs. In lampreys cleavage is holoblastic, the entire egg dividing to form smaller cells or micromeres at the animal pole and macromeres at the vegetative pole. In hagfish, elasmobranchs and teleosts development is meroblastic (Figure 8.7). Here cleavage at the animal pole leads to a blastoderm or cap of cells. The blastoderm overgrows the yolk (epiboly) eventually enclosing it to form a gastrula, a hollow sphere of cells containing yolk with a small opening into the perivitelline space – the blastopore. The embryonic axis is laid down in relation to the dorsal lip of the blastopore and the neurula stage forms with the future head, spinal cord and body musculature soon visible. After a

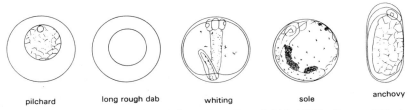

pilchard    long rough dab    whiting    sole    anchovy

**Figure 8.6** Eggs of various species. Redrawn from Russell, F.S. (1976). *Eggs and Planktonic Stages of Marine Fishes*. Academic Press, London.

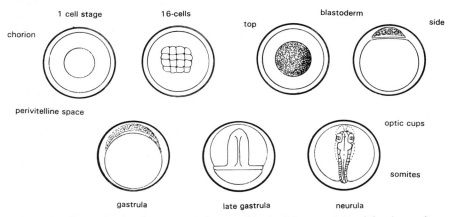

**Figure 8.7** Stages in development of the meroblastic fish egg of *Fundulus heteroclitus*. Redrawn from Lagler *et al.* (1962).

time, the tail region grows away from the neurula and coils round inside the perivitelline space. The optic cups (the future eyes) and heart are the first organs to be identified easily. The heart starts to beat well before hatching and in some species the eyes become pigmented and functional and the embyonic fish often wriggles and rotates within the chorion. The prelude to hatching is a softening of the chorion as a result of enzymes secreted by glands on the head. Some of the chorionic material is probably utilised as a nutrient by the embryo before hatching. The embryo then breaks free from the chorion to become a larva.

There has been considerable interest of late in the problem of storing fish spermatozoa, unfertilised eggs and embryos for aquaculture. Free gametes or whole gonads can be stored for one or two days at temperatures of 1–5°C, without much loss of fertilisability. Long-term storage is only possible for spermatozoa, which are held in antifreeze agents, such as glycerol or dimethyl sulphoxide and stored in liquid air or nitrogen. Carp, cod, herring and salmon spermatozoa have been frozen for many months in this way.

## 8.6 Parental care

### 8.6.1 Ovoviviparity

Retention of the eggs within the body of the female for subsequent development is necessarily preceded by internal fertilisation, but even where fertilisation is external, some species then take the eggs into their bodies to protect them. The male marine catfish (*Galeichthys felis*) for example, takes up to 50 eggs (20 mm in diameter) into its mouth and broods them until hatching, and even shelters the young for 2 weeks afterwards. Similarly, the male in seahorses and pipefish (Syngnathiformes) takes the fertilised eggs into special pouches or marsupia, whence they emerge as they hatch. Curiously enough, in the related *Solenostomus* the female lays eggs, which are fertilised, and she then takes them into a marsupium formed by the pelvic fins.

In ovoviviparous fishes the eggs develop in the 'uterus' – the oviduct of elasmobranchs and the ovary of teleosts. During the 'gestation' period – note that many of the terms are borrowed from mammalian reproduction – the eggs hatch within the mother and are eventually born alive as larvae or juveniles. The gestation period can last from a day or two in small tropical teleosts to one or two years in sharks. During gestation the eggs are protected from predation (although if the mother is eaten all the young are lost!) and live in physiologically regulated surroundings. At birth the young are larger than their oviparous counterparts and have better locomotor powers to avoid predators, and to feed on a wider range of food organisms. Presumably such advantages compensate for a lower reproductive rate.

In ovoviviparous species there are two reproductive strategies; the eggs may merely develop on their endogenous yolk reserves so that the young are born at a lower weight than the original egg as in *Sebastes*; or, more commonly, the developing young take in nutrients from the uterus, either by mouth from uterine secretions, as in the electric ray *Torpedo*, or via the thread-like extensions of the uterine wall (trophonemata) which pass into the oesophagus of the young through its spiracles, as in some sting rays, and the butterfly ray, *Pteroplatea* (Figure 8.9). In these cases the new born pup may be 50 times the weight of the unfertilised egg. In lamnoid sharks, the embryos develop a precocious dentition after their yolk reserves are used, and generally only a single young shark in each uterus survives pregnancy, reaching a size over 1.0 m. This huge size at birth is because the survivors have fed within the uterus on the unfertilised eggs (oophagy) or on their smaller siblings.

### 8.6.2 Viviparity

In viviparous species, an intimate connection is established between the maternal and embryonic circulation akin to that in the mammalian placenta. In the small cyprinodont *Heterandria formosa*, for example, the embryos are retained in the ovarian follicles, where a maternal capillary network is in intimate contact with the external surface of the embryo, and nourishes it. In the four-eyed fish *Anableps*, the placenta is derived from a large expansion of the pericardial wall, whilst in the small cyprinodont *Jenynsia*, the placenta links the maternal circulation with the branchial capillaries of the embryo. But it is in hammerhead and carcharhinid sharks that placental structure is best known (see Section 8.8.2).

### 8.6.3 Nest building and brooding

Some littoral species of bullheads (Cottidae), blennies (Blenniidae) and gobies (Gobiidae) protect their eggs which are simply attached to the substratum, but many species from a range of taxonomic groups build nests in which the eggs are guarded, and sometimes ventilated, by one or both parents. These include the blenny *Ictalurus*, the stickleback *Gasterosteus* (p. 300), centrarchid sunfish, cichlids, the bowfin (*Amia*) and the lungfishes *Protopterus* and *Lepidosiren* (but not *Neoceratodus*). The Siamese fighting fish, *Betta splendens*, makes a nest of bubbles which may help to aerate the eggs as well as protecting them. Fertilisation is almost always external in species with paternal care and internal with maternal care, an exception being the sculpin, *Artedius harringtoni*, in which the male guards the eggs when they are laid some time after internal fertilisation. Other species have brood pouches for carrying the eggs as in sea horses (*Hippocampus*) and pipefish (*Syngnathus*). The marine ariid catfish, the apogonid cardinal fish (*Apogon semilineatus*) and some tilapias are mouth brooders. *Tachysurus barbus* (see Figure 8.13, p. 187) incubates its eggs intestinally and *Platystachus cotylephorus* over its abdomen. The bitterling (*Rhodeus amarus*) lays its eggs within the gills of a freshwater mussel. These modes of parental care have not been adopted by elasmobranchs as a reproductive strategy since internal fertilisation is universal. Elasmobranchs are much more commonly ovoviviparous than teleosts. All types of parental care are associated with small numbers of larger young, often with a fairly long reproductive period, and are therefore examples of K-selection.

## 8.7 Agnatha

The eggs or sperm from the single gonad of lampreys and hagfishes are shed into the body cavity and thence into the water through abdominal pores. In the large anadromous sea lamprey (*Petromyzon marinus*) the fecundity ranges from 24 000 to 236 000 but is only 400–9000 in smaller non-parasitic species. The eggs are laid in the stream bed and hatch in about two weeks as small pro-ammocoete larvae. They soon change into active ammocoete stages that burrow into silt banks and filter-feed. After a period of up to five years, longer in parasitic species, they metamorphose into adults at a length of 12 cm or more. Some species descend to the sea to hitch rides on, and rasp blood from, other fishes, but the non-parasitic brook lamprey (*Lampetra planeri*) remains in freshwater.

The mortality of the ammocoete stages is relatively low judging by the decrease in population size of different year classes. A protracted ammocoete stage and delay of metamorphosis is probably favourable to the parasitic species where an adult existence is hazardous if host fishes of suitable size are scarce. The ammocoete is unique among fish larvae in having such a long sheltered existence and microphagous mode of feeding. Because of their size at metamorphosis and slow maturation rate, ammocoetes have the characteristics of K-selected organisms. However, their high fecunity and unlimited food supply, at least in the pre-adult stage of parasitic species, suggest a typical r-selected species. This demonstrates the difficulty of neatly compartmentalising theories. As the philosopher J.L. Austin sagaciously remarked in another connection, '. . . it is essential, here as elsewhere, to abandon old habits of *Gleichschaltung*, the deeply ingrained worship of tidy looking dichotomies'.

Hagfish are probably functionally dioecious, although their gonads pass through an hermaphroditic phase. They fit better as K-selected organisms and tend to be deep-sea forms where the environment is rather stable but food-limited. They lay linked batches of five or six large ellipsoidal eggs, 14–25 mm long, on the sea bed, or perhaps within their burrows. After two months or more, these hatch as juveniles. Curiously, although the eggs of *Bdellostoma* and the Australian *Geotria* are well known, the eggs of the European *Myxine* have very rarely been found. Holmgren found only 131 during a 20-year search and almost all of these were undeveloped. The impecunious and optimistic reader will be disappointed to learn that the prize offered in 1865 by the Copenhagen Academy of Sciences for a description of the embryology of *Myxine* remained unclaimed, but has now been withdrawn!

## 8.8 Elasmobranchiomorpha and *Latimeria*

### 8.8.1 Reproduction

In elasmobranchs and holocephalans (chimaeras) fertilisation is internal, the inner elements of the pelvic fins, the claspers, being rolled up to form a tube with overlapping edges. The claspers transfer sperm to the oviduct of the female. After fertilisation the eggs may be laid on the sea bed or retained within the uterus. Oviparity is confined to the chimaeras, some families of skates (Rajoidea) and four families of sharks – Port Jackson sharks (Heterodontidae), whale sharks (Rhincodontidae), some of the carpet sharks (Orectolobidae) and all but one species of the dogfish (Scyliorhinidae). Eggs within their horny cases range in size from over a centimetre in dogfishes to 30 × 15 cm in the whale shark (Figure 8.8) and 25-cm long in the chimaeroid *Callorhynchus*. The egg capsule is horny and may be equipped with tendrils to attach it to weed or stone. By elasmobranch criteria, fecundity is quite high. Some species of skates (*Raja*) lay 100 eggs or so. The eggs incubate for several months or even

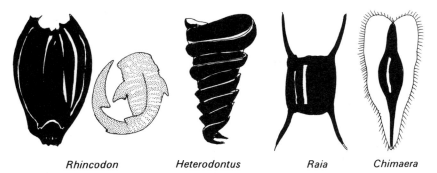

Rhincodon            Heterodontus            Raia            Chimaera

**Figure 8.8** Egg cases of oviparous elasmobranchiomorphs (not to same scale). After Linaweaver and Backus (1970) *The Natural History of Sharks*, Lippencott, Philadelphia.; Dean (1906) *Chimaeroid Fishes and their Development*, Carnegie Inst. Washington, **32**; and Daniel (1922) *The Elasmobranch Fishes*, University of California Press.

longer. Recorded incubation times are 2.5–3 months in a carpet shark (*Chiloscyllium griseus*), 4.5–8 months in *Raja* species, 6–8 months in *Scyliorhinus* and 9–12 months in the Port Jackson shark (*Heterodontus*) and the chimaera *Hydrolagus colliei*.

### 8.8.2 Ovoviviparity and viviparity

Most sharks and rays with parental care are ovoviviparous (aplacental), the embryos depending on their yolk reserves or eating other eggs in the uterus (oophagy) as in the mackerel shark (*Lamna nasus*), and *Odontaspis*, the sand shark. *Lamna* and the thresher shark (*Alopias vulpes*) have a gestation period of about two years and the young (five individuals or less) are born at a length of about 60 cm. The ovoviviparous rays depend on uterine milk rich in protein and lipid, and produced by processes very similar to those seen in the production of mammalian breast milk. Such rays lack placentae, but thread-like trophonemata extend from the uterine wall into the mouth and gill chambers producing the milk and acting as respiratory membranes, as in the sting ray (*Dasyatis violacea*), the eagle ray (*Myliobatis bovina*) and the butterfly ray (*Pteroplatea*) (Figure 8.9).

The viviparous (placental) elasmobranchs are all either hammerhead sharks (Sphyrnidae) (Figure 8.9) or belong to the Carcharhinidae, the grey

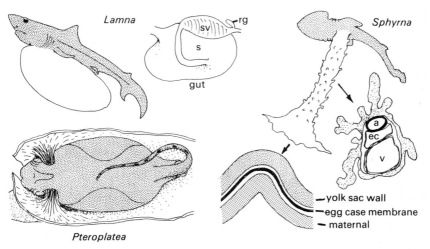

**Figure 8.9** Elasmobranch embryonic nutrition. Upper left: egg-eating embryo of porbeagle (*Lamna*) showing greatly enlarged stomach (seen in detail on right). Lower left: embryo of butterfly ray (*Pteroplatea*) showing trophonemata entering spiracles. Right: placenta of the hammerhead (*Sphyrna*). The embryo is attached to the placenta by a cord bearing absorptive processes, seen in section to the right, a, artery; v, vein; ec, extra-embryonic coelom; rg, rectal gland; s, stomach; sv, spiral valve. After Alcock (1892) *Ann. Mag. Nat. Hist.*, **10**, 1; Schlernitzauer and Gilbert (1966) *J. Morphol.*, **120**, 219 and Shann (1923) *Proc. Zool. Soc. Lond.*, **11**, 161.

sharks; the smooth dogfish (*Mustelus*) is a member of this family. In *Sphyrna*, a placental cord forms bearing fine processes which aid in the absorption of the maternal secretions. In some species such as *Mustelus canis*, the embryo depends upon its yolk for the first three months, and only then does a placenta develop, to nourish the embryo for the remaining seven or eight months of pregnancy.

Such sharks often come close inshore to give birth in shallow water nursery areas (often in bays) where adults that might eat the young do not normally occur. For example, at least eight species of carcharinids and hammerheads use Cleveland Bay in North Queensland as a nursery area.

Oviparous sharks and rays tend to be bottom dwellers of inshore waters and of relatively small size and with young usually less than 30 cm long at birth. The larger sharks and rays are viviparous and produce young mostly 30–70 cm in length. The largest oviparous shark, the whale shark, has a relatively small egg case measuring 15 × 30 cm, while the largest viviparous species, the basking shark, produces young 150–180 cm long. *Lamna nasus* produces young about 70 cm long and weighing about 10 kg, whereas the blue shark (*Prionace glauca*) has a maximum litter size of 54, the pups being 31–47 cm long and with a mean weight of 0.14 kg. As one might expect, there is a relative trade-off between numbers and size in species that have evolved parental care.

The genus *Mustelus* is of particular interest in the way that different species have adopted different reproductive strategies. While *Mustelus laevis* and *M. canis* are viviparous, although the placenta only develops after a period of ovoviviparity in the latter species, *M. vulgaris* and *M. antarcticus* are ovovivparous.

### 8.8.3 Latimeria

The coelacanth *Latimeria* incubates very large eggs (like tennis balls) in the oviduct for a considerable period. The embryos, 20–30 in number, have large yolk sacs which can also be seen in some fossil coelacanths (Figure 8.10). The adults live to an age of 11 years or more.

## 8.9 Growth

To study growth rates, we need some means of determining age. In well-studied species such as the spiny dogfish (*Squalus acanthias*) and the skate (*Raja clavata*), transverse sections of the spines or vertebral centra, respectively, show rings that correspond to changing seasonal patterns of growth. These annual rings are used to age the fish. In tropical or sub-tropical waters where there is less seasonality, the annual rings may be less evident, or absent.

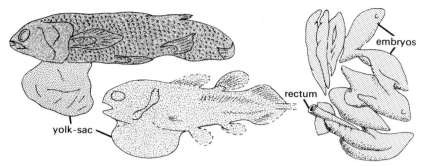

**Figure 8.10** Coelacanth embryos. Above: *Latimeria* embryo from oviduct, shown schematically on right. Below: *Rhabdoderma* (Upper Carboniferous)—note yolk sacs in fossil and recent embryos. After Smith *et al.* (1975) *Science*, **190**, 1105 and Schultz (1977) *Evolutionary Biology*, **10**, 277.

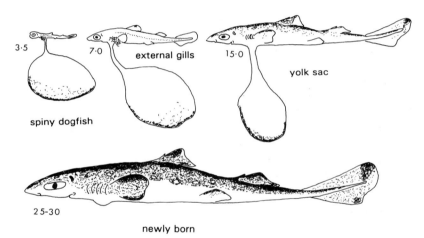

**Figure 8.11** Stages in the uterine development of the young of the ovoviparous dogfish *Squalus acanthias*; young removed from the uterus. Lengths in centimetres. Redrawn from Hishaw, F.L. and Albert, A. (1947). *Biol. Bull. Woods Hole*, **92**, 187.

Holden studied growth of spiny dogfish using a modification of the von Bertalanffy growth equation on the assumption that the growth curves of the embryos can be extrapolated to give those of free-living individuals (Figure 8.11). The derived equation is:

$$l_{t+T}/L^{\infty} = 1 - \exp(-KT) \qquad (8.1)$$

where $l_{t+T}$ is the length at birth, $L^{\infty}$ is the maximum observed length, $T$ is the length of gestation and $K$ is a constant. In European waters, $l_{t+T}$ is 27.5 cm, $L^{\infty}$ is 108 cm and the gestation period two years. Thus $27.5/108 = 1 - \exp(-2K)$ or $K = 0.15$. This derived growth constant may be compared with the value 0.11 from actual growth data of free-living individuals. In general, values of $K$ seem to be 0.1–0.2 for sharks and 0.2–0.3 for rays but an exception is found in the seven species of smooth hounds (*Mustelus*) where $K$ ranges from 0.22 to 0.53 for males and 0.21 to 0.36 for females. In this genus, maturity is reached up to four years after birth, a much faster rate of growth than the spiny dogfish which takes 10 years to mature.

Elasmobranchs generally have low rates of reproduction, slow growth and a long time to maturation compared with teleosts. The period for 50% to mature (the generation time) is probably 7–13 years in sharks and 5–6 years in rays. These K-selected species are very susceptible to overfishing and many of the fisheries are not sustainable as the result of a low rate of recruitment.

## 8.10  Teleosts

Teleost life histories are diverse. Although parental care is widespread, viviparity is only well represented in two major groups, the ophidioids and cyprinodontoids. The majority of teleosts are oviparous and much more fecund than elasmobranchs. The incubation period is short and the newly hatched larvae have a relatively short period with high mortality before they metamorphose; they are therefore r-selected species. There are always exceptions to such generalisations, the three-year drift of the leptocephalus larvae of European eels (*Anguilla anguilla*) across the Atlantic being an example.

There are no particular distinguishing characteristics of freshwater compared with marine eggs and larvae although they have different osmotic problems. Freshwater eggs are more likely to be larger (1.0–2.0 mm in diameter) and to be attached to weed or stones and some are laid in nests or brooded in various ways. Although a few species, such as gouramis and grass carp produce buoyant eggs, non-floating kinds may be best suited to freshwater, to prevent excessive drift in rivers and streams. Marine eggs are usually smaller (about 1.0 mm in diameter) and almost always buoyant and liberated into the pelagic zone. The herring and capelin (*Mallotus villosus*) are exceptions, laying sticky demersal eggs that adhere to weed or gravel on the sea bed or in the intertidal zone. The mummichog (*Fundulus heteroclitus*), Atlantic silverside (*Menidia menidia*) and Californian grunion (*Leuresthes tenuis*) lay their eggs on various substrata in the intertidal zone and spawning is usually linked with tidal lunar cycles (see p. 313).

Marine eggs can easily be arranged to be buoyant since the solute concentrations within the yolk can be maintained osmotically below that of the ambient sea water, and hence less dense. There is usually a massive

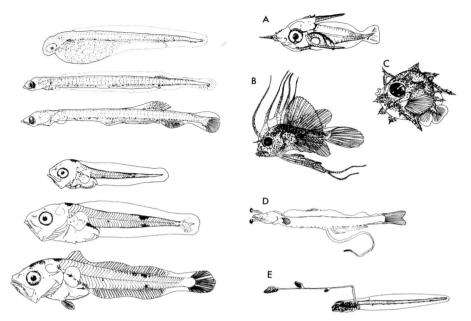

**Figure 8.12** Teleost larvae. Left above: three stages in larval development of the northern anchovy (*Engraulis mordax*). Left below: three stages in development of the hake (*Merluccius productus*). Right: more unusual larvae (A) *Holocentrus vexillarius*; (B) *Lophius piscatorius*; (C) *Ranzania laevis*; (D) *Myctophum aurolaternatum*; (E) *Carapus acus*. From Blaxter (1988).

uptake of water at ovulation; the fertilised eggs of plaice, for example, are over 90% water. The specific gravity of the whole egg is then kept below that of seawater although the yolk material itself is of higher specific gravity. A globule of low specific gravity oil may also help flotation. In freshwater, solute concentrations within the egg are of course higher than in the surrounding water, so buoyant eggs require large oil globules.

The larvae which hatch from teleost eggs are rather various in form, as seen in Figure 8.12, and often quite unlike the adult, e.g. leptocephali (some over 1 m long!) having curious forms with bizarre shapes. Unfortunately, there is insufficient space to cover the full range of teleost reproductive habits; all that can be done is to outline a few case histories, and to recommend the reader to seek further in the byways of teleost reproductive adaptations.

## 8.10.1 Freshwater species

Freshwater environments include the stable and mainly non-seasonal great tropical lakes, the great lakes of more temperate regions where conditions are more seasonal with ice-cover in the winter, the highly seasonal flood plains of rivers like the Amazon and Nile and the many rivers of temperate regions where there is a strong seasonality of temperature and water flow. Life-history adaptations are found to all these conditions.

In the African great lakes such as Lake George and Lake Victoria, ripe individuals of most kinds of cichlid fishes exist at any time of year, although there may be seasonal peaks in the number of spawning individuals. The dominant cichlid fauna forms 'flocks' of over 300 species in Lake Malawi, over 200 in Lake Victoria and 120 in Lake Tanganyika. For the most part these cichlids produce small egg batches that are guarded on the substratum or mouth-brooded, usually by the female. The substratum-guarders are usually monogamous and the parents form pair bonds. The mouth-brooders are polygamous and the parents separate soon after mating. In the flood-plain rivers, the entire life of some fish communities may be geared to the rains. Flooding not only greatly increases the available habitat, it also releases nutrients that evoke blooms of phyto-plankton and an increase in microzooplanktonic food organisms. Many of the larger fishes, especially the ostariophysans (carp, roach, etc.) spawn just before or during the flood, while others spawn in grass swamps at the edge of the advancing flood. For many tropical fish, the high-water period is also the main feeding and growing season, when they build up fat stores to carry them through the dry season.

In some genera of the oviparous killifishes (Cyprinodontidae) there are 'annual' species with a life history adapted to exploit the temporary pools that appear each year in the tropical and sub-tropical regions of South America and Africa. At the beginning of the rainy season, killifish eggs buried in the mud hatch quickly and the young grow rapidly on the abundant (but temporary) food supply and are ready to spawn in six to eight weeks. They produce drought-resistant eggs with thick chorions which can go through an insect-like suspension of growth and development in a diapause phase, although there are also 'escape' eggs that avoid diapause.

In the viviparous killifishes (Poeciliidae), fertilisation is internal. The male is smaller than the female (unlike the oviparous species) and some of the anal fin rays are elongated to form an intromittent organ, the gonopodium (Figure 8.13). In the mosquito-fish, *Gambusia affinis*, so called for its great liking for the larvae and pupae of *Anopheles*, the male is about 3.5 cm and the female 6.5–8.0-cm long. Broods of 43–205 young are produced every 21–28 days. The mosquito-fish occurs naturally in south-eastern parts of the United States but has been introduced to many other parts of the world to help in the control of malaria.

If frequent enough, the repeated broods of viviparous killifishes may make them almost as fecund as their oviparous relatives. In some poeciliid species, sperm is stored in the ovarian wall and the embryos gestate within intact egg follicles in a process called 'superfoetation'. By a successive

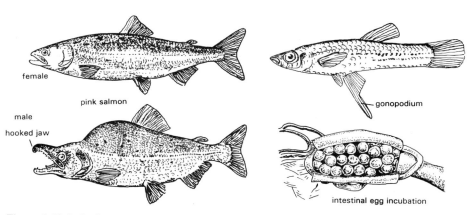

**Figure 8.13** Left: female and male pink or humpback salmon (*Oncorhynchus gorbuscha*) showing hooked jaw and hump of the male at spawning. Right, top: male intromittent organ (gonopodium) of *Gambusia* (mosquito fish). Right, bottom: intestinal egg incubation in *Tachysurus barbus*. Redrawn from Nikolsky (1963).

ripening and fertilisation of the egg batches, up to nine broods, each in its own state of development, can be gestated in the ovary, for example in *Poeciliopsis retropinna*, *P. elongatus* and *Heterandria formosa*. Viviparity in killifishes and other small species is largely confined to tropical and subtropical environments where there is less risk of marked seasonal fluctuations. In North America, the killifishes most at home in more temperate conditions belong to the oviparous genus *Fundulus*.

Finally, salmonids deserve special mention because of their varied life histories. The Atlantic salmon (*Salmo salar*) (p. 309), the Pacific salmon (*Oncorhynchus*) and their relatives – the trouts *Salmo* sp., the rainbow trout (until recently called *Salmo gairdneri*, but now *Oncorhynchus mykiss*), charr (*Salvelinus*), grayling (*Thymallus*) and whitefish (*Coregonus*) – are all of considerable economic importance. Salmonids comprise both landlocked as well as anadromous species and are characterised by their large size for freshwater fish, the adults usually growing to at least 1–2 kg and in some salmon 25 kg or more. Because of their size and the excellent quality of their flesh they are much sought after for food and sport. Salmon, rainbow trout and charr, in particular, are farmed in cages and raceways in both the sea and freshwater (p. 321) and many efforts have been made to transplant and establish them in new habitats, for example in the southern hemisphere. The salmonids are oviparous, producing a few thousand rather large eggs, 5–7 mm in diameter. The male develops secondary sexual characters at spawning, the hooked jaw of the Atlantic and pink salmon being especially conspicuous (Figure 8.13). Salmonid eggs are laid in the shallow gravel beds of fast-running streams and take several weeks to hatch. The young of the Atlantic salmon hatch as fairly well-developed alevins about 20 mm long, with red-pigmented blood and a large yolk sac. They pass through a parr stage and migrate to the sea as smolts at a length of 10–15 cm, usually after one or two years in freshwater. The typical development of a salmonid, the rainbow trout, is shown in Figure 8.14.

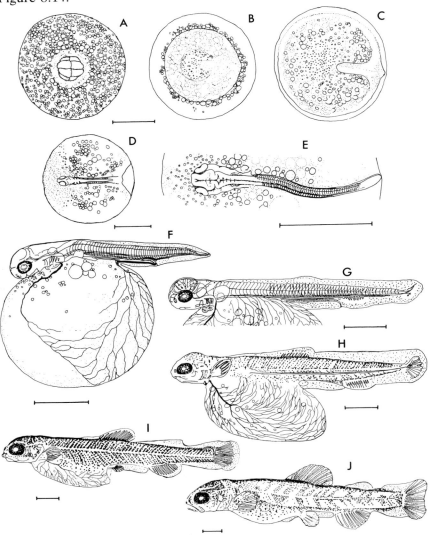

**Figure 8.14** Development of the rainbow trout *Onchorhynchus ÿkiss*. 0 (A) 8-Blastomeres. (B) Early embryo apparent, one-third epiboly. (C) 0–5 Somites, one-half epiboly. (D) Otic placodes, three-fourths epiboly. (E) Caudal bud with 10–20 somites, total somites 51–58, heart beating. (F) Posterior cardinal veins formed, choroid of eye pigmented. (G) Near hatching, pelvic fins develop. (H) Hatched alevin, first anal and dorsal fin rays. (I, J) Later alevin stages as yolk is resorbed. Scale bars 2 mm long. Redrawn from Vernier, J.-M. (1969) *Ann. Embryol. Morphol.*, **2**, 495–520.

### 8.10.2 Marine species

*Distribution.* The spawning grounds of commercially important marine species are well known, especially in temperate latitudes, from the distribution of the fishing fleets and sampling of their catch. Data from zooplankton research cruises on the distribution of eggs and larvae can also be back-plotted to identify spawning grounds and can be used to measure mortality rates and assess paths of dispersal. Much less is known about the spawning behaviour of the adult fish. At one time it was thought that spawning was a rather hit-or-miss process in which large numbers of females and males milled around together in mid-water. There are doubtless many exceptions to such a scheme. We now know that the haddock, for example, goes through a quite elaborate courtship involving visual displays by the male (Figure 8.15) and vocalisation. The male of the coral reef angelfish (*Pygoplites diacanthus*) produces a vortex ring of the eggs and sperm by flexing its tail, presumably improving the success of spawning. In the herring (a demersal spawner) males and females interact, often on a one-to-one basis, and the female will only release its eggs onto a suitable substratum. Some species such as the sea horse are monogamous, staying with the same mate for life.

About three-quarters (9000 out of 12 000) teleost species produce buoyant eggs so that the most common habitat for the eggs and larvae is the pelagic zone. Here the young stages comprise that part of the zooplankton called the ichthyoplankton. They have little control of their

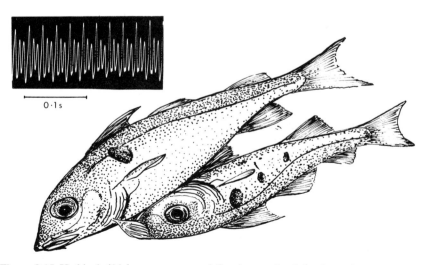

**Figure 8.15** Haddock (*Melanogrammus aeglefinus*) spawning behaviour; the male is below. Inset: audiogram of humming sound made by the male as it leads the female. Redrawn from Hawkins, A.D. *et al.* (1967) *Nature, Lond.*, **215**, 923.

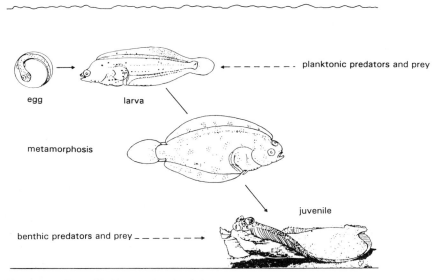

planktonic predators and prey

egg                    larva

metamorphosis

juvenile

benthic predators and prey _ _ _ _ _ _ →

**Figure 8.16** Early life history of plaice (*Pleuronectes platessa*), a typical flatfish which spends its larval life in the plankton and settles on to the sea bed at metamorphosis. Drawing by Dr. R.N. Gibson.

own destiny; apart from the older larvae making limited diel vertical migrations, dispersal is mainly under the control of the currents. As they grow towards and through metamorphosis and become independent, there is a tendency for the juveniles to collect on inshore nursery ground or, in the case of flatfish, to settle on the sea bed (Figure 8.16).

As almost all fish are poikilotherms, the rate of development depends on temperature (Figure 8.3). In tropical regions, not only is development fast but the generation time is short and spawning almost year-round. The eggs of coral fishes require only a day or so to hatch and larval life lasts a few days. Many families of coral fishes such as the sea basses (Serranidae), snappers (Lutianidae), red mullet (Mullidae) and butterfly fishes (Chaetodontidae) lay pelagic eggs but some gobies, blennies and damselfishes lay non-buoyant eggs, which may be scattered over the bottom or guarded by one or both parents. Total fecundities may range from tens of thousands to millions of eggs. Mortality is high and only small numbers return to the reef as juveniles, and presumably even fewer disperse to other areas. An unusual characteristic is found in eggs and larvae of the sharpnose puffer fish (*Canthigaster valentini*) which are unpalatable to reef fish predators.

In the open ocean of the tropics and sub-tropics, epipelagic teleosts such as flying fishes (Exocoetidae), sauries (Scomberesocidae), sailfishes and marlins (Istiophoridae), tunas and ocean sunfishes (Molidae) produce numerous buoyant eggs as do the species in the mesopelagic and bathypelagic zones. Species at both these levels are presumed to spawn where they live, and the eggs float upwards towards the surface. The young stages then join the ichthyoplankton for a time before moving down again to the adult habitat.

Both in the open ocean and in coastal waters, species diversity drops markedly between sub-tropical and temperate latitudes. Of the 111 species of British fishes, 68 lay floating eggs. The remainder produce grounded eggs (sand eels, blennies and clingfishes) are nest-building (wrasse, stickleback), viviparous (*Sebastes*) or have brood pouches (pipefishes). Such adaptations no doubt reduce mortality in inshore areas where the eggs could be stranded or damaged by wave action.

At high latitudes the inshore waters are near freezing and often covered with ice. Many species lay large eggs on the sea floor. The Antarctic notothenioids (ice fish) have eggs 2–5 mm in diameter and fecundities of 2500–12 000. Growth is slow and maturity is not reached for several years. Fish in these waters must exploit the short growth season with a short-lived larva or become less dependent on seasonality by producing large demersal eggs. These hatch into bottom-feeding larvae with substantial yolk reserves that allow them to grow quite large by endogenous nutrition, increasing the range of prey available. Polar fishes are typically K-selected with their large eggs, low fecundity and delayed maturity.

*Larval ecology.* The high commercial value of food fishes which are mainly marine (p. 318), has led to a great input of research over the last hundred years. This research has been aimed at clarifying the factors that control the recruitment of the young stages to the fishable stock. A full understanding of the ecological mechanisms underlying recruitment would help in the scientific management of stocks and enable the prediction of yields to the fishing fleets. Most of the high-yielding species are r-selected with high fecundity. Mortality rates of both eggs and larvae are of the order of 5–30% of the population per day. At one time it was thought that this mortality was mainly caused by starvation, the larvae being unable to find adequate microzooplanktonic food, with a particular critical period at the end of the yolk-sac stage. More recently, it has become fashionable to consider that predation is of equal or greater importance as a mechanism of mortality. Careful experiments have shown that the mortality rates of eggs are similar to those of larvae (the eggs cannot die of starvation) and that there is little evidence of a critical period at the completion of yolk resorption. Furthermore, experiments on rearing larvae of species like cod in shore-based tanks show very high survival rates in the absence of predators.

Estimation of natural mortality is difficult. Of late, fishery biologists have been investigating marine food chains in which fish larvae are involved and measuring predation rates under experimental conditions. It is thought that the transparency of eggs and younger larvae and the disruptive camouflage of their melanophores reduces their conspicuousness to predators, at least under some conditions of illumination and background. It is important for survival that growth be rapid to reduce the spectrum of predators; generally speaking, larvae have to be less than half the size of their predators to be vulnerable. As larvae grow they become more conspicuous but their ability to escape predatory attack is improved. Changes of behaviour such as schooling or settling on the sea bed are also likely to reduce predation pressure.

That is not to say that feeding (p. 155) and starvation do not play a significant role in these high mortalities. Larvae of species like herring and plaice can survive for about a week without food when they are small and plaice can withstand starvation for as long as three weeks as they approach metamorphosis. Starving larvae reach a point where they are still alive but are too weak to feed if food becomes available (the so-called point-of-no-return (PNR) or 'ecological death'). Considerable efforts have been made over the past decade to categorise the nutritional status or condition of sea-caught larvae, to ascertain their chances of survival. To this end, condition factors (weight divided by the cube of the length), body dimensions, organ histology, fat content and DNA:RNA ratios have been used in a wide range of species including anchovy, cod, herring, sardine and many species of flatfish.

One of the most interesting findings is that microzooplankton density in the sea is rarely adequate to sustain survival and growth unless the food is distributed patchily, e.g. at hydrographic 'fronts' or discontinuities (p. 155). This has led to the concept of Lasker years, named after a distinguished marine biologist from the USA. It was found in the California current that good brood survival was associated with calm weather when such fronts (and patches) could be built up and maintained, so providing circumstantial evidence for the advantage of patchiness.

The importance of a match between larvae and their microzooplanktonic food has led to the match–mismatch hypothesis of larval survival expounded by Cushing (1990). The main production of larvae should be geared to the production of their food. Since the youngest larvae have very small mouths, it is essential that their food, such as copepod eggs and nauplii, is available when the larvae are very small. Batch spawning may help in matching larvae to their food but, of course, the numbers of larvae will then be less than with a one-off spawning. As well as a match with the food, a mismatch with the predators is also advantageous for survival. Thus larvae have to be considered both as predators and prey (p. 304) and this leads to the concept of cohort competition. The best way to survive is to grow fast, to enable larger food items to be eaten and larger predators to be avoided. If the potential prey can 'outgrow' the potential predator it has a greater chance of survival. Biologists put a high premium on such fast growth to yield a good brood and high recruitment.

*Growth.* Fishery biologists are much interested in growth because (obviously) it is a major component of the fishery equations that are used to calculate the yields at different levels of exploitation. It is not only numbers of fish that are important (resulting from recruitment and the number of fish surviving from year to year) but also the addition to the biomass of stock caused by the growth of individual fish.

Growth is enhanced by a good food supply and high temperature. In high latitudes it is thus seasonal and overwintering fish may grow slowly or not at all. Growth can depend on intra- or interspecific competition and may thus be density-dependent. A considerable breakthrough in refining the fishery equations was made in the 1940s by applying the von Bertalanffy growth equation (p. 183) to the growth of commercially important teleosts as part of the study of fish population dynamics.

A key requirement in estimating growth is to be able to age individual fish. In higher latitudes with seasonality the overwintering growth checks are reflected in the growth patterns of a number of tissues, especially the scales and otoliths, giving rings equivalent to the age in years (Figure 8.17). In larvae, it is possible to remove the otoliths and, after suitable treatment, observe *daily* growth rings. This fairly recent finding has enabled larval growth and mortality rates to be measured, giving much greater insight

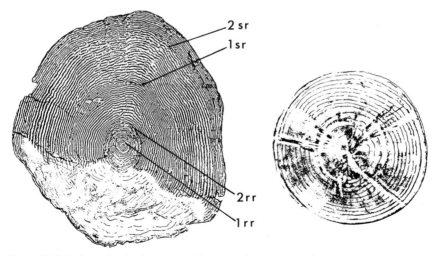

**Figure 8.17** Left: scale of a four-year-old salmon (two years in the river, rr: river ring; two years at sea; sr: sea ring). Redrawn from Jones (1959) *The Salmon*. New Naturalist series, Collins. Right: Otolith of herring larva with 12 daily rings. Redrawn from photograph by Dr. A. Geffen.

into the ecological pressures on populations of larvae. In low latitudes the lack of seasonality makes these techniques less applicable, or even useless.

## Bibliography

Bailey, K.M. and Houde, E.D. (1989) Predation on eggs and larvae of marine fishes and the recruitment problem. *Advances in Marine Biology*, **25**, 1–83.

Balon, E.K. (ed) (1985) *Early life history of fishes: new developmental, ecological and evolutionary perspectives*. W. Junk, Dordrecht, 280 pp.

Blaxter, J.H.S. (1969) Development: eggs and larvae, in *Fish Physiology*, **III** (Hoar, W.S. and Randall, D.J. eds) Academic Press, New York, pp. 177–252.

Blaxter, J.H.S. (1988) Pattern and variety in development, in *Fish Physiology* **11A** (Hoar, W.S. and Randall, D.J. eds) Academic Press, San Diego, pp. 1–58.

Cushing, D.H. (1990) Plankton production and year-class strength in fish populations: an update of the match/mismatch hypothesis. *Advances in Marine Biology*, **26**, 249–293.

Fenton. G.E., Short, S.A. and Ritz, D. (1991) Age determination of orange roughy *Hoplostethus atlanticus* (Pisces: Trachichthydae) using $^{210}$Pb:$^{226}$Ra disequilibria. *Marine Biology*, **109**, 197–202.

Hoar, W.S. (1969) Reproduction, in *Fish Physiology* **III** (Hoar, W.S. and Randall, D.J. eds) Academic Press, New York, pp. 1–72.

Holden, M.J. (1974) Problems in the rational exploitation of elasmobranch populations and some suggested solutions, in *Sea Fisheries Research* (Harden-Jones, F.R. ed), Elek Science, London, pp. 117–137.

Kamler, E. (1992) *Early Life History of Fish; an Energetics Approach*. Chapman & Hall, London.

Keenleyside, M.H.A. (ed) (1991) *Cichlid Fishes, Behaviour, Ecology and Evolution*. Chapman & Hall, London.

Lagler, K.F., Bardach, J.E. and Miller, R.R. (1962) *Ichthyology*. John Wiley, New York.

MacArthur, R.H. and Wilson, E.O. (1967) *Theory of Island Biogeography*. Princeton University Press, Princeton.

McDowall, R.M. (1988) *Diadromy in fishes*. Croom Helm, London.

Meyer, A. and Lydeard, C. (1993) The evolution of copulatory organs, internal fertilization, placentae and viviparity in killifishes (Cyprinodontiformes) inferred from a DNA phylogeny of the tyrosine kinase gene *X-src*. *Proceedings of the Royal Society of London, B*, **254**, 153–162.

Nikolsky, G.V. (1963) *The Ecology of Fishes*. Academic Press, London.

Potts, G.W. and Wootton, R.J. (1984) *Fish Reproduction: Strategies and Tactics*, Academic Press.

Purdom, C.E. (1993) *Genetics and Fish Breeding*, Chapman & Hall, London.

Robertson, D.R., Reinboth, R. and Bruce, R.W. (1982) Gonochorism, protogynous sex change and spawning in three species of sparisomatinine parrot fishes from the Western Indian Ocean. *Bulletin of Marine Science*, **32**, 868–879.

Simpfendorfer, C.A. and Millward, N.E. (1993) Utilisation of a tropical bay as a nursery area by sharks of the families Carcharinidae and Sphyrnidae. *Environmental Biology of Fishes*, **37**, 337–345.

Wourms, J.P., Grove, B.D. and Lombardi, J. (1988) The maternal embryonic relationship in viviparous fishes, in *Fish Physiology* **11A** (Hoar, W.S. and Randall, D.J. eds) Academic Press, San Diego, pp. 1–134.

# 9  Endocrine systems

Fish endocrine organs and the hormones they secrete are particularly interesting because different groups of fishes show different stages in the evolution of the endocrine systems better known in terrestrial vertebrates. Studies on fishes, moreover, have significantly added to our knowledge of endocrine systems in higher vertebrates. For example, it was in teleosts that Scharrer more than 60 years ago first demonstrated the link between brain neurons sending hormonal material down their axons, and the region of the pituitary where they are stored and later released. In elasmobranchs and teleosts there is an endocrine organ lacking in other vertebrates (the uropophysis), and perhaps most interesting of all, recent sequence studies of fish hormones have provided insights into the evolutionary relationships of different vertebrate peptide hormone families.

## 9.1 Origins

In protochordates like amphioxus or ascidian tunicates, there are no direct equivalents of the endocrine organs of fishes, but there are intriguing morphological and biochemical hints of precursors to those of vertebrates. For instance, the amphioxus endostyle (which secretes mucus for filter feeding) takes up iodine and couples it to tyrosine, then forming thyroxine just as in the vertebrate thyroid. There is therefore a direct link with the mucus-secreting endostyle of the lamprey ammocoete larva which gives rise to the adult thyroid on metamorphosis. Again, the tunicate neural gland and its ciliated duct open to the pharynx are reminiscent of the way that the anterior part of the vertebrate pituitary (the adenohypophysis, see p. 200) forms during embryonic development from an upgrowth of the roof of the pharynx (Rathke's pouch).

The difficulty has been to find in these protochordate 'anlagen' of vertebrate endocrine organs, equivalents of the hormones they produce in vertebrates, or if such are found, to know what role they may perform. For example, we have really no idea what thyroxine may do in amphioxus, or what may be the function of the cells containing hormonal peptides (shown by immunocytochemical studies) in the ascidian neural gland. It is, of course, striking that these peptides in the neural gland are labelled by antisera to vertebrate peptides, including prolactin and secretin, found in the vertebrate anterior lobe, so supporting the idea that the neural gland is homologous to the anterior lobe only of the vertebrate pituitary. What *has* emerged from such immunocytochemical studies though, is that there is in both protochordates and vertebrates similar peptide hormone activity in gut cells and in brain neurons.

### 9.1.1 The brain–gut axis

The development of immunocytochemical methods for hormonal peptides has shown that such peptides as those of the gastrin/cholecystokinin family are not only found in the endocrine cells of the gut, where they were first discovered (by less-sensitive techniques), but are also present in neurons of the peripheral and central nervous system. There is the same duality in nervous system/gut neuropeptide location in most invertebrates (molluscs, arthropods and annelids, for example), but in cnidaria neuropeptides (as for example FRMFamide) are only present in the nervous system, and this is why it is supposed that the neuropeptides originally arose in the nervous system and only later became products of gut cells. So the concept of a brain–gut axis for neurohormonal peptides arose; the same peptide hormones present in the gut and the brain may have had their origin in neurons of the ancestral invertebrate central nervous system. Obviously, although the structure of these peptides in neurons and gut cells in higher invertebrates and in fishes may be the same, at different sites they are likely to have very different functions and target organs. For example, the remarkable modulator neurons of nerve 0 (the nervus terminalis) in the teleost *Colisia* (p. 286) contain gonadotrophin releasing hormone that presumably had modulating functions on neuronal activity in the central nervous system, whilst the same GnRH is also released by the neuro-hypophysis of the pituitary to act on the gonads. Similarly, neuropeptide Y which is a member of the pancreatic polypeptide family found in most regions of the teleost brain, is also seen in the hypophyseal nerve fibres contacting different types of secretory cells in the adenohypophysis (see below) presumably modulating release of such hormones as gonadotrophin and melanotrophin.

## 9.2 The endocrine organs of fishes

The main endocrine organs in a teleost fish, with the hormones they are known or supposed to secrete, are shown schematically in Figure 9.1. To determine the presence and functions of the hormones secreted has sometimes been rather difficult, for although a variety of approaches has been adopted (e.g. immunocytochemistry and radioimmunoassay; bio-assays by injection of extracts from the organs into fish and other vertebrates such as frogs and mice; injection of higher vertebrate

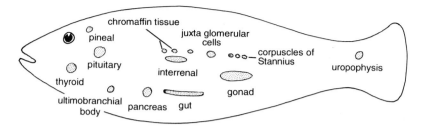

**Figure 9.1** The endocrine organs of a teleost fish.

hormones into fishes themselves; and the effects of surgical removal or chemical block of the organs), clear results have not always been obtained. Thus injection of extracts from the teleost caudal neurosecretory organ (uropophysis) into rats produces vasoactive effects, whilst in the fish itself, the extracts have osmoregulatory effects. Even if a well-characterised mammalian hormone, like gastrin, is injected into a fish, it is hardly surprising that similar effects to those produced in mammals may not be obtained, given the possible changes in receptor structure, and changes in the hormone structure itself that have occurred between fish and mammals.

Table 9.1 shows what is known of the functions of hormones in fish. Most hormones are named from their function in mammals, so that their piscine homologues are often inappropriately named. For example, although prolactin was named for its stimulatory action on mammalian mammary glands, in teleosts (and indeed, it later became evident, in higher vertebrates) PRL has a remarkably wide spectrum of action, being involved in osmoregulation, reproduction, growth, lipid metabolism and steroid synergism. As neurons containing prolactin-like material are revealed immunocytochemically in various brain regions in fishes, we can also add neurotransmitter action to this remarkably varied list.

Two endocrine organs of fish consist in part of neuron cell bodies in the central nervous system: the uropophysis and the pituitary. We begin this brief survey of fish endocrine organs with the uropophysis, which is in some ways a much simpler version of the pituitary.

## 9.3 The uropophysis

In elasmobranchs and teleosts, large neurosecretory neurons at the caudal tip of the spinal cord send axons ventrally to end in palisades along capillary walls. The neurons are more condensed in teleosts, and send their axons down a uropophyseal stalk, exactly analogous to the hypophyseal stalk of the pituitary, so that in some teleosts there is a conspicuous ventral swelling at the end of the cord (Figure 9.2). The whole arrangement is

**Table 9.1** The endocrine organs, their hormones and hormonal actions in teleosts

| Site (Hormones) | Target organ | Effect(s) |
|---|---|---|
| *Uropophysis* | | |
| Urotensins I–IV | Kidney, gills | Changes salt–water balance |
| *Pituitary* | | |
| Pars distalis | | |
| Prolactin | Several | Many |
| Adrenocorticotrophic hormone | Interrenal | Stimulation of cortisol |
| Somatostatin | Pituitary | Inhibition of growth hormone release |
| Growth hormone | Several | Stimulation of growth |
| Thyroid stimulating hormone | Thyroid | Stimulation of thyroxine |
| Gonadotrophic hormone | Gonads | Stimulation of gonads |
| Pars intermedia | | |
| Melanophore stimulating hormone | Pigment cells | Darkening |
| Neurohypophysis | | |
| Arginine vasopressin | Blood vessels | Increases blood pressure |
| Isotocin, mesotocin | Blood vessels | Constricts gill blood vessels, systemic vasodilation |
| *Thyroid* | | |
| Thyroxine | Many | Various, but not calorigenic |
| *Ultimobranchial* | | |
| Calcitonin | Gills, kidney | Regulation of $Ca^{2+}$ metabolism |
| *Corpuscles of Stannius* | | |
| Hypocalcin | Gills | $Ca^{2+}$ homeostasis |
| *Pancreas* | | |
| Insulin | All cells | Increases glucose permeability |
| Glucagon | All cells | Glycogen and lipid metabolised |
| *Gut* | | |
| Cholecystokinin, vasoactive intestinal peptide | Gut muscle and secretory cells | Controls gut motility and secretory activity |
| *Chromaffin tissue* | | |
| Adrenaline | Circulation | Gill vasodilation, systemic vasoconstriction, |
| Noradrenaline | | increases heart rate, increases blood glucose |
| *Interrenal* | | |
| Corticosteroids | Gills, kidney | Stress response, osmoregulation |
| *Juxtaglomerular cells of kidney* | | |
| Renin | Kidney, chromaffin tissue | Osmoregulation in euryhaline fishes, blood pressure regulation |
| *Gonads* | | |
| Androgens and oestrogens | Many, including brain | Reproductive status and behaviour also of other fish (as pheromones) |

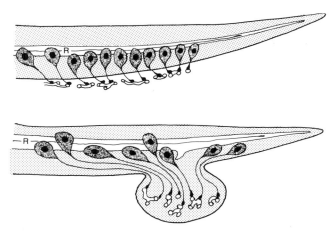

**Figure 9.2** The uropophysis of the elasmobranch (above) and the teleost. Note association of swollen axon terminals of the neurosecretory neurons with blood vessels, and Reissner's fibre (R) in the central canal of the cord. After Fridberg and Bern (1968). *Biol. Rev.*, **43**, 175.

similar to that in the neurohypophysis of the pituitary, albeit on a simpler scale, and naturally led to studies of the properties and structure of the hormones it was presumed to secrete. A sustained effort by Lederis and his colleagues at Calgary, processing large numbers of uropophyses mainly of suckers (*Catostomus commersoni*), showed that two principal peptides were released, urotensins I and II, together with less well-known urotensins III and IV. These were distinguished by their effects on mammalian preparations. Urotensin I has hypotensive actions, urotensin II has smooth-muscle stimulating actions, urotensin III influences sodium transport and urotensin IV has osmoregulatory activity.

In fish, urotensin II stimulates the smooth muscle of the reproductive tracts. It also is involved in sodium exchange. Urotensin I stimulates skin chloride cell secretion in the goby (*Gillichthys*). Not much is known of the functions in fishes of urotensins III and IV. It thus seems clear that the original suggestion by Enami that the teleost uropophyseal hormones are concerned in osmoregulation was right, despite initial failures to confirm his original experiments. Nothing is yet known of the functions of the elasmobranch uropophysis.

## 9.4 The pituitary

The pituitary is the most complex endocrine gland in the body. It controls the secretory activity of three other endocrine glands as well as producing hormones of its own (for example, melanocyte stimulating hormones, MSH) that act directly on effector tissues, and it is in addition the chief link between the nervous and endocrine systems. The basic structure of the pituitary (hypophysis) is essentially the same in all vertebrates; it consists of two parts of different structure, function and embryological origin. The nervous part (pars nervosa or neurohypophysis) is a downgrowth from the floor of the diencephalon under the hypothalamus (p. 276), and the epithelial part (the adenohypophysis) is an upgrowth from the roof of the pharynx, arising as Rathke's pouch in development. In less-advanced teleosts like the milkfish (*Chanos*) and the tarpon (*Megalops*) the connection with the pharynx is still retained. Between these two regions, there is a complex system of blood vessels. In lampreys and hagfish, these vessels are more simply arranged than in other craniates, for with the exception of these two groups, the vascular link between the two divisions of the fish pituitary (as in higher vertebrates) forms a portal system transporting blood and hormones from the neurohypophysis to the distal part of the adenohypophysis. Although the basic plan of the pituitary is the same in all fishes, there are differences in the arrangement and relative sizes of the different pituitary regions, as seen in Figure 9.3. The different divisions of the adenohypophysis were first named in tetrapods, and those in fishes have been homologised with them, based upon the same

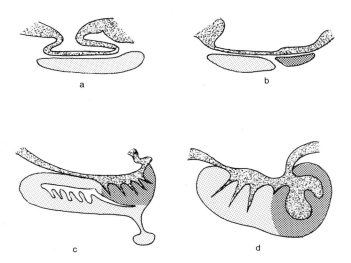

**Figure 9.3** Schematic mid-sagittal sections of the pituitary in different fish. (a) hagfish; (b) lampreys; (c) elasmobranchs; (d) teleosts. Random shading: nervous tissue; light regular stipple: pars distalis of the adenohypophysis; heavier regular stipple: pars intermedia of the adenohypophysis. After Ball and Baker (1969) in *Fish Physiology*, **2** (Hoar and Randall eds), p. 1.

hormonal content of the secretory cell types (reflected in their differential staining reactions). So, for example, the pars distalis is characterised by cells secreting thyroid stimulating hormone, prolactin, adrenocortico-trophic hormone and growth hormone.

How did this complex organ arise in chordate evolution? Although as we have seen, the ascidian neural gland is in some respects a possible precursor of the adenohypophysial part of the pituitary, no trace of the neurohypophysis is to be found in the ascidian brain. In amphioxus, the infundibular organ in the ventral part of the 'brain' vesicle has been suggested as equivalent to the neurohypophysis, and the development of Hatschek's pit on the left side of the head suggests that it may be equivalent to the adenohypophysis. Several immunocytochemical studies on amphioxus have been undertaken in an attempt to demonstrate the presence of vertebrate neuropeptide hormones in these structures, but the results so far have not been entirely convincing, and their functional role remains enigmatic. One difficulty, which seems likely soon to be resolved by recent continuing work on the expression of homeobox genes (p. 274) in amphioxus, is that homology remains to be established between the different regions of the amphioxus cerebral vesicle, and those of the vertebrate brain. So although amphioxus seems to show hints of the origin of the chordate pituitary, it is in hagfish and lampreys that we first see the pituitary clearly.

### 9.4.1 The pituitary in hagfish and lampreys

As seen in Figure 9.3, in lampreys, the adenohypophysis is divided into three regions, termed the anterior and posterior pars distalis and the posterior pars intermedia (although the homology of these regions with those in other chordates is uncertain). They are linked to the neuro-hypophysis by an extensive capillary network, but there does not seem to be the same portal system carrying neurosecretory material from the neurohypophysis as is seen in 'higher' fishes. In hagfish, the two parts of the pituitary are separated by a thick sheet of connective tissue. In other words, it seems that in both lampreys and hagfish, the nervous control of hormonal release by the adenohypophysis (which is such a striking feature of pituitary function in other chordates), is absent. The neurohypophysis receives neurosecretory axons from the pre-optic nucleus and hypo-thalamus. Not a great deal is known of pituitary function in hagfish, whilst in lampreys, hypophysectomy has provided evidence for a melanophore expansion hormone (the lamprey becomes pale), and less convincingly, for a gonadotropic hormone in the adenohypophysis. In the neurohypophysis, arginine vasotocin (see Table 9.2) has been demonstrated, but it is debatable what its role may be.

**Table 9.2** The amino acid sequences of the arginine vasotocin family found in the neurohypophysis of fishes. After Heller (1974) *Gen. Comp. Endocrinol.*, **22**, 315

|  | 1 | 2 | 3 | 4 | 5 | 6 | 7 | 8 | 9 |
|---|---|---|---|---|---|---|---|---|---|
| Arginine vasotocin | Cys | Tyr | Ile | Gln | Asn | Cys | Pro | Arg | Gly(NH$_2$) |
| Mesotocin | Cys | Tyr | Ile | Gln | Asn | Cys | Pro | Ile | Gly(NH$_2$) |
| Isotocin | Cys | Tyr | Ile | Ser | Asn | Cys | Pro | Ile | Gly(NH$_2$) |
| Glumitocin | Cys | Tyr | Ile | Ser | Asn | Cys | Pro | Gln | Gly(NH$_2$) |
| Valitocin | Cys | Tyr | Ile | Gln | Asn | Cys | Pro | Val | Gly(NH$_2$) |
| Aspartocin | Cys | Tyr | Ile | Asn | Asn | Cys | Pro | Leu | Gly(NH$_2$) |

Note that, of the nine amino acids, those in positions 4 and 8 differ.

### 9.4.2 The pituitary in elasmobranchomorpha

In elasmobranchs and holocephalans, the pars intermedia is very large and closely interdigitated with the neurohypophysis, and the pars distalis is elongated with an unique ventral lobe in the floor of the chondrocranium linked to the rest of the pars distalis by a long stalk (Figure 9.3). In holocephali, the equivalent of the ventral lobe is a group of follicles entirely separate in the adult from the rest of the pituitary. So the pituitary does not look much like the pituitary of other fishes. Unlike lampreys and hagfish, there is an abundant and complex vascular bed including a portal system between neuro- and adenohypophysis, and palisades of neurosecretory terminals from the pre-optic nuclei lie closely adjacent to the capillaries, in just the same manner as in the teleost uropophysis. In addition to this 'indirect' hormonal link some nerve fibres pass directly to innervate cells in the adenohypophysis. Because of the cartilaginous chondrocranium, experimental hypophysectomies are easier in elasmobranchiomorphs than in bony fishes (although in salmonids, an approach via the orbit has been successful), so that it is possible to remove different regions of the adenohypophysis separately to examine the effects produced. Thus, for example, removal of the ventral lobe in dogfish has shown that various reproductive hormones are secreted there, such as follicle stimulating hormone and gonadotropin releasing hormone, whilst incubation of ventral lobe extract with thyroids increases thyroxine release indicating secretion of thyroid stimulating hormone. Other experiments have shown that melanophore stimulating hormone is secreted by the intermediate lobe, acting to expand the melanophores and darken the skin, and prolactin by the distal lobe, acting (in the euryhaline sting ray *Dasyatis violacea*) to control plasma osmolarity and sodium and urea retention. Although the ventral lobe is not directly linked to the neurohypophyseal hypothalamic neurosecretory axons, the control hormones they release reach the ventral lobe via the general circulation to regulate the secretion of the ventral lobe hormones.

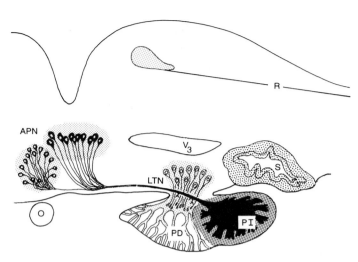

**Figure 9.4** Sagittal section of the pituitary of the eel, *Anguilla*, showing the nerve tracts from the anterior preoptic nucleus (APN, divided into parts with large and small neurosecretory neurons), and from the lateral tuberal nucleus (LTN). The infundibular organ dorsally secretes Reissner's fibre (R) which passes down the spinal cord. $V_3$: third ventricle; S: saccus vasculosus; O: optic chiasma; PD: pars distalis; PI: pars intermedia. After Knowles and Vollrath (1966), *Phil. Trans. Roy. Soc. B*, **250**, 311; and Olivereau (1967) *Z. Zellforsch. Mikr. Anat.*, **80**, 286.

In the neurohypophysis, there may be several hormones of the arginine vasotocin family, differing in their amino acid sequence, as well as arginine vasotocin itself (Table 9.2). Not much is yet known of the function of these hormones in elasmobranchs, though when injected into rats they have antidiuretic, lactational and oxytocic (parturition-inducing) effects. Remarkably enough, even within the same species of elasmobranch, different individuals may have different members of the other arginine vasotocin family hormones! Evidently, 'random' amino acid substitutions can take place so long as the functional region(s) of the hormone remain conserved, and it is certainly very striking that in another hormone, relaxin (p. 213) Callard and his colleagues (1989) emphasise that on present data, there has been almost as much amino acid substitution and sequence change *within* the elasmobranchs, as there has been *between* elasmobranchs and mammals!

### 9.4.3 The teleost pituitary

This differs from that of other fishes because the neurohypophysis ramifies into all the regions of the adenohypophysis (Figures 9.3 and 9.4), sending at least two kinds of axons into intimate contact with the secretory cells of

the adenohypophysis. The range of hormones produced by the adenohypophysis is similar to that in elasmobranchs, but in some areas, owing to
the commercial importance of teleosts such as eels and salmonids, more is
known of their actions. For example, in eels and in salmon, growth
hormone from the pars distalis acts on receptors in the liver inducing the
liver to secrete insulin-like growth factors (or somatomedins), which
directly stimulate chondrogenesis. Since hypophysectomy in salmon
reduces receptor number, the pituitary itself regulates the hepatic growth
hormone receptors.

Growth hormone has many actions, e.g. osmoregulatory, adrenocortico-
trophic activity (thus being potentially diabetogenic), thyrotropic, and
possibly immunological and reproductive effects. Hence the present
considerable interest and effort in producing transgenic salmonids using
growth hormone transgenes to obtain larger fish might perhaps be better
devoted to the use of insulin-like growth factor-I transgenes. In contrast to
elasmobranchs, teleost melanophores are innervated by autonomic nerve
fibres, and although in some species, melanophore stimulating hormone is
present, and darkens the skin if injected; in others it has no effect, perhaps
in the latter being overridden by the nervous control mechanism.

The neurohypophyseal hormones of teleosts include arginine vasotocin
(as in other fishes), and two different members of this family, isotocin and
mesotocin (Table 9.2). The role of these is still unclear, though arginine
vasotocin probably has a variety of effects including peripheral vasoconstriction and oviducal smooth-muscle contraction.

## 9.5 The thyroid

In contrast to the pituitary, the thyroid is of simple design, and the
hormones it produces are of the same relatively simple structure in all
vertebrates. The evidence for the evolutionary origin of the thyroid is
clear. Not only is there a direct morphological link between the
protochordate endostyle and that of the lamprey ammocoete, but also the
thyroid hormones are found in protochordates. Even more convincing, the
thyroid stimulating hormone (TSH) and goitrogens exert the same effects
on the iodine-binding cells of the protochordate endostyle as they do on
vertebrate thyroid follicle cells. However, we do not know what the
thyroid hormones of protochordates do.

In adult lampreys the thyroid gland is a series of follicles scattered along
the pharyngeal floor. It arises in ontogeny from the larval ammocoete
endostyle which opens into the floor of the pharynx (Figure 1.11). In
lampreys and in all fishes, the thyroid consists of a series of unicellular
follicles, which are usually more scattered than in higher vertebrates
(Figure 9.4); these take up iodine and produce the thyroid hormones. Both

the ammocoete endostyle and the adult thyroid follicles take up iodine, and iodinate tyrosine adding one or two iodine atoms to make mono- or diiodotyrosine. These then condense in pairs, losing an alanine residue (Figure 9.5) to make the two thyroid hormones triiodothyronine ($T_3$) and thyroxine ($T_4$). At the periphery of the thyroid follicles, these hormones are bound to the glycoprotein thyroglobulin which has a similar amino acid composition to the thyroglobulins of higher vertebrates. The hormones are unbound by protease hydrolysis of the thyroglobulin within the follicle cell, and then released into the circulation.

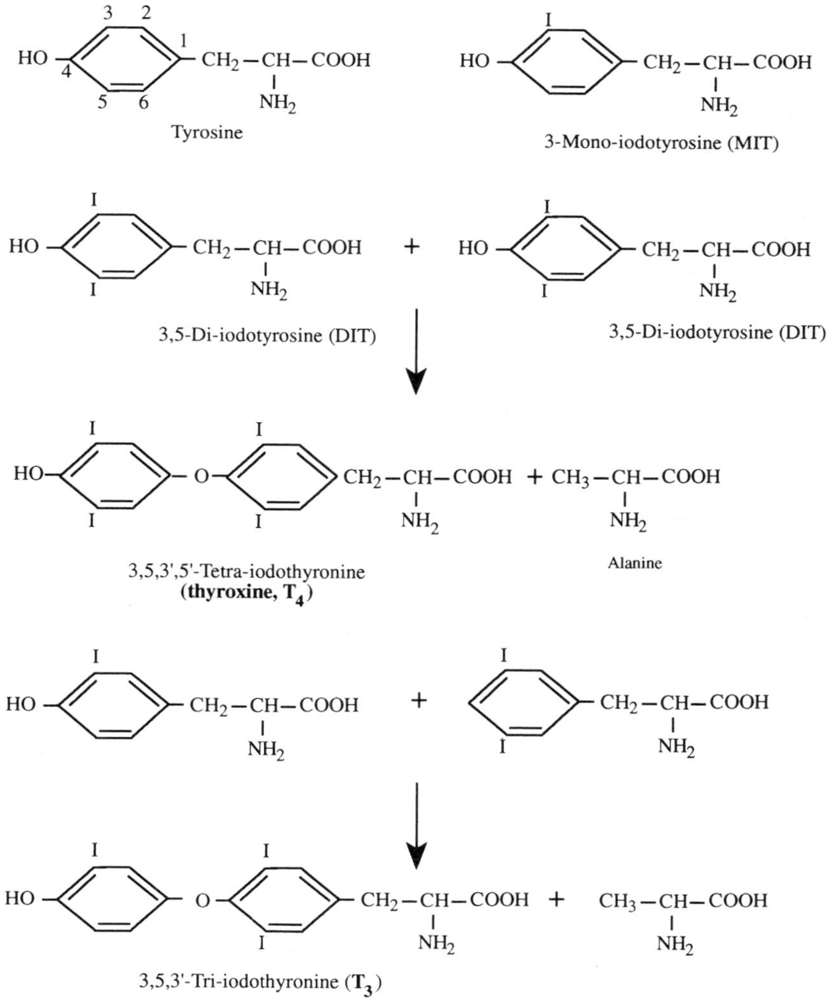

**Figure 9.5** The derivation of the two thyroid hormones, triiodothyronine and thyroxine from tyrosine. After Barrington (1964).

Thyroxine levels in ammocoete blood are around 8.0 μg%, compared to 0.5 μg% in adult lampreys, but there is no evidence for any effects of thyroxine in metamorphosis, nor clear evidence for control of the thyroid by TSH.

What is the function of the thyroid hormones in gnathostome fishes, and how is their release controlled? Curiously enough, although there is good evidence for TSH thyroid *regulation* by the pituitary, in fishes the role of the thyroid hormones is rather confusing, and different from that in other vertebrates.

In birds and mammals, the thyroid hormones are calorigenic, stimulating oxidative metabolism, but this function has not been shown in poikilotherms like fishes and amphibia. In amphibia, thyroxine is necessary for metamorphosis from tadpole to adult, but it does not seem to be necessary for the (admittedly less drastic) metamorphosis of the fish larva to the juvenile. However, it certainly plays a central role in the later transformation of the salmon parr to the smolt, prior to the latter going downriver to the sea. Smoltification (as it is inelegantly called) involves all kinds of changes in pigmentation, silveriness, carbohydrate and lipid metabolism, and in osmoregulatory ability. Silvering of the skin increases when parr are treated with thyroxine (since there is increased deposition of reflecting guanine platelets in the scales; see p. 249). Blood thyroxine levels in *Oncorhynchus* species reach a peak at the time of the new moon in the spring, before the downstream migration, and although recent work has shown that this peak does not synchronise with peaks in tissue concentration, it provides an excellent 'timer' for release of young salmon from hatcheries. In brown and rainbow trout (*Salmo trutta* and *Oncorhynchus mykiss*), $T_3$ is essential for seawater adaptation. Pituitary hormones such as cortisol, prolactin, growth hormone and insulin-like growth factor-I are also involved in smoltification and in osmoregulation. The present situation seems best summarised (as Matty points out) by the view that the fish thyroid aids adaptation of the fish to environmental changes such as temperature and osmotic stress, and to the rapid internal changes during growth and sexual maturation. One interesting observation which has yet to be fitted into any scheme of fish thyroid function, is that thyroxine levels are high in the yolk of teleost eggs (derived from the maternal blood supply to the ovary), and may be important in regulating development.

## 9.6 The ultimobranchial gland

Although lacking in hagfish and lampreys, in other fishes paired or unpaired outgrowths from the last branchial pouch migrate during development to lie over the pericardium. Originally described in elasmobranchiomorpha, where the tissue is follicular like the thyroid, in teleosts it is solid; but in both groups, the ultimobranchial gland contains high concentrations of the straight-chain peptide calcitonin. In mammals, calcitonin is secreted by the parathyroid and is a potent hypocalcaemic factor, inhibiting bone resorption. Because teleost bone is virtually acellular and elasmobranchiomorphs are almost entirely cartilaginous (p. 15), what calcitonin does in such fishes is unclear. Experimental studies have shown in teleosts that calcitonin increases renal output of $Ca^{2+}$, and increased efflux but decreased influx of $Ca^{2+}$ across the gills. Further, in female salmon, serum calcitonin levels rise until spawning when there is a dramatic decline. Thus it seems that calcitonin in fishes is involved in $Ca^{2+}$ metabolism, together with the activity of the corpuscles of Stannius, and prolactin produced by the pituitary, even if the interrelation of these different factors remains to be worked out.

An interesting finding by Girgis and her colleagues (1983) is that a calcitonin very closely similar to human calcitonin occurs in the nervous system of amphioxus and the hagfish (*Myxine*), where it presumably functions as a neurotransmitter. These workers suggest that this human calcitonin-like molecule was the parent brain peptide from which the later members of the calcitonin family found in the ultimobranchial bodies and in the human thyroid were derived.

## 9.7 The corpuscles of Stannius

These are small spherical bodies lying on, or embedded in, the kidneys of bony fishes, first described by Stannius in 1839. Each is well vascularised and innervated, and its cells show the fine structure typical for protein-secreting cells. Removal of the corpuscles produces an immediate rise in plasma $Ca^{2+}$, whilst conversely, injection of extracts of the corpuscles into normal eels results in a rapid fall in plasma $Ca^{2+}$. The hypocalcaemic hormone of the corpuscles, hypocalcin, may or may not be the same as a partially characterised (possibly identical) hormone teleocalcin found in salmonids. In any event, the corpuscles are important in $Ca^{2+}$ homeostasis in bony fish, and may control the ratio of $Ca^{2+}$ to $Na^+$ and $K^+$ in the plasma.

## 9.8 The gastro-entero-pancreatic endocrine system

### 9.8.1 The pancreas

The structure and location of the endocrine pancreas varies in different groups of fishes (Figure 9.6). In some teleosts, such as *Cottus* and *Lophius*, relatively large lumps of endocrine pancreatic tissue (Brockmann bodies)

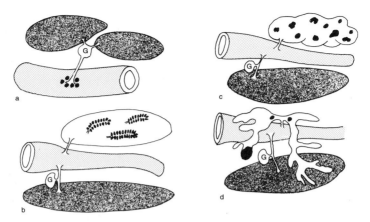

**Figure 9.6** The pancreas in different fishes. (a) *Myxine*; (b) Elasmobranch; (c) Compact teleost type (*Anguilla*); (d) Ramified teleost type. Islet tissue black, liver dark random shading, gut: light stipple. G: gallbladder. After Epple (1969) *Fish Physiology*, **2**, 275.

can be isolated, and their hormones extracted, to see the effects of pancreatomy. The endocrine portion of the pancreatic tissue contains various secretory cell types staining differentially with dyes like par-aldehyde-fuchsin, in a similar way to those in the mammalian pancreas, and producing similar hormones.

In mammals, the level of blood glucose is controlled by two pancreatic hormones, insulin and glucagon. Glucose enters cells only in the presence of insulin, which thus lowers blood glucose levels; glucagon raises blood glucose levels by stimulating glucose production from stored glycogen. So blood glucose levels result from the balancing of two hormones with opposite actions. Not surprisingly, the two are produced by different cell types in the pancreas. In fishes as in mammals, the A cells produce glucagon, the B cells insulin, and the D cells somatostatin (which in mammals inhibits release of growth hormone by the pituitary). Holo-cephalans have, in addition, unique X cells making up half of all their islet cells.

As well as the three pancreatic hormones mentioned above, immuno-cytochemistry has shown cells containing other regulatory peptides of the neuropeptide Y family, like pancreatic polypeptide, neuropeptide Y itself and peptide YY. Some of these may be co-localised in the same cells as the more familiar pancreatic hormones (e.g. pancreatic polypeptide and glucagon). A similar peptide Y, but without a terminal amide (as in Table 9.2) has been found in the angler fish (*Lophius*) pancreas. In addition, two regulatory proteins of the RF-amide family have also been demonstrated, sometimes co-localised (e.g. with glucagon or neuropeptide Y). In mammals, neuropeptide Y and RF-amides are only in the nervous system, peptide YY in the nervous system and gut, and only pancreatic polypeptide has so far been found in the pancreas.

What is the function of the different hormones of the fish pancreas? In lampreys, there is good evidence that insulin regulates blood sugar, but why blood insulin levels vary seasonally is not understood. In the holocephalan *Hydrolagus*, treatment with glucagon evokes a rapid but transient hyperglycaemia, whilst insulin evokes hypoglycaemia, as it does in elasmobranchs, where insulin is probably also involved in protein and lipid metabolism. In many bony fishes, however, glucose homeostasis may not be a major role of insulin, which seems more to be concerned with amino acid metabolism. Somatostatin in mammals inhibits insulin and glucagon secretion, and in hagfish and *Anguilla* inhibits insulin release, but its effects on glucagon release have not been examined in fishes.

### 9.8.2 Gut hormones

The many different polypeptide hormones released from the gut cells themselves and the 'neurohormonal' substances released from the auto-nomic nerve fibres innervating the gut, have two main functions: (1) to control secretory activity, and (2) to control gut motility. Some also act on other organs in the body. There is a strikingly long list of regulatory polypeptides recognised in the fish gut, belonging to two main families, gastrin/cholecystokinin, and secretin/glucagon/vasoactive intestinal peptide. There are also many others, such as substance P (found also in the dorsal horn of the spinal cord, see p. 266), bombesin, enkephalin, somatostatin and neurotensin, and in elasmobranchs, rectin controlling $Cl^-$ secretion by the rectal gland. The functions of most of these are still poorly known.

The peptides of the cholecystokinin (CCK) family share the highly evolutionarily conserved common C-terminal pentapeptide amide sequence (-Gly-Trp-Met-Asp-Phe-$NH_2$). It seems that CCK was the 'original' parent of the family, and has long been involved in stimulating digestive enzyme secretion (as it does in ascidians), only later taking over the roles of stimulating gall-bladder contraction, and stimulation of pancreatic enzyme secretion, as these new target organs appeared. In bony fishes, CCK from cells in the intestine inhibits acid secretion by the oxyntic cells of the stomach, which are stimulated to secrete by bombesin from cells in the stomach as the meal enters. In mammals, acid secretion is evoked by gastrin release, but in fishes this peptide has apparently not yet diverged from CCK.

Gut motility in fishes is controlled by the autonomic nervous system via the cholinergic, adrenergic and aminergic supply to the gut musculature, but this control is modulated by gut hormones. Thus bombesin greatly potentiates the effects of acetylcholine, and substance P and enkephalin stimulate gut muscle contractions.

### 9.9  Chromaffin tissue and the interrenals

In higher vertebrates, the adrenals lie next to the kidneys, and consist of an inner medulla and outer cortex, producing different hormones (the medullary catecholamines adrenaline and noradrenaline, and cortical corticosteroids) so that they are really two endocrine glands in one. In fishes, however, they are separate: the medulla represented by chromaffin cells innervated by spinal autonomic fibres, and the interrenals by yellowish patches or lumps near the kidneys and posterior cardinal veins (Figure 9.7).

#### 9.9.1  Chromaffin tissue

The release of adrenaline and noradrenaline into the blood from chromaffin tissue in response to stress is rapid and dramatic. Handling trout, for example, causes levels in the circulation to rise several hundred times. This release not only increases heart beat amplitude, and gill vessel

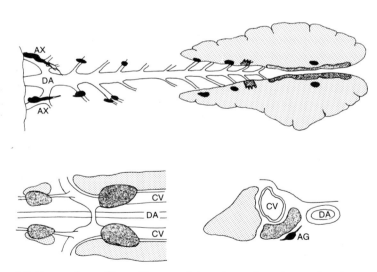

**Figure 9.7** Above, chromaffin and interrenal tissue in an elasmobranch (*Raia*). Below left, a teleost, *Anguilla*, anterior. Kidneys, light stipple; interrenals, dark random shading; chromaffin tissue, black. Below right, transverse section of kidney and interrenal. DA: dorsal aorta; CV: cardinal veins; AG: autonomic ganglion; AX: axillary body. After Vincent and Curtis (1927) *J. Anat. Physiol.*, **62**, 110, and Vivien (1958) *Traité de Zoologie*, (P.P. Grassé, ed) **13**, pt 2.1470 Masson, Paris.

vasodilation (hence lowered vascular resistance), as well as systemic vasoconstriction, but also induces hypoglycaemia, changes in gill ionic permeability and sometimes increased lipolysis. These varied stress responses under the control of the autonomic nervous system, like the stress responses of mammals, evidently fit the fish to withstand disturbance by external factors, and they are of obvious importance in aquaculture where monitoring circulating catecholamines has enabled stressful procedures to be avoided.

### 9.9.2 Interrenals

The interrenals produce corticosteroids (such as cortisol) under the regulation of corticotrophin (adrenocorticotrophic hormone, ACTH) from the adenohypophysis. These hormones have several functions. In elasmobranchs, a unique corticosteroid is involved in control of secretion by the rectal gland, thus the interrenal may be involved in osmoregulation, as it certainly is in teleosts, where blood cortisol levels rise when euryhaline fish are transferred to seawater, and where cortisol injections alter gill ionic and water permeability. In teleosts, stress such as handling or cold shock elevates blood cortisol levels, acting synergistically to the immediate catecholamine response, and it also changes during the reproductive cycle.

### 9.9.3 Kidney hormones and the renin–angiotensin system

In mammals, the juxtaglomerular cells of the kidney secrete renin which, in the blood, acts on the plasma precursor of the polypeptide angiotensin. Angiotensin causes a rise in blood pressure, and stimulates corticosteroid secretion by the adrenals resulting in kidney sodium retention. In fishes, juxtaglomerular cells have not been found in agnatha and elasmobranchs, but they are seen in other fish groups, including *Latimeria*, and renin pressor activity is seen when kidney extracts are incubated with the blood and then tested in mammals. However, what renin does in fishes and how its release is regulated are not yet clear, though the system probably plays some part in water and ion regulation, at least in euryhaline fishes. In lungfish, the renin–angiotensin system seems to be important in the normal control of blood pressure.

## 9.10 Gonadal hormones

### 9.10.1 Elasmobranchs

As we might expect, with their wide repertoire of reproductive strategies (p. 180), the hormonal control of elasmobranch reproduction has attracted much attention. Elasmobranchs do provide excellent models for the regulation of reproduction in higher vertebrates.

The development of the gonads and sexual behaviour, and their seasonal changes, are controlled by gonadotrophin releasing hormone from the neurons of the pre-optic nuclei, released into the systemic circulation from the neuropophysis, which then regulates gonadotrophic hormone secretion by the ventral lobe of the adenohypophysis. Strikingly, gonadotrophin releasing hormone is present in the neurons of the terminal nerve (nerve 0) in elasmobranchs, which as in teleosts (p. 286) is linked to many brain regions, including those controlling sexual behaviour and physiology.

Gonadotrophic hormone evokes increased androgen and oestrogen production from the gonads. These steroid hormones (testosterone and oestradiol-17$\beta$) are found in the serum, and in the ovaries and testes. The relatively slight differences in the structure of the different steroids (seen in Figure 9.8) testify to the specificity of the receptors for which they are designed. Oestradiol-17$\beta$ is produced in the ovary by cells of the follicle walls, whilst progesterone is produced by the corpora lutea (post-ovulatory follicles). In rays, progesterone is involved in the sequence of events of vitellogenesis, egg capsule formation and oviposition, whilst oestradiol is required for the development and maintainance of the reproductive tract. Ovarian steroid secretion signals the proper timing of egg laying or

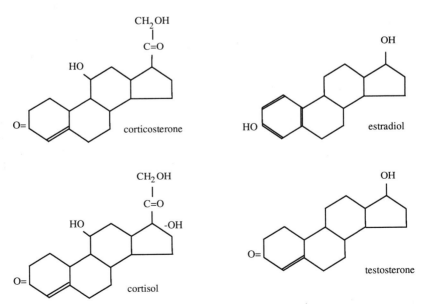

**Figure 9.8** The structure of two gonadal steroid hormones (right), compared with two interrenal steroids. After Barrington (1964).

parturition. Not all gonadal hormones in elasmobranchs are steroids however. The ovaries in viviparous sharks like the sand tiger (*Odontaspis*) and the oviparous ray (*Raja erinacea*) contain the peptide hormone relaxin, a member of the insulin-like growth factor hormone family, which all share the α- and β-chains linked by disulphide bridges familiar in the insulin molecule itself. Relaxin acts on the lower part of the female reproductive tract causing it to expand to permit the exit of the egg case or foetus, and as in mammals, the greatest effect is seen after priming by oestradiol. It also seems to regulate uterine contractions in conjunction with progesterone.

### 9.10.2 Teleosts

In general terms, the gonadal hormones of teleosts are similar, and play similar roles to those of elasmobranchs, although since relatively few teleosts are viviparous, control of the female reproductive tract is less striking than in elasmobranchs. More is known in teleosts about the sites of origin of hormones in the gonads, and of the hormonal control of behaviour, including sex changes. These last are of considerable importance in aquaculture, where ripening of fish by pituitary injection (not necessarily from the same species) enables fish farmers to produce fry at all seasons of the year. But it is the production of single-sex batches of fish (see p. 321) that has stimulated much fairly recent interest in teleost endocrinology, chiefly in salmonids and tilapias. The aim is to farm only the sex with the better conversion efficiency and growth rate. In *Tilapia aurea*, all male stocks are produced by treating the young newly hatched fry with methyl testosterone. Male tilapias are more appreciated than females, but with salmonids the reverse is the case, and all-female stocks are produced by various oestradiol treatments.

The effects of androgens and oestrogens on the secondary sexual characters of elasmobranchs is not dramatic (increase in clasper size in immature males, for example), but in teleosts, there may be striking 'nuptial' colour changes and elongation of intromittent organs (as in the bitterling (*Rhodeus*) which lays its eggs in freshwater lamellibranchs). There may also be behavioural changes of an equally striking kind in such reproductive activities as nest-building, and male parental behaviour. There are, however, quite different effects of castration and steroid injection in different species, and it seems that different pathways regulate sexual behaviour in different fishes.

### 9.11 The pineal

The pineal and parapineal arise as dorsal evaginations from the roof of the diencephalon (see p. 276), and although the parapineal is present in adult lampreys, in other fishes it disappears or is much reduced during ontogeny. The presence of receptor cells, pigment and associated nerve fibres in connection with the posterior commissure makes it quite clear that the pineal has a sensory function, but as in mammals, there are dense-cored secretory vesicles in the sensory cells, and the pineal is obviously a photosensitive endocrine organ. As in higher vertebrates, in lampreys and teleosts (little is known of the elasmobranchiomorph pineal) the pineal contains the hormone melatonin and its precursor serotonin, both of which show marked diurnal changes, rising at night. In trout plasma, melatonin levels rise at night to around 150 pg ml$^{-1}$, twice the level seen during the day.

What is the significance of these marked diurnal changes in pineal and plasma levels? In goldfish, removal of the pineal abolishes or changes diurnal variations in liver glycogen and in other fishes, the usual circadian rhythm of colour change is eliminated by pinealectomy. On the whole, it seems that melatonin secretion by the pineal provides the link between photoperiod and hypothalamic–pituitary function, and between photo-period and seasonal gonadal development.

### 9.12 Origin and evolution of fish hormones

Many questions have arisen about the origin and evolution of fish hormones. For example, when and from where did the different hormones arise in phylogeny, and what were their original functions? Does information about the amino acid sequences of hormonal peptides in living forms make it possible to reconstruct the structure of ancestral hormones? Naturally, such questions are linked to similar questions about the evolution of hormonal receptors and their associated intracellular signalling G-proteins.

#### 9.12.1 Origins

Since microorganisms produce substances similar to vertebrate hormones (e.g. insulin and human chorionic gonadotrophin) one possibility is that invertebrates, fishes and man, inherited hormones from early ancestral microorganisms. But this implies that the genes for the precursors of such hormones were conserved to an extraordinary degree, over $10^9$ years in the case of human chorionic gonadotrophin! Perhaps more likely, the microorganisms may have gained these genes much later by DNA recombination from higher animals, instead of passing them on to vertebrates from early in evolution.

Still, fishes and other vertebrates have evidently inherited some hormones from invertebrates, a good example being insulin, found in vertebrates and (in neurons) of several invertebrates. Insulin is supposed to have arisen from an ancestral proinsulin-like protein, perhaps a serine protease, whose original functions may have been in food processing and digestion. Hagfish insulin differs from mammalian (pig) insulin in almost 40% of the 51 amino acid residues, and apart from the 'invariant' regions of the molecule which stabilise the insulin monomer and dimer, hagfish insulin is not very like the insulins of other fishes. So we might suppose that the insulin molecule inherited by hagfish from invertebrates underwent changes from the ancestral precursor, giving rise in later fish and higher vertebrate groups to the insulin superfamily of related peptides (insulin, insulin-like growth factors-I and -II, and relaxin).

However, living hagfish whose insulin we can sequence, have undergone a long evolution from the ancestral forms speculatively placed in evolutionary trees like that of Figure 1.1. Like other aspects of their organisation, the structure and functions of their hormones today presumably exhibit both 'ancient' and 'modern' components. One might reasonably suppose that the highly conserved region was the active site, and that random neutral substitutions took place in the amino acids in other positions. If it is assumed that such neutral changes took place at a constant rate (i.e. that the molecular clock ran steadily), then the degree of similarity would indicate the time elapsed between the two hormones in different animals. Unfortunately, despite the evidence for a constant rate of neutral changes for certain proteins like albumin, it is far from clear that a constant rate is necessarily correct for hormonal peptides that are under stringent functional constraint. It is not always easy to reconcile phylo-genetic trees for hormones based on sequence comparisons with those based on the properties of the hormones. Figure 9.9 shows such a comparison for the calcium-regulating hormone calcitonin. Note that an ancient gene duplication is held to have occurred at the base of the fishes, and that a more recent duplication took place in teleosts. There are two calcitonin-like molecules in birds, supporting the idea of a primitive duplication, but the difficulty is that the artiodactyls seem to have been separated for a remarkably long time from the rest of the vertebrates, and a more recent rapid evolution of the artiodactyl calcitonin molecule seems more probable. We still need more sequence and pharmacological data before a less-speculative tree can be produced.

Despite doubts that may be felt about the constant rate of neutral sequence changes, nevertheless, the striking differences between elasmo-branchiomorph and other fish hormones of the same families (e.g. the neuropituitary hormones of the arginine vasotocin family) clearly indicates a long separate history, even if we are unsure when the divergences took place.

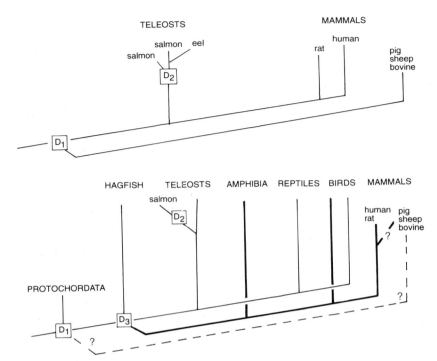

**Figure 9.9** Evolution of calcitonins according to 'molecular' data (above) and to data on various properties of the calcitonins in different groups. Gene duplications are assumed to have taken place at $D_1$–$D_3$. Modified from Fontaine (1985) in *Evolutionary biology of primitive fishes*. NATO ASI series A, **103**, 413. Plenum Press, New York.

## 9.12.2  Changes in function

A striking feature of vertebrate peptide hormones, as we have already seen, is the existence of families of hormones sharing sequence homology to a considerable degree (e.g. Table 9.2), but which nevertheless may have quite different functions. It seems clear that the members of these families have arisen by gene duplication, followed by subsequent divergence resulting from substitutions and deletions. An example of a relatively recent duplication is given by work on the two forms of growth hormone in the chum salmon (*Oncorhynchus keta*) where the two forms have conserved a core of invariant amino acid residues (presumably representing the active site) and differ little elsewhere. Interestingly, in the same fish, the growth hormone variants share with prolactin a 24% homology along the 188 amino acid chain, *but* there is no identical conserved core. In other words, it looks as if a more ancient gene duplication took place and that subsequent changes led to different active sites and the evolution of two hormones with different active sites, but still retaining nearly a quarter of the ancestral amino acid sequence.

Change in function need not necessarily, however, involve changes in the active sites, but may involve recruitment of new receptors and new target organs. For example, cholecystokinin appears in ascidian tunicates, where it is involved in control of intestinal enzyme secretion (as well as being widespread in brain neurons). In hagfish, intestinal enzyme secretion is again its function, but in other fishes, it has recruited new target organs, being involved in control of stomach acid secretion as well as gall bladder secretion.

Like the hormones themselves, the receptors probably belong to families associated with the different families of related hormones, and like them also, these related receptor types probably arose by gene duplication and subsequent modification to change their specificity and sensitivity. In teleosts and the South American lungfish *Lepidosiren*, arginine vasopressin injection evokes blood pressure increase and diuresis, whereas in mammals the equivalent hormone is antidiuretic. This puzzling difference seems most easily explained by the existence of two related receptors, one set in the pre-glomerular circulation, and the other in the peripheral vasculature. In fishes, the latter seems to be more sensitive or much more abundant, hence arginine vasopressin produces a pressure diuresis. In mammals, by contrast, the receptors of the peripheral circulation are either less sensitive or they are much less abundant than the pre-glomerular receptors, hence the result of injection of the hormone is antidiuretic.

## Bibliography

Barrington, E.J.W. (1964) *Hormones and Evolution*, English Universities Press, London, pp. 154.
Bern, H. (1990) The "New" Endocrinology: Its scope and its impact. *American Zoologist*, **30**, 887–885.
Bern, H.A. and Nishioka, R.S. (1993) Aspects of salmonid endocrinology: the Known and the Unknown. *Bulletin of the Faculty of Fisheries Hokkaido University*, **44**, 55–67.
Callard, I.P., Klosterman, L.L., Sorbera, L.A., Fileti, L.A. and Reese, J.C. (1989) Endocrine regulation of reproduction in Elasmobranchs: Archetype for terrestrial vertebrates. *Journal of Experimental Zoology*, supplement **2**, 12–22.
Cimini, V., Van Noorden, S.V. and Nardini, V. (1989) Endocrine regulation of reproduction in Elasmobranchs: Archetype for terrestrial vertebrates. *Journal of Experimental Zoology*, supplement **2**, 146–157.
Clark, N.B., Norris, D.O. and Peter, R.E. (organisers) (1983) Evolution of endocrine systems in lower vertebrates, a symposium honoring Professor Aubrey Gorbman. *American Zoologist*, **23**, 595–748.
Falcon, J., Thibault, C., Begay, V., Zachmann, A. and Colin, J.-P. (1992) Regulation of the rhythmic melatonin secretion by fish pineal photoreceptor cells, in *Rhythms in Fishes* (Ali, M.A. ed), NATO ASI series A, **236**, 167–198.
Fredriksson, G., Fenaux, R. and Ericson, L.E. (1989) Distribution of peroxidase and iodination activity in the endostyles of *Oikopleura albicans* and *Oikopleura longicauda* (Appendicularia, Chordata). *Cell and Tissue research*, **255**, 505–510.

Foreman, R.E., Gorbman, A., Dodd, J.M. and Olsson, R. (eds) (1985) *Evolutionary Biology of Primitive Fishes*. NATO ASI series A, **103**, Plenum Press, New York and London, pp. 463.

Girgis, S.I., Galan Galan, F., Arnett, T.R., Rogers, R.M., Bone, Q., Ravazzola, M. and MacIntyre, I. (1980) Immunoreactive human calcitonin-like molecule in the nervous systems of protochordates and a cyclostome, *Myxine*. *Journal of Endocrinology*, **87**, 375–382.

Hoar, W.S., and Randall, D.J. (eds) (1969) The endocrine system, in *Fish Physiology*, **2**, Academic Press, New York and London, pp. 446.

Hoar, W.S., Randall, D.J. and Donaldson, E.M. (eds) (1983) Reproduction. Endocrine tissues and hormones. *Fish Physiology*, **IXA**. Academic Press, New York and London.

Ince, B.W. (1983) Pancreatic control of metabolism, in *Control Processes in Fish Physiology*, (ed Rankin, J.C., Pitcher, T.J., and Duggan, R.) Croom Helm, London, pp. 89–102.

Jönsson, A.-C. (1993) Co-localization of peptides in the Brockmann bodies of the cod (*Gadus morhua*) and the rainbow trout (*Oncorhynchus mykiss*). *Cell and Tissue Research*, **273**, 547–555.

Matty, A.J. (1985) *Fish Endocrinology*. Croom Helm, London and Sydney, pp. 267.

Meier, (1992) Circadian basis for neuroendocrine regulation, in *Rhythms In Fishes* (Ali, M.A. ed), NATO ASI series A, **236**, 109–126.

Thorpe, A., and Thorndyke, M.C. (1975) The endostyle in relation to iodine binding. *Symposium of the Zoological Society of London (1975)*, **36**, 159–177.

# 10   Sensory systems and communication

Fishes have more senses than ourselves, for they have an elaborate lateralis system to detect vibrations, and in addition, a fair number of species have modified part of this lateralis system for electroreception. The sensory systems of fishes which show such remarkable and fascinating adapations is the subject of this chapter.

## 10.1 Proprioception

Compared with terrestrial animals, fish are poorly equipped with proprioceptors. The muscle spindles and Golgi tendon organs of terrestrial vertebrates, monitoring the length and tension of muscles, and hence very important in controlling posture, seem to be absent in fishes (although there are reports of spindles in some teleost jaw muscles) and proprioceptors associated with locomotor muscles are known only in *Myxine* and rays. Although ray proprioceptors (in parallel with the slow locomotor muscle fibres of the pectoral and pelvic fins) are much simpler than spindles, their responses to ramp stretches are very similar to those of spindles, and seem able to provide the same kind of information for locomotor control as do muscle spindles. In sharks, there are no proprioceptors amongst the muscles, but coiled corpuscular pressure receptors under the skin act as proprioceptors. Probably the paucity of proprioceptors, except in special cases like the barbel of the mullet and the anterior fin rays of the gurnard, is related to the relatively insignificant role of gravity for aquatic organisms of similar density to water and to the damping of movements by the medium.

## 10.2 The acoustico-lateralis system

Moving bodies in water produce changing fields of pressures and hence pressure gradients around themselves. Their mechanoreceptors must obtain information from these fields as well as be able to perceive mechanical stimuli from adjacent organisms such as conspecifics, predators and food. Different parts of the acoustico-lateralis system can respond to sounds and gravity as well as to linear and angular accelerations of the fish's own body. Some of the abilities seem to have appeared late in chordate phylogeny: amphioxus has no analogue of the system although one is present in certain tunicates and the epidermal head lines of some cephalopods contain ciliated cells and are sensitive to weak movements of the surrounding water.

In fishes the inner ear (or membranous labyrinth) is present close to the brain on either side of the head. It has an upper part comprising orthogonally arranged semicircular canals and a lower part comprising the utriculus, sacculus and lagena (Figure 10.1). The swellings, or ampullae, of the semicircular canals contain a sensory crista (Figure 10.2) equipped with a cupula, the cristae responding to angular accelerations as the fish turns. The three parts of the lower labyrinth each contain a sensory macula covered by an otolith membrane (Figure 10.2). In teleosts this membrane invests a fairly massive single calcification, the otolith, but in elasmobranchs the calcifications consist of more diffuse otoconia and in the bottom-living *Rhina*, sand grains are even present. The maculae respond to sound, gravity and to linear accelerations of the fish's body. The sacculus is enlarged in most species except for the clupeoids which have an enlarged utriculus. In the holostean bowfin (*Amia calva*) and some cypriniform species, the lagena is enlarged. The other component of the acoustico-lateralis system, the lateral line (Figure 10.2), consists of lines of sense organs over the head and body as well free sense organs distributed more randomly.

The basic sensory cell in all parts of the acoustico-lateralis system is a

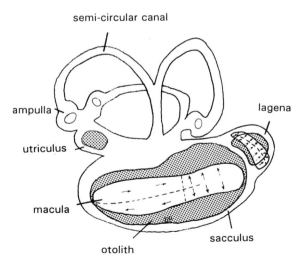

semi-circular canal

ampulla

lagena

utriculus

macula

otolith

sacculus

**Figure 10.1** Diagram of the labryinth of a typical teleost. The stippled areas show the extent of the otoliths overlying the sensory macullae. The polarity of the 'fields' of hair cells is shown by arrows in the sacculus and lagena.

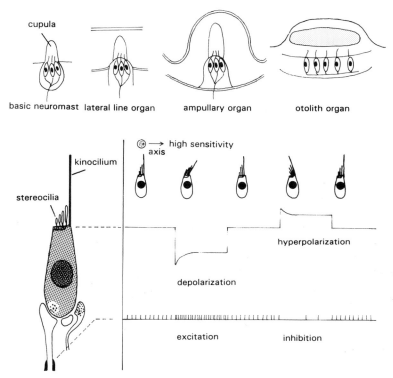

**Figure 10.2** Top line from left, diagram showing the structure of the basic neuromast organ found on the surface of the skin and modifications where it is incorporated into the lateral line canal, in the cristae of the ampullary organ of the semicircular canals and in the maculae of the utriculus, sacculus and lagena. The lower part of the figure shows a single hair cell on the left and on the right how displacement of the hairs along the axis of sensitivity leads to excitation or inhibition of the hair cell.

displacement detector (Figure 10.2). This consists of a dually innervated cell with an external bundle of hairs projecting into a gelatinous cupula. There is one long hair, the kinocilium, which has the typical 9 + 2 filament internal array of tubules found in all cilia. To one side is a 'staircase' of much shorter microvilli or stereovilli (misleadingly called 'stereocilia' for they do not have internal tubules) so that the hair cell is morphologically asymmetrical. The hair cells not only synapse with afferent nerve fibres carrying information to the brain, but they also receive inhibitory efferent fibres whose action 'switches off' the cells. The hair cells have directional properties; if displacement of the cupula causes the stereocilia to bend towards the kinocilium, the hair cell becomes depolarized, causing excitation; if the stereocilia are bent away from the kinocilium, the cell become hyperpolarized with an inhibitory effect (Figure 10.2). If the cupula bends the hair bundle at right angles to the kinocilium–stereocilia axis, there is no response; in between, the sensitivity of the response follows a cosine law. The sensory hair cells are grouped into neuromast organs comprising a few to many thousands of hair cells. There has been much work by scanning electronmicroscopists looking at groups or 'fields' of hair cells and plotting their 'polarity' or axis of higher sensitivity. Usually the hair cells in certain areas of a neuromast organ are all polarised along the same axis, but adjacent pairs of hair cells have opposing polarity.

The neuromast is the basic displacement receptor in all the components of the acoustico-lateralis system – as free neuromast organs on the body surface, as lateral line neuromast organs buried in pits, grooves or canals, in large fields (cristae) in the ampullary organs of the semicircular canals (superior part of the labyrinth of the inner ear) and in the maculae of the sacculus, utriculus and lagena of the inferior part of the inner ear (Figure 10.2). In these neuromast organs, the basic cupulae coalesce to follow the shape of the field of hair cells and may be cylindrical, flat vertical ribbon-like plates, or in the case of the maculae, quite extensive membranes containing the calcareous otolith.

To appreciate in more detail how the neuromasts respond to mechanical stimuli we need to understand the nature of stimuli produced by sound sources. Water is very incompressible (14 000 times less than air at 1 atmosphere) and sound travels at $1500 \text{ m s}^{-1}$ compared with air at $300 \text{ m s}^{-1}$. A sound source produces two types of stimulus – a back-and-forth motion of the particles in the medium (particle displacement) and a sinusoidal change in pressure (sound pressure). Their amplitudes drop at different rates depending on the distance from the source. Particle displacement is more important near the source and sound pressure at a distance from the source. The region close to the source that is important for particle displacement is called the 'near-field' (its radius from the source being the wavelength of the sound divided by six) and the more distant region where sound pressure is important is the 'far-field'. The near-field is thus more important for low frequencies; for example at 100 Hz, the wavelength is 15 m and the near-field 2.5 m; at 1000 Hz, the wavelength is 1.5 m and the near-field only 0.25 m.

The implication of this is clear; to increase sensitivity, especially at high frequencies at some distance from a sound source, sense organs are needed that respond to sound pressure. We shall see on page 226 how teleosts have evolved mechanisms that couple pressure-sensitive structures (which are always compressible gas-filled spaces such as swimbladders or otic bullae) to displacement receivers (the neuromast organs) by pressure–movement transduction.

Many larval fishes hatch having only free neuromast organs (each containing a few hair cells) scattered over the head and trunk. The lateral line develops later in ontogeny, but before metamorphosis. The adult base

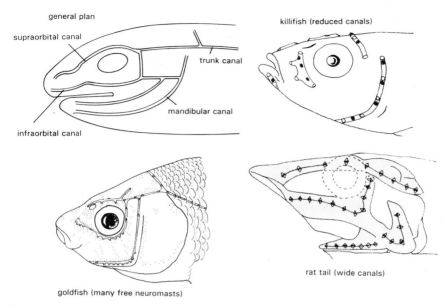

**Figure 10.3** Top left, the base plan of the lateral line canals of a teleost showing modifications in different groups. The goldfish retains the basic plan with large numbers of free neuromasts over the head and body. The killifish has its canals and number of neuromast organs much reduced. The rat tail, a fish of deep water, has wide and deep canals covering much of the head (redrawn from Blaxter, 1987 and Puzdrowski, 1989).

plan (Figure 10.3) consists of a supra- and infra-orbital canal, a mandibular canal and a trunk canal, on each side, the two sides being connected by a commissure. There are many modifications of this base plan – in some species, the canals are reduced in number and size and probably missing in some monognathid eels; in other species, the canals are not interconnected or they become greatly enlarged (Figure 10.3). Some species, such as the goldfish, have as many as 1000 free neuromasts on the head and six to nine on *each* scale! Neuromast organs lie along the lateral line canals, their cupulae partially blocking the lumen, and with the axis of the hair-cell polarity along the axis of the canal. The arrangement strongly suggests that the receptors (except in the clupeoids) are laid out to produce a long baseline. In holocephalans and macrourids this baseline extends to the tip of a long thread-like tail, and in many other species the lines of neuromast organs continue on to the caudal fin.

Obviously fish can detect very localised sources of vibration close to the body. For example, Antarctic fish such as *Pagothenia borchgrevinki* and the piper (*Hyporhamphus ihi*) locate and feed on small crustaceans (whose limb movements produce vibrations) very close to the body or head. It has, however, been shown that the whole body of a fish vibrates in a sound field if the sound source is some distance from the fish. Although at some point along the length of the fish its body and the water are vibrating at the same amplitude (i.e. there is no relative movement), at points nearer the source the body is vibrating less than the water particles, and at points further from the source the body is vibrating more than the water particles. The long sensory baseline is obviously useful in this situation to give the maximum differences along the body. Within the lateral line canals the cupulae will move relative to the wall of the canal as a result of frictional forces caused by the velocity of the water in the canal differing from the velocity of the wall itself. A flow within the canal is required for this, provided for by pores from the canal to the exterior but some fishes have membranes in place of pores. The sensitivity of the neuromast organs is indeed remarkable: they can respond to any displacements of less than 1 nm. This must present problems for detecting sound sources because of the noise produced by the fish's own swimming movements. Enclosure of the neuromasts in the lateral line may afford some protection in this respect, at the expense of reducing sensitivity. In addition, the efferent (inhibitory) innervation of the hair cells may reduce their sensitivity during active swimming. In fact, fish not only produce noisy flow round their bodies but they also generate trails of vortices in their wake, which last for several seconds.

Some elongate fishes appear to adapt their locomotor style to retain the functioning of the lateral line. *Aphanopus*, the trichiurid scabbard fish, sculls itself along by a minute forked tail (see Figure 3.18), keeping the body rigid and flexing the body in rapid anguilliform locomotion only when within visual range of the prey. The related *Trichiurus* moves by undulating its long dorsal fin so reducing disturbance to the ventrally placed lateral line. The sculpin listens for vibrations of buried prey by placing its lower jaw on the substratum, presumably helping to stimulate the large mandibular neuromasts.

Apart from detecting the movements of prey close to the body, the lateral line has other functions. Some freshwater species, such as the topminnow (*Aplocheilus lineatus*) and the butterfly fish (*Pantodon buchholzi*) use their lateral lines to detect surface waves created by insects at the air–water interface and so home in on their prey. The blind Mexican cave fish (*Astyanax mexicanus*) analyses distortions of its own flow field to give itself information about surrounding obstacles. Fish such as the brook trout (*Salvelinus fontinalis*) shelter behind obstacles in a water current to prevent themselves being swept downstream, so saving energy. Such entrainment is less effective in the dark after section of the lateral line nerves. The lateral line is also implicated in intra-specific behaviour such as schooling. Although blindfolded saithe (*Pollachius virens*) can follow changes in course and velocity of their neighbours, they cannot school if the posterior lateral line nerves are also cut. If danger threatens, the male fighting fish (*Betta splendens*) attracts its young by vibrating movements of the pectoral fins close to the water surface. The young return to the male for protection from up to 40 cm, even in the dark.

It has been shown that rapidly swimming fish such as herring generate a very rapid pressure pulse which might be useful in signalling to its neighbours. Although the lateral line does not normally respond to sound pressure, pressure–displacement transduction can occur in some instances. For example, a gas-filled swimbladder pulsating in a sound field could stimulate nearby neuromasts by displacement movements through the surrounding tissues. One of the best researched pressure–displacement transduction mechanisms occurs in the otic bulla system of clupeoid fishes (see p. 227). Here a gas-filled structure is coupled to both the inner ear and head lateral line canals, allowing sound pressure to stimulate neuromast organs in both parts of the acoustico-lateralis system.

## 10.3 Sound reception

In the sense that the lateral line can perceive stimuli from sound sources under some circumstances it may be said to be involved in hearing. The inner ear is, however, much more versatile than the lateral line in terms of both sensitivity and frequency response. If the head of a fish vibrates in a sound field, the calcareous otoliths overlying the maculae (Figure 10.1) make smaller movements than the surrounding tissues since their density is higher. This causes the hairs of the hair cells to bend, so firing the hair cells if their polarity is appropriate to the direction of the vibration. The hairs act as a spring to return the otolith to its resting position but it will vibrate at the same frequency as that of the sound stimulus.

In the cristae of the semicircular canals, the hair cells all have their polarity along the same axis so that excitation occurs with fluid motion in one direction and inhibition in the opposite direction. The maculae, however, are divided into 'fields' of hair cells (Figure 10.2). In each field hair cells polarised in one direction are accompanied by adjacent hair cells polarised in the opposite direction. There is some evidence that the cells are broadly tuned to particular bands of frequencies. In those fishes without accessory hearing structures such as the swimbladder or otic bullae, the inner ear perceives only the particle displacement aspect of the sound stimulus. Fish like cod can perceive the direction of the stimulus as well as appreciate its intensity and frequency. In fact, the cod and many other fishes have extensions of the swimbladder close to the back of the skull so giving some opportunity for pulsations of the swimbladder (in response to sound pressure) to stimulate the inner ear via the bones of the skull. What the swimbladder and other gas-filled structures do is to enhance the particle displacement aspect of the sound stimulus by transducing the sound pressure to particle displacement.

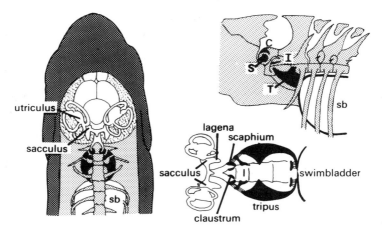

**Figure 10.4** Ostariophysan ear ossicles. Left: dorsal view of goldfish; Right: detailed view of ossicles. S, scaphium; C, claustrum; T, tripus; I, intercalarium; sb, swimbladder. After v. Frisch (1936) *Biol. Revs.*, **38**, 110.

Pressure to displacement transduction should thus improve both sensitivity and the range of frequency that is perceptible. In the Ostariophysi, a chain of ossicles links the sacculus to the swimbladder. These ossicles were described by Weber as long ago as 1820 and are still called the Weberian apparatus (Figure 10.4). As the swimbladder pulsates in a sound pressure field, displacements of the swimbladder wall rock the tripus, and this movement is transferred to the claustrum via the intercalarium and scaphium. The claustrum is coupled to a sinus containing perilymph adjacent to the saccular maculae. The analogy with the ossicles of the mammalian ear is intriguing.

In the clupeoids such as anchovy, herring and sprat, an otic bulla on either side of the head also acts as a pressure–displacement transduction mechanism, in this case very close to the utriculus of the inner ear (Figure 10.5). The bullae contain gas in the lower part separated from perilymph in the upper part by an elastic bulla membrane. The gas-filled part is also connected to the swimbladder by extremely fine gas ducts about 8 μm in diameter. During the passage of sound, the gas in the bulla changes in volume in sympathy with the changes in sound pressure. The bulla membrane vibrates at the same frequency, forcing perilymph in and out of a fenestra or orifice in the upper wall of the bulla. The fenestra is adjacent to the utriculus and the shear of perilymph over the external surface of the utriculus stimulates the macula. (In the sprat there is even a very fine elastic ligament joining the bulla membrane to the utricular wall which may help in sensing changes of depth – hydrostatic pressure.) Nearby there is another membrane, the lateral recess membrane in the lateral wall of the skull, that also moves in sympathy with the bulla membrane. The lateral

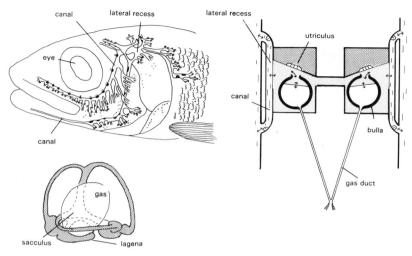

**Figure 10.5** Gas-filled bullae and inner ears in mormyrids and clupeids. Bottom left: mormyrid ear showing gas bulla applied to the sacculus (the bulla has a rete supplying it with gas, not shown). Right: schematic diagram of gas-filled bullae (in dorsal view) of a clupeoid; clear areas – gas, fine stippling – perilymph, coarse stippling – endolymph, short lines – seawater in lateral line canals; top left: side view of a herring head showing the complex lateral recess. Adapted from Stipetic (1939) *Z. vergl. Physiol.*, **26**, 740; Denton and Gray (1979) and Blaxter *et al.* (1983).

recess membrane is at the back of the sac on the external surface of the head from which all the lateral line canals radiate (Figure 10.5). Thus changes of sound pressure stimulate not only the displacement-sensitive maculae in the utriculus but also the neuromast organs of the head lateral line, a condition unique to the clupeoid fishes.

There is a further arrangement in the clupeoids that makes this elaborate acoustico-lateralis system independent of hydrostatic pressure. The fine gas ducts connecting the bullae to the swimbladder allow the swimbladder gas to act as a source or sink of gas as the fish move up and down in the water. When the fish move down, gas is pulled forward to the bulla and when the fish move up, gas is pushed back into the swimbladder. The time constant for this is the order of 30 s which prevents changes of the higher frequency sound pressure affecting the mechanism. Without this connection to the swimbladder, the bulla membrane might burst during a big change of depth (and remember that the hydrostatic pressure increases by 1 atm every 10 m!); the bulla membrane also tends to remain in its flat resting condition where it is more responsive to sound pressure. The clupeoids make very extensive vertical migrations (see p. 307) so that the compensation mechanism described above is essential.

Analogous gas-filled structures occur in the mormyrids. Sacs of gas closely adjacent to the labyrinth become isolated from the swimbladder during ontogeny and when functional are equipped with rete mirabile to supply them with gas (see Figure 10.5). Anabantid fish have an accessory respiratory chamber above the gills (see p. 111) coupled to the inner ear by a thin membrane. Neither of these freshwater fishes make extensive vertical migrations and do not require a rapid compensation mechanism for hydrostatic pressure change, as do the clupeoids.

The auditory performance of fishes is usually shown as an audiogram (Figure 10.6). Note that the non-ostariophysans have narrower frequency responses and lower sensitivity than ostariophysans, with the herring somewhere in between (but with an especially good low frequency sensitivity). There is something of a mystery in the very high-frequency responses of ostariophysans and clupeoids, one a predominantly freshwater,

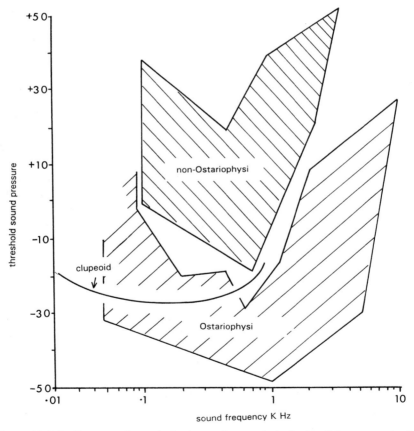

**Figure 10.6** Audiograms of ostariophysine and non-ostariophysine fish compared with a clupeoid. The shaded areas enclose a number of audiograms of different species in each group. Threshold scale is logarithmic, in decibels re 1 μ bar.

shallow living group, the other a marine group able to live at greatly varying depths. The latter may be able to hear the high-frequency echolocation clicks of hunting marine mammals. There have also been suggestions that some fish species can themselves echolocate but this remains unproven. High-frequency responses also help fish to respond quickly to short sharp sounds, probably aiding in escape from predators or responding to the movements of neighbours in a school (p. 306).

## 10.4 Sound production

Fishes in 50 different families produce sounds in various ways, by rasping spines and fin rays, or by burping, farting or gulping air. Some of these mechanisms may involve the swimbladder acting as a resonator. In some fishes like the gadoids, special drumming muscles vibrate the swimbladder wall causing it to act as an internal loudspeaker.

Most sounds are involved in social behaviour within a species and less commonly communication between species. Gurnards (Triglidae) grunt when disturbed by a predator but also display visually, perhaps helping them to escape predation. Sounds are relatively common during courtship, as for example in a gadid, the haddock (*Melanogrammus aeglefinus*). In the batrachoid toadfish (*Opsanus tau*) the male occupies areas of the sea bed and makes so-called boat whistle calls apparently as a means of delineating its territory, or attracting mates, since the rate of calling often increases when a female approaches.

The role of sound production in schooling is far from clear. It is likely that schoolmates of species such as herring respond to hydrodynamically created sounds from their neighbours (p. 303) and surface feeding fish can home in on the disturbances produced by prey struggling at the surface (p. 224).

## 10.5 Electroreceptors and electric organs

Electrical signals arise from various sources: inanimate electric fields are caused by geomagnetism and tides, motional electric fields by fish swimming through the Earth's magnetic field, animate electric fields by gill movements and other relative movements of parts of the body, and animate electric fields by specialised electric organs evolved from muscle and (in one instance) from nerves. Electroreceptors are able to pick up the small steady d.c. fields generated by living prey and are found in lampreys, all elasmobranchiomorph fishes, in a variety of teleosts, in sturgeons and *Polyodon* and in Dipnoi; they probably also occur in *Latimeria*. Electroreceptors of a different kind, responding to high frequencies (and insensitive to d.c. fields) are found in several families of freshwater teleost – Gymnarchidae, Gymnotidae and Mormyridae – that also have electric organs producing electric fields used in prey detection and social signalling.

### 10.5.1 Ampullary (tonic) receptors

These consist of a canal leading from the surface of the skin to an ampulla in whose wall a group of sensory cells is embedded (Figure 10.7). The apex of each cell bears microvilli and/or a single kinocilium protruding into the lumen of the ampulla, which is full of low-resistance jelly. Since the canal wall has a high resistance (30–100 times that of the nerve myelin sheath) the canals act as ideal 'submarine cables' ending in open circuit at the ampullary end of the canal. The ampullary organs are tonic receptors giving a long-lasting response to very low frequency or d.c. stimuli. Although they are modified neuromasts, the sensory cells appear to have no efferent innervation.

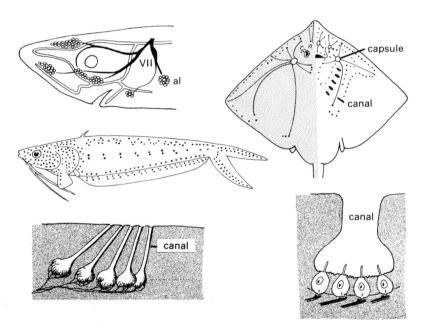

**Figure 10.7** Ampullary electroreceptors. Upper left: groups of ampullae (al) on head of a shark. Mid-left: distribution of ampullary organs on the catfish (*Kryptopterus*). Bottom left: groups of ampullary organs from shark snout. Upper right: upper and underside of ray showing openings of ampullary organs, and capsules into which canals collect. Only a few canals are indicated. Bottom right: Ampullary organ from sturgeon snout, the sensory cells are ciliated. After Murray (1967) in: *Lateral Line Detectors* (P. Cahn, ed), Indiana Univ Press; Jorgensen (1980) *Acta Zool. Stockholm*, **61**, 87; and Bennett (1971) *Fish Physiology*, **5**, 347.

Ampullary receptors (the ampullae of Lorenzini) occur around the head in sharks and over the upper and lower surface of the 'wings' of rays (Figure 10.7). The receptors are distributed in clusters surrounded by a connective-tissue capsule. In the weakly electric freshwater gymnarchids, gymnotids and mormyrids and in the siluriform catfish *Ictalurus* (*Amieurus*) and *Kryptopterus*, the ampullary organs are arranged in longitudinal rows or scattered over the head and body at high density (Figure 10.7). The canals are much shorter in freshwater than marine fishes and the ampullae even open directly to the surface in the polypterid *Erpetoichthys*.

The widespread distribution of the ampullae allows the fish to compare electric potentials over a considerable area, no doubt improving their ability to detect the direction of the source of an electric stimulus (just as the long baselines of the lateral line canals assist in directional perception). It is interesting that the elasmobranchiomorph ampullae are collected into capsules, for it means that while the openings of the canals cover a wide area of the body, the sensory cells are all located close together and as near isopotential as possible, so allowing a greater ability to detect small potential differences over the body surface.

### 10.5.2 Tuberous (phasic) receptors

A second type of electroreceptor is also found in the weakly electric gymnotids and mormyrids (Figure 10.8). These tuberous organs, varying somewhat from group to group, consist of epidermal capsules with no connection to the exterior. The sensory cells have closely packed microvilli at their apices but no kinocilia. The receptors are phasic, giving brief responses to high-frequency electric stimuli. Although there is no canal, there is presumably a low-resistance pathway to the surface of the skin. The receptors adapt rapidly and are well suited for the reception of high-frequency discharges from the electric organs.

### 10.5.3 Electric organs

With the exception of the freshwater sternarchids, where the electric organs are derived from modified nerve fibres, fish electric organs are modified striated muscle fibres and consist of stacks of flattened cells innervated on one side (Figure 10.8). This arrangement, like batteries in series, sums the small electric potentials from membrane depolarisations so giving rise to much larger external potentials. In a few species such as the marine ray (*Torpedo*), the freshwater eel (*Electrophorus*) and the catfish (*Malapterus*), the voltage generated (500 V in *Electrophorus*) is sufficient to stun prey and certainly acts as an effective defence against predators. Gumboots and rubber gloves are *de rigueur* for handling such fishes!

Lissmann in 1963 was the first to show that the much smaller discharges

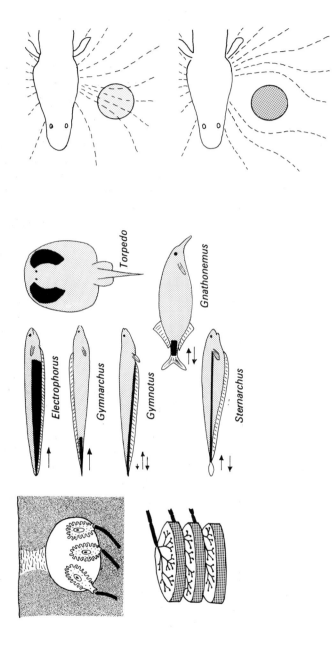

**Figure 10.8**  Active electroreception and electric organs. Upper left: tuberous mormyrid receptor (note large surface area of receptor cells within cavity). Lower left: stacked modified muscle fibres and their innervation in the electric organ of *Torpedo*. Centre: positions of electric organs in different fishes – the arrows indicate directions of current flow. After Bennett (1971) *Fish Physiology*, **5**, 347, and Szamier and Wachtel (1970) *J. Ultrastruct. Res.*, **30**, 450. Right: lines of current flow around *Gymnarchus* with a poor conductor (above) and a good conductor (below). Redrawn from Lissmann (1963).

of many other species functioned as an electrolocation mechanism akin to echolocation. Weakly electric fish can distinguish objects of different impedance from the medium by the distortions they produce in the electric field generated by their electric organs (Figure 10.8). Analyses of the signals is simplified if the fish keeps its body rigid and, indeed, these fish scull around their environment by undulating their dorsal or ventral fins.

It is possible to pick up and record the signals produced under different conditions. Some species of mormyrids and gymnotids increase the frequency of their electric pulses when stimulated but in some situations the discharges cease altogether, presumably as a 'camouflage' against electroreceptive predators. Other gymnotids and *Gymnarchus* emit pulses at the extraordinarily high frequency of 2000 Hz. Many of these species are found in tropical rivers and lakes and it is likely that species recognition is involved, each species having a characteristic discharge. When the fish are present at high density, there is a considerable problem of interference between the electric discharges of individuals close to one another. Species producing a steady frequency of discharge are found to change the frequency whenever another fish with a frequency within 1–2% comes within range. This jamming avoidance response (JAR) always operates to increase the difference between the frequency of the intruder and the fish's own frequency, allowing a number of such fishes to operate successfully when in close proximity.

The marine rays are also weakly electric with caudal electric organs. They evidently do not use these for electrolocation (the discharges are infrequent and irregular) and so, perhaps, use the electric discharges for intra-specific recognition. The bizarre electric organ of another marine group, the stargazers, also presents a puzzle. *Astroscopus*, with its electric organ modified from its eye muscles, is an ambush predator lying in wait for small prey on the sea bed. The electric discharges are insufficient to stun the prey and are not used in echolocation. During feeding the electric organs emit a burst of high-frequency pulses for 150–300 ms followed by a train of discrete pulses lasting about 1 s. The duration of the burst, which occurs as the mouth opens, is correlated with the length of the prey. Can this be a signal to other stargazers about the size of available prey?

## 10.6 Vision and photophores

Very few fishes live in total darkness. In water, light is attenuated by both absorption and scattering and falls off with depth at a logarithmic rate. In the clearest oceanic water the specially adapted eyes of deep-sea fish can still operate down to about 1100 m in daylight and perhaps 600 m at night. Bearing in mind that the ocean trenches are as deep as 10000 m and the *average* depth of the ocean is about 4000 m there are obviously vast spaces of water with no light penetration from the surface. Here, the natural darkness is relieved by flashes and glows of bioluminescence from the photophores of both fish and invertebrates. Only in caves is the visual system useless and there some 40 species of blind fishes are known. Another characteristic of water is that it absorbs different colours of light selectively, some wavelengths being transmitted more readily than others. The wavelength ($\lambda$) that is best transmitted, $\lambda_{max}$, is 470–480 nm in the deep ocean, 500–530 nm near the coast and 550–560 nm or longer in many freshwaters. This change from blue towards yellow is caused by the presence of yellow pigments, mainly breakdown products of chlorophyll and humic acids, in freshwater and in the run-off near the coast. This has important implications for the fish eye in terms of its performance in different coloured light (p. 244).

Fish eyes show many more variations than terrestrial animals because of the variety of light regimes in which they live. Fish also have an inflexible neck which may require modifications of the eye to ensure a wide binocular visual field in which there is overlap of the images of the outside world in each eye. Near the surface, because of refraction, fish see the whole horizon above the surface compressed into a solid of angle of about 98°, called Snell's window (Figure 10.9). Refraction presents problems for fish feeding above the surface like the archer-fish (*Toxotes*) as well as those avoiding predatory diving birds, because they have to make visual corrections for distorted images.

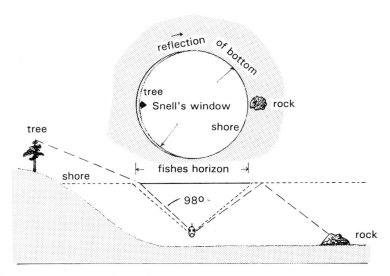

**Figure 10.9** Diagram of Snell's window. In calm water a fish will see the whole horizon above the surface subtended by an angle of about 98°. Redrawn from Walls (1963).

   Although the eyes of hagfishes are reduced and lie buried below the skin, those of adult lampreys are like those of gnathostomes, and in essentials all fish eyes (Figure 10.10) are built to the same design seen in all vertebrates.

## 10.6.1 Optics

Light passing into the eye is brought to a focus on the retina where the light-sensitive visual cells, the rods and cones, are situated. Because the retina is 'inverted', with the visual cells at the periphery of the eye, far from the lens, light rays pass through the associated nerve cell layers of the retina (the bipolar and ganglion cell layers) *before* reaching the visual cells. These layers are very transparent to light in order not to interfere with vision. In some fishes there are reflecting layers in the retina or chorioid (see Figure 10.10) that cause incident light to pass back again through the visual cells. The refractive index of the cornea and ocular fluids is similar to that of water. Refraction takes place almost entirely in the lens which is usually almost spherical with a short focal length (in teleosts about 2.55 × radius of the lens, or Matthiessen's ratio). There is little spherical or chromatic aberration; try looking at a piece of graph paper in a shallow dish full of water through a glass marble and a fresh fish lens, and you will be astonished at the quality of the fish lens! This is achieved by the refractive index varying across the diameter of the lens so that the rays follow a curved path (Figure 10.10). A number of fishes, for example the yellow perch (*Perca flavescens*), have yellow corneas and lenses which act as filters cutting off the shorter wavelengths and so reducing chromatic aberration.

   There is still some doubt how fishes accommodate to examine close or distant objects and whether at rest the eye is focused close to, or far from, the cornea. While most vertebrates accommodate by changing the radius of curvature of the lens, and so its focal length, fishes accommodate by changing the distance between the lens and the retina by moving the lens back and forth along the optical axis. The teleost lens is moved relative to the retina by the retractor lentis muscle, usually towards the tail of the fish. This is surprising because it might be expected that it would move the lens towards the midline of the fish, into the eyeball and towards the centre of the retina. However, the retina is not always concentric with the lens and it has been suggested that the resting eye is short-sighted anteriorly and far-sighted laterally. As the lens is moved posteriorly the lateral view changes very little but the fish would become able to focus distant objects anteriorly.

   Lampreys accommodate by an extraordinary and unique mechanism, for the cornea is attached to an external circular muscle (of myotomal derivation) that contracts to flatten the cornea, so pushing the lens towards

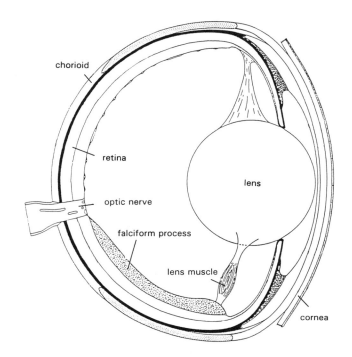

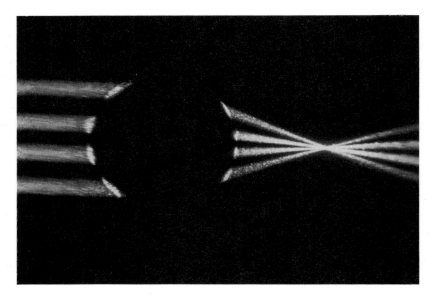

**Figure 10.10** Top, diagram of a section through the eye of a fish. Bottom, excised cichlid lens focusing four argon laser beams (photograph by kind permission of Dr. R. Fernald, Stanford University.)

the retina to focus distant objects. In elasmobranchs, the accepted view was that the lens was moved towards the retina by contraction of a protractor muscle, but histological examination of the supposed muscle has not demonstrated muscle fibres and electrical stimulation experiments have yielded negative results so it is at present unclear whether elasmobranchs can accommodate. In some sting rays, the distance between the retina and the lens varies round the eye because the lens is not quite spherical. Perhaps different regions of the retina are used to focus near and far objects so that accommodation is a static rather than a dynamic process. No-one who has been in the water close to an inquisitive pelagic shark like *Lamna* would doubt that some elasmobranch eyes can be used to inspect prospective prey!

In order to get overlapping fields for binocular vision the posterior part of the eye often plays an important role because it is on the posterior retina that images of the world to the front of the fish are brought to a focus. Binocular vision is important especially for determining the range of objects around the fish, which must move the whole front part of the body to inspect its surroundings (except in the unusual fish *Lepidogalaxias salamandroides* which has a flexible neck). Different parts of the retina must be usable for different purposes. Unlike our eyes, in which we monitor a rather limited part of our surroundings at any one time, a fish can probably see some parts of the environment with binocular vision and other parts, to the side, with lateral vision using each eye independently. Some fishes, like the mudskipper (*Periophthalmus*) have protuberant eyes which can move independently just like those of chamaeleons. A forward-directed optic axis is sometimes found in which objects to the front of the head are brought to a focus on a specialised part of the retina in the posterior part of the eye (p. 242). It is likely that this part of the retina is often most important for behaviour such as feeding where binocularity improves the judgement of distance, while the anterior part and centre of the retina monitor to the side and rear of the fish for the approach of predators or the presence of conspecifics.

### 10.6.2 *Tubular eyes*

The teleost eye has been strikingly modified in many mesopelagic fishes. Eleven families, including the hatchet fish (sternoptychids) and giganturids, have evolved tubular eyes pointing upwards and forwards with their optical axes more or less parallel (Figure 10.11). These tubular eyes were first supposed to act as telescopes but the distance between the lens and the main retina is the same as the normal fish eye. Such eyes have a number of advantages: they allow the fish to achieve good binocular vision in one direction; they allow the eye to have a large lens, with good light-collecting properties, without taking up too much space in the head; some are also

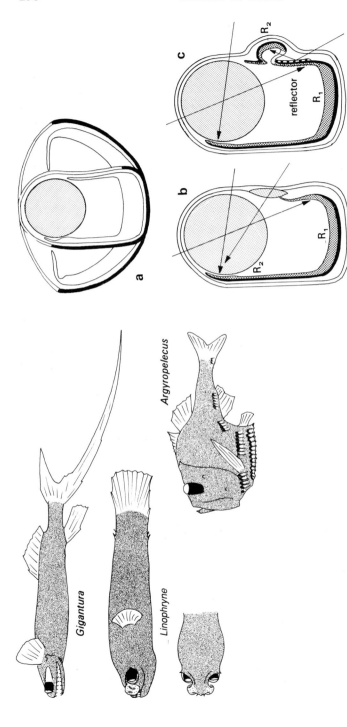

**Figure 10.11** Left: mesopelagic fishes with tubular eyes. Note that *Linophryne* looks through its transparent olfactory organ. Right: modification for lateral vision. (a) Tubular eye of the sternoptychid *Argyropelecus* superimposed upon normal fish eye; (b) *Scopelarchus* showing accessory scleroid lens; (c) the opisthoproctid *Dolichopteryx* with reflector and accessory globe. $R_1$, $R_2$: main and accessory retinae. After Munk (1966) *Dana Rpt.*, **70**, 1; and Locket (1977) *Handbk. Sensory Physiol.*, **7** (pt.5) (Crescitelli, F. ed), 67.

positioned so that the fish can look predominantly upwards to see their prey in silhouette against the vertically down-welling light (p. 251). One problem of a tubular eye is that the peripheral parts of the retina cannot be brought into focus because they are too near the lens; they are probably only useful for unfocused movement detection. Some species have developed an accessory retina or accessory refracting devices to obtain focused images from light entering the eye outside the main axis (Figure 10.11). The argentinoid *Dolichopteryx* has an accessory retina illuminated by lateral light which has not passed through the lens but through the side of the eye, then being reflected by the argentea into the accessory retina. Scopelarchids and evermannellids have no accessory retina but lateral light is focused on to the main retina by accessory refracting lens pads and ocular folds. Evidently inefficient lateral vision is a high price to pay for efficient binocular vision, hence the elaborate modifications of the tubular eyes.

### 10.6.3 Aerial vision

Quite different modifications have arisen in teleosts which need to see in air. The intertidal mudskipper, *Periophthalmus*, has a flattened lens for vision in air but vision is probably poor under water. The 'four-eyed' fish, *Anableps*, swims at the water surface with the upper part of each eye exposed to the air (Figure 10.12). The lens is elliptical with the longer axis

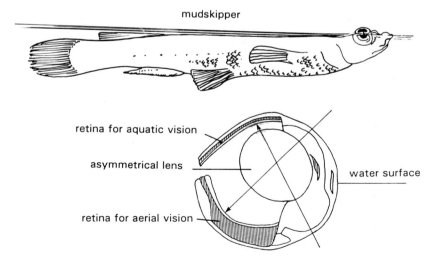

**Figure 10.12** View of *Anableps* at the water surface showing how the asymmetrical lens allows it to focus objects above and below the surface on to the retina. Adapted from Walls (1963).

refracting light from below the surface and the short axis refracting light from above the surface. In air, the corneas of all fishes become refractive making the fish short-sighted. Flying fish and some shore-living clinids avoid this by have flat corneal 'windows' that reduce the extent of refraction and make vision in air more satisfactory.

### 10.6.4 Reflecting tapeta

In most elasmobranchs, in Holocephali, sturgeons, *Polypterus*, the lungfish *Neoceratodus* and in *Latimeria*, light passing through the retina is reflected back by a tapetum at the back of the eye (Figure 10.13). When illuminated, the eyes of these fishes shine, as do those of cats; the eyes of deep-sea squaloid sharks shine with a superb greenish-blue colour when hauled aboard ship. The tapetum is a specular reflector, consisting of layers of reflecting cells in the chorioid layer packed with thin platelets of guanine, like those found in the teleost scale. The platelets are arranged at suitable angles to reflect light back into the retina along the long axis of the visual cells (Figure 10.13). Reflectors of this kind are adjusted to reflect certain wavebands by appropriate thickness and spacing of the guanine crystals; those of deep-sea sharks reflect best at wavelengths that penetrate into deep water.

Teleosts also show eyeshine, but the reflecting layer is instead usually in the retina, often consisting of tiny reflecting particles, and light is backscattered in a more diffuse manner than with specular reflectors. Chorioidal tapeta are quite rare, only being found in midwater fishes like myctophids and the castor oil fish, *Ruvettus*.

However constructed, these reflectors all reflect light back through the

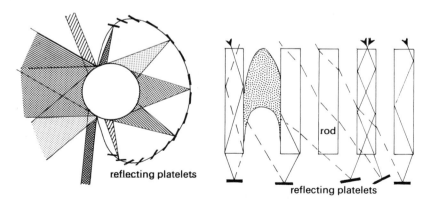

**Figure 10.13** Reflecting tapeta. Left: orientation of reflecting platelets in *Squalus* tapetum. Right: Paths of light rays channelled by rods and reflected by platelets in sturgeon tapetum. Pigment cell stippled. After Denton and Nicol (1964), and Nicol (1969) *Contr. Mar. Sci.*, **14**, 5.

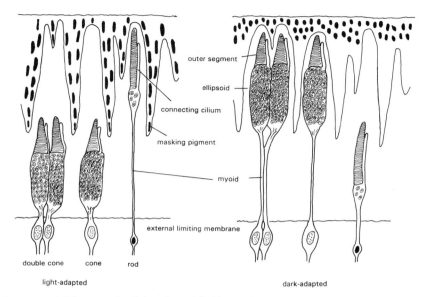

outer segment

ellipsoid

connecting cilium

masking pigment

myoid

external limiting membrane

double cone    cone    rod

light-adapted                                              dark-adapted

**Figure 10.14** Diagram of a light-adapted (left) and dark-adapted retina (right) of a typical teleost. Note the changes in position of the retinal masking pigment and the cones and rods.

visual cells, so enhancing sensitivity. It has also been suggested that tapeta reduce 'noise' from the spontaneous breakdown of visual pigment since fishes with tapeta need less visual pigment for a given sensitivity. The inevitable snag is, however, that some spreading of the reflected light occurs and this may reduce the perception of detail (acuity, see p. 247) since sharp boundaries in the image become more diffuse as more visual cells are stimulated by the reflected light.

Reflecting layers in the eye have another potential disadvantage, for they could make fish very conspicuous when illuminated by daylight. Many sharks can occlude the tapetum by migration of black masking pigment over the tapetal surface on transfer to light, a process that takes 60–90 min; unmasking in the dark takes a somewhat shorter time. Masking pigment movement is also found in teleosts but here it is related to the movements of the visual cells, the rods and cones (Figure 10.14).

### 10.6.5 The receptors

Most fishes have a duplex retina (like ours), i.e. one containing both rods and cones, but some sharks such as *Centrophorus*, chimaerids and rays and some deep-sea teleosts have a pure rod retina. In many larval fishes there is a pure cone retina at hatching and the rods develop progressively during ontogeny.

Both rods and cones share a common plan: an outer segment containing visual (photosensitive) pigment, an ellipsoid packed with mitochondria, an extensible myoid or foot-piece and a nuclear region (Figure 10.14). The rods and cones are held in position by an external limiting membrane. Typically rods are longer and thinner with cylindrical outer segments and ellipsoids while in cones the outer segment is conical, and the ellipsoid rather bulbous. The ellipsoids and outer segments are joined by an eccentrically placed connecting cilium with the characteristic nine filaments. Cones are often present as pairs and generally the visual cells are disposed in a regular mosaic when viewed in tangential section (Figure 10.15). The rods vastly exceed the cones in number in the adult eye.

The duplex retinas of many fishes have regions of specialisation. For example, an area temporalis consists of a patch of closely packed cones (with few, or no rods) appropriately placed to receive light along the main axis of feeding, where high acuity would be an advantage. In pelagic feeders like herring and horse mackerel, looking upwards and forwards to feed, the area is postero-ventral on the retina. In horizontally feeding fish

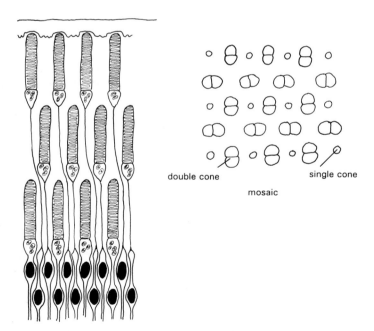

double cone                     single cone

mosaic

**Figure 10.15** Left: tiered rod retina of a deep sea fish (redrawn from Locket, 1977, in Nicol, 1989). Right: retinal mosaic of the perch. Adapted from Engström (1963) *Acta Zool. Stockholm*, **44**, 179.

such as the sailfish (*Istiophorus*), it is posterior on the retina, and in bottom feeders like the sea bream (*Sparus*), the area is postero-dorsal on the retina. The amphibious mudskipper (*Periophthalmus*) has a horizontal band of cones placed to perceive objects near ground level where both food and predators might be present. A fovea or depression with a high density of cones (as in the reader's eye) is present in sea horses (*Hippocampus*) and pipefish (*Syngnathus*).

The rods and cones are less distinct morphologically in lampreys and the retina has degenerated in the hagfish. While double cones are extremely common in most fishes, triple cones sometimes occur in the brown trout and quadruple cones in the minnow. In some deep-sea fishes and the cusk eel the retina is composed of a number of tiers of rods (Figure 10.15), presumably arranged to absorb all the limited light that enters the eye. *Diretmus* has two photosensitive pigments, which may confer colour vision, and a stable yellow pigment, perhaps acting as a yellow filter (p. 254).

Both rods and cones are connected to the brain by a complex 'wiring' system via bipolar cells and ganglion cells whose axons pass along the optic nerve to the optic lobes. The rods contain much more visual pigment (of one type, see p. 244) than the cones and are thus more sensitive. Their sensitivity is enhanced by the convergence of many rods to one bipolar cell, a system called summation. The combined effect of light on a number of rods can then be additive, helping to fire the bipolar cell. The rods respond in dim light but with poor acuity (because of summation) and without conferring any sense of colour. The cones, on the other hand, are divided into 'populations', each with a different visual pigment (p. 246); they do not summate at the bipolar cells and confer high-acuity colour vision (where this has been rigorously tested) in brighter light.

In teleosts there is a switch in the eye from dim light to bright light. This light or dark adaptation is accompanied by retinomotor or photomechanical movements of the visual cells and melanin masking pigment (Figure 10.14), akin to the occlusion of the tapetum. In dim light, the masking pigment is well retracted towards the outside of the retina, the rod myoids are short and the cone myoids are long, so that the rod outer segments are near to the external limiting membrane and the cones are not impeding this penetration of light to the rods. As the eye light-adapts, the cone myoids shorten, bringing their outer segments towards the external limiting membrane; the rod myoids lengthen, taking their outer segments away from the cones and into the advancing masking pigment, which protects them from the brighter light. This process takes 20–30 min and occurs around dusk and dawn. In larval fishes, before the rods develop, and in adult fish without cones, retinomotor movements do not take place during the transfer from dark to light.

These intricate adaptations of the fish retina are not found in terrestrial animals (although to a limited extent in Amphibia) where rapid changes in pupil diameter play an important role in controlling the amount of light entering the eye. In elasmobranchs (which usually have rod-dominated retinas) variation of pupil diameter is important in species that may be subjected to changes of light intensity, but in deep-sea selachians and chimaerids, the pupils are immobile. With some exceptions, the teleosts have relatively immobile pupils. Most teleosts seem to have iris muscles or some form of sphincter but the variations of pupil diameter in different light levels are nothing like so dramatic as in elasmobranchs.

### 10.6.6 Visual pigments

Visual pigments consist of a protein (opsin) linked to an aldehyde of vitamin $A_1$ or vitamin $A_2$. The $A_1$ pigments are sometimes called rhodopsins, the $A_2$ pigments porphyropsins. Depending on its precise chemistry, each pigment can be shown to have a characteristic light absorption as measured by a spectrophotometer. The pigments are usually referred to by the wavelength of light at which such absorption is maximal (the $\lambda_{max}$). Because the pigments are photosensitive and break down in light they are most sensitive to light of wavelengths near the $\lambda_{max}$. As they break down chemically, the pigments bleach and only regenerate to their original chemical state in the dark, a process that may take several hours to complete. The visual pigments are located in the outer segments of both rods and cones and after bleaching some of the breakdown products migrate into the epithelium surrounding the visual cells where part of the regeneration process occurs.

There has been much interest in whether fishes from different environments – the deep sea, coastal waters or freshwater – have visual pigments matched to the wavelengths of light that penetrate the water best. Certainly deep-sea fishes have rod pigments (chrysopsins, related to rhodopsin) with a $\lambda_{max}$ of 470–480 nm matched to the clear blue water in which they live (Figure 10.16). Coastal fishes with a $\lambda_{max}$ of 490–515 nm and freshwater fishes of $\lambda_{max}$ of 500–545 nm, although with visual pigments shifted in the appropriate direction, are well offset from the maximum light-transmission characteristics of their environments. At the expense of sensitivity, this is thought to enhance the contrast between objects such as food, and the background.

When fish migrate into water of different spectral quality, as do eels and salmon, their rod pigments change. As Pacific salmon move from the sea into freshwater for the spawning run, the $\lambda_{max}$ changes from 503 to 527 nm – the rhodopsin being gradually replaced by a porphyropsin more suited to the yellower freshwater environment. The sea lamprey (*Petromyzon marinus*) changes from a porphyropsin-dominated system when migrating upstream to a rhodopsin system migrating downstream. In the freshwater juvenile of the eel, there are a pair of rod pigments

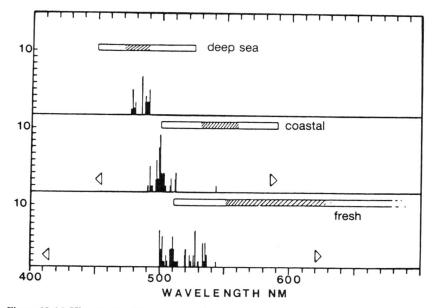

**Figure 10.16** Histograms of fishes containing single-rod visual pigments (maximum absorption, $\lambda_{max}$). Horizontal bars show the possible range of visual pigment $\lambda_{max}$ that would give maximal sensitivity in each water type; shaded areas show the most likely range. Open triangles indicate $\lambda_{max}$ of the most blue- and red-sensitive visual pigments measured in the cones of fishes inhabiting that water type (exclusive of ultraviolet-sensitive cones). From Lythgoe (1980).

absorbing maximally at 501 and 523 nm but as the eyes enlarge and the eel returns to the sea, the 523-nm pigment disappears and a new pigment develops with a $\lambda_{max}$ of 482 nm.

A dual rhodopsin–porphyropsin system is found in some inshore and freshwater fishes. The relative proportions vary with the individual, its age, with the spectral nature of the environment and with season and water temperature. Sometimes these variations seem to correlate with changes in the spectral quality of the water, but not always.

### 10.6.7 Colour vision

Much early controversy raged over whether fishes had colour vision, well described by Walls in his classic book *The Vertebrate Eye*. Some earlier workers trained fish to make behavioural choices of various sorts between different colours but failed to control the brightness (greyness) of the colours. More enlightened experiments, in which not only absolute brightness (as determined by a brightness meter) but also the subjective brightness to the fish (p. 247) were controlled, showed that some teleosts do indeed have colour vision. Whether sharks do is less certain.

Colour vision was thought to be mediated via the cones but it is only in the last 20 years or so that sufficiently refined optical equipment has been developed to check the mechanisms. The microspectrophotometer is a sophisticated instrument that can measure the spectral characteristics of individual cones. Bearing in mind the low density of the cone pigments and the fact that the pigment-containing outer segment is ≤5 μm in diameter, this is certainly a considerable feat. In the goldfish and several other species, three populations of cones each with a pigment of characteristic $\lambda_{max}$ can be identified (Figure 10.17) to provide a trichromatic mechanism for colour vision. Some fishes, such as the weever (*Trachinus*), are only dichromatic with blue- and green-sensitive cones and some cyprinids are tetrachromatic.

It is not surprising that colour vision has evolved; the brilliant colours of many fishes from shallow water, especially coral reef fishes, are an indication of colour vision, which also enhances contrast, making prey and perhaps predators more conspicuous.

Quite recently, experiments have shown that some cyprinids, trout and *Anableps* can see ultraviolet (UV) light. In *Tribolodon hakonensis* (related to dace), goldfish and the roach (*Rutilus rutilus*), small single cones are sensitive to UV light and (a necessary prerequisite) the cornea and lens

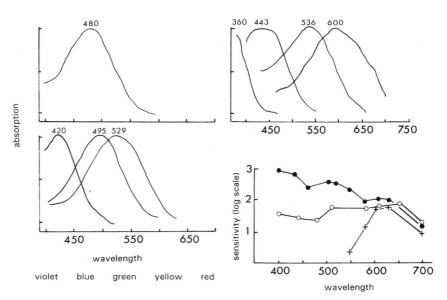

**Figure 10.17** Absorbance spectra of visual pigments obtained by microspectrophotometry. Top left: from the rods of a deep-sea fish *Searsia koefoedi*. Top right: from the freshwater brown trout (*Salmo trutta*) (note the pigment absorbing in the violet end of the spectrum at 360 nm). Bottom left: from the coastal marine dragonet (*Callionymus lyra*). Data redrawn from various sources in Douglas and Djamgoz (1990). Bottom right: Photopic spectral sensitivity curves of the rudd (*Scardinius erythrophthalmus*) obtained using three types of light-dependent behaviour criteria. Redrawn from Muntz (1975).

also transmit UV light down to 350 nm. How these fish make use of the UV information is uncertain. The goldfish also responds behaviourally to the plane of polarized light, maintaining a fixed orientation to the e-vector and rainbow trout can be trained to swim to a refuge using the e-vector as an orientation mechanism. There is some evidence of a link between UV and polarised light perception.

### 10.6.8 Sensitivity and acuity

The larger the eye, the better its light-collecting ability and so its sensitivity, but as explained on p. 243 the sensitivity of vision also depends on the density of the visual pigment and the extent to which the visual cells summate. Adaptations to increase sensitivity such as summation and the reflection of light back through the retina by a tapetum tend to fuzz sharp boundaries and so impair acuity – the perception of detail.

The equivalent of the audiogram is the spectral sensitivity curve, the relationship between sensitivity to light and wavelength (Figure 10.17). Spectral sensitivity curves can be obtained by experiments in which the threshold for light-dependent behaviour is measured at a number of wavelengths. These curves may vary with the behaviour criteria used but sometimes the curves obtained can be matched to the absorption characteristics of the rod visual pigment, if the fish is dark-adapted, or to the combined absorption characteristics of the cone pigments if it is light-adapted.

During the transfer from light (photopic vision) to dark (scotopic vision) there is not only an increase in sensitivity to light but a change in the response to different wavelengths – the Purkinje shift – the eye becoming more sensitive to light at the blue end of the spectrum (Figure 10.18). This should not be confused with colour vision. The fish may, indeed, switch from colour to monochromatic vision as they dark-adapt but they may only see colours as shades of grey when both light- and dark-adapted. We do not always know.

In nature fish may be exposed to changes of ambient illumination ranging over 10–12 logarithmic units. It is the dark–light adaptation mechanism with the populations of cones and rods with different intrinsic thresholds that allows their eyes to function over such a wide range. In many fishes, light-dependent behaviours such as feeding, spawning and schooling become reduced in intensity as the illumination falls (Figure 10.19) and cease altogether when these fishes become dark-adapted. The function of the rods in dark-adapted fishes is not evident. In some they may still allow limited feeding, avoidance of predators, or schooling (perhaps operating in conjunction with the other senses); they may be involved in controlling activity, maintaining a certain depth depending on the amount of down-welling light or in the control of vertical migration by the appreciation of an optimum light intensity (p. 307) that may be tracked at dusk and dawn.

| **Dark-adapted** (Rods) | **Light-adapted** (Cones) |
|---|---|
| High sensitivity | Low sensitivity |
| (Summation) | (No summation) |
| Poor acuity | Good acuity |
| Monochromatic vision | Colour vision |

←Retinomotor movements→

←Purkinje Shift→

**Figure 10.18** Summary of the physiological changes taking place as fish light- and dark-adapt.

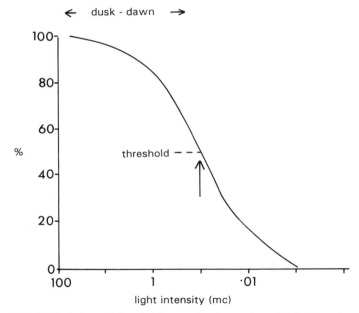

**Figure 10.19** Typical sigmoid threshold curve for the waning of light-dependent behaviour such as feeding or schooling, as the light intensity falls. The intensity range of the dusk-dawn period is 100–0.1 mc.

Many of the features of fish vision (and indeed those of most vertebrates) are set out in Figure 10.18 in rather a simplistic way. However, it should never be assumed that animals have evolved mechanisms that always fall into neat categories. For example, there is some limited evidence that a few highly sensitive, low-threshold, cones continue to function near the absolute visual threshold, and so some fishes may retain limited colour vision even when dark-adapted.

## 10.7 The pineal body

This organ, known to be light-sensitive, is present as a lobe on the upper surface of the forebrain (p. 277). In trout and tuna, for example, it lies below a transparent window but in other species, access to light is more questionable. In the trout the organ is most responsive to light of about 500 nm, suggesting a rod-like visual mechanism. It has been suggested that the pineal acts as a dusk detector in its tonic mode and as a shadow detector in a dynamic mode. The hormone melatonin is secreted by the pineal, causing expansion of the melanophores, and darkening of the skin in the dark (p. 214). The rhythmical secretion of melatonin is implicated in the timing of circadian rhythms, synchronising cycles of activity with light cycles in the environment. In man, melatonin plays a role in sleep patterns and melatonin secretion at the wrong time is a cause of jet lag.

## 10.8 Camouflage

Many benthic fishes are perfectly camouflaged against the substratum by their chromatophores. It is only too easy to step on well-nigh invisible stonefish (*Erosa*) when walking on coral reefs, and remarkable matches to their backgrounds are made by flatfishes such as *Pleuronectes*. Not only are these fishes coloured like the substratum, but many have fringed margins and projections making their bodies 'unfishlike' in outline. But the hardest camouflage problems are faced by pelagic fish. Countershading is common (inverted in the upside-down freshwater catfishes like *Synodontis*), but there are other much more effective and ingenious camouflage methods in pelagic marine fishes.

### 10.8.1 Camouflage by reflection

Many fishes living in the upper layers of the ocean are silvery, because of silvery scales (herring) or underlying silvery layers (mackerel). The silveriness results from light reflection by organised stacks of thin guanine crystals, separated by sheets of cytoplasm, as in the tapetal reflectors. Such layers of material of alternating high- and low-refractive indices operate as very efficient reflectors. Since the wavelength reflected depends upon the optical thickness (refractive index × thickness) of the layers, the highest reflectivity being when the optical thickness is $0.25\lambda$, such reflectors can be (and are) 'tuned' to reflect particular wavelengths.

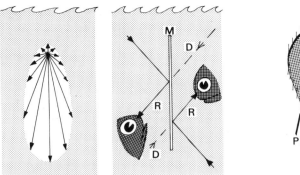

**Figure 10.20** Light in the sea and camouflage with silvery surfaces. Left: distribution of radiance in the sea. The length of the arrows indicates relative radiance in a given direction from their origins. This polar diagram of radiance distribution remains constant with depth, although light intensity changes. Middle: predatory fishes looking at a perfect mirror (M) cannot distinguish between reflected rays (R) and direct rays (D), hence the mirror is invisible. Right: a herring has reflecting platelets (PL) on its sides inclined slightly upwards, thus compensating for less than perfect reflection by reflecting light slightly brighter than if oriented vertically as in middle diagram. Note narrow keel and dark pigment dorsally. After Denton (1970) *Phil. Trans. Roy. Soc. Lond., B*, **258**, 285.

The polar diagram of light in the sea (Figure 10.20) is symmetrical except just at the surface, and so fishes can camouflage themselves by reflecting light, as Denton (1970) showed.

Consider a flat mirror vertical in the water. An observer (a predatory fish) looking at the mirror obliquely from any angle (Figure 10.20) will see light reflected from the mirror but will be unable to distinguish this from light which would have reached its eyes if the mirror were not there, and hence will be unable to detect the presence of the mirror. Although fish are not flat-sided, they can arrange that the reflecting platelets are vertical (Figure 10.20) so achieving the same result as if the sides of the fish were flat. The bottom of the fish will be hardest to camouflage in this way, because if looked at directly from below, it will be silhouetted against down-welling light.

Two solutions to this problem are possible: either the fish can make the ventral region compressed and knife-like, as do herring and sprat (Figure 10.20, right), or it can use photophores to generate its own light to shine down to mimic the ambient down-welling light.

### 10.8.2 Luminescence and photophores

Light organs of different kinds are an intriguing and important feature of many mesoplelagic and deep-sea fishes, and are even found in some fishes living in shallow water, such as the midshipman (*Porichthys*), though no freshwater fishes have them. Samples of fishes collected off Bermuda and in the South Atlantic showed that some 70% of all species caught had light organs, and systematic surveys have shown that around 10–15% of all marine fish genera contain luminous species. Some species use lights as lures – barbels and fishing rods with luminous tips are found in several families (Figure 10.21), whilst others have light organs in the mouth – or like the flashlight fish (*Photoblepharon*) use them as headlights to illuminate their prey. Most fishes, however, use their photophores for signalling to other members of the same species, or for camouflage. In the upper 1000 m of the ocean down-welling light will silhouette fishes looked at from below (which is why paralepids, for example, adopt a 45° head-up attitude to seek their prey). Many fish surmount this difficulty by shining light downwards to match natural down-welling daylight (see Figure 10.24 below) and so make themselves invisible from below. Since it is also necessary to match the background when viewed *obliquely* as well as directly from below, it is hardly surprising that the most complex photophores and photophore arrays have been developed for ventral camouflage.

Fish photophores are very diverse, and their structure can only be touched on here. There are two basic types: those in which light is produced by special photocytes, and those where symbiotic luminous bacteria are cultured in special sacs. So far, only one fish, the angler *Linophryne*, has been discovered which has both kinds.

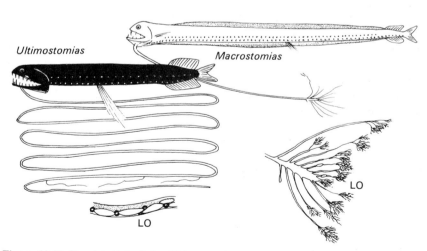

**Figure 10.21** Stomiatoid barbels. *Ultimostomias* has light organs (LO) aligned along the barbel (below) whilst *Macrostomias* has luminous organs (LO) at the tip. Note ventrally directed camouflage photophores along body. After Beebe (1930) *Copeia* **1933**, 160.

### 10.8.3 Bacterial photophores

Bacterial light organs are either linked to the gut, or are open to the sea; the species of *Photobacterium* found in these light organs are obligate symbionts which infect the chambers during larval life. In angler fishes, analysis of their 16S rRNA genes has shown that they are specific to each host fish, and different from the free-living species. They glow continuously, so the fish can only 'switch' them off by masking them with a shutter, or by rotating them into a black-lined pocket. The bacteria are in colossal numbers in these light organs. In the spectacularly luminescent flashlight fish *Photoblepharon*, the light organs contain $10^{10}$ bacteria $cm^{-3}$! *Photoblepharon* lurks in reef caves in the tropics during the day, flashing intermittently (by blinking the shutter over its light organs); at night it emerges to hunt copepods by the continuous light shone forwards. Some other bacterial light organs are very much dimmer, and were only discovered by a dark-adapted observer, for example, the peri-anal organs of *Chlorophthalmus*; these may be used as cues for schooling.

A number of fishes with bacterial light organs use them to illuminate the ventral surface, presumably for camouflage. Externally, there are no special modifications, but internally, there are remarkable specialisations. The light organs are diverticula of the gut in such fishes, and are surrounded dorsally and laterally by a connective-tissue reflecting layer. Light therefore emerges downwards from the light organ, and is refracted by translucent ventral muscles before passing out of the ventral region of the fish (Figure 10.22). In the extraordinary *Opisthoproctus* (Figure 10.22), a light organ near the anus is enclosed in black epithelium except anteriorly, where it shines into a long hyaline ventral light guide surrounded dorsally by reflecting platelets. The bottom of the fish is completely flat and light emerges evenly over the whole of this flattened sole. *Opisthoproctus* lives in the upper mesopelagic zone, and it seems certain that this bizarre arrangement must be used for ventral camouflage, but we do not yet know how it is tuned to cope with being seen obliquely from below.

### 10.8.4 Photophores with intrinsic light production

Perhaps partly because of problems of infection and maintenance, fishes with bacterial light organs have four at most and usually only one or two; hence they may economically make the same organ serve several purposes like the flashlight fishes, where the organs are used to illuminate prey, for schooling, and for sexual communication. Fish with photophores where the light is generated intrinsically, on the other hand, often have many organs, and so can use different ones for different purposes. In lantern fishes (myctophids), for example, ventral series of photophores are used for

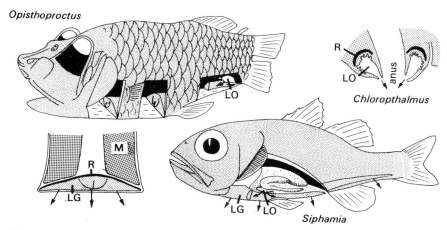

**Figure 10.22** Bacterial light organs. Left: *Opisthoproctus* has a rectal light organ (LO) which transmits to a flattened light guide (LG) backed by a reflector (R) that emits light along the 'sole' of the fish, seen in transverse section below. M: myotomal muscles. Right: anal light organs (LO) of *Chlorophthalmus* open from the rectum and are backed by reflectors (R). Bottom: the cardinal fish *Siphamia* has a light organ (LO) opening from the gut and illuminating a light guide (LG) which emits along the ventral surface. After Herring (1977), *Nature*, **267**, 788; Iwai (1971) *Jap. J. Ichthyol.*, **18**, 125 and Somiya (1977) *Experientia*, **33**, 906.

camouflage whilst lateral ones (differently patterned in the two sexes) are evidently used for intra-specific signalling.

Sometimes different photophores emit light of different colours. This kind of photophore is often rather like an eye, for the photocytes may be backed with a reflecting layer, and capped with a lens. They are richly innervated and under the control of the autonomic nervous system, the transmitter being adrenaline or noradrenaline. By far the most complicated photophores so far studied are those of the sternoptychid hatchetfish, *Argyropelecus* (Figure 10.11). These are arranged in groups of ventrally directed tubes along the lower part of the fish (Figure 10.23). In each group of photophores, light is produced in a dorsal chamber, lined with black pigment apart from a series of small ventral windows into the photophores (Figure 10.23). Light passes through these windows, and then through a colour filter transmitting at 485 nm, and then enters the wedge-shaped photophore. This is lined with a reflective guanine layer, and the flat external surface is covered with a half-silvered mirror, again made from guanine crystals. The result of this rather complicated design is that light entering the photophores from the dorsal chamber will leave it in a particular pattern differing in intensity at different angles. Because the inner reflecting surfaces of the photophores are curved, each photophore will distribute its light in a wide arc. Denton and his colleagues who worked out the way in which these photophores operated, set up living hatchetfishes

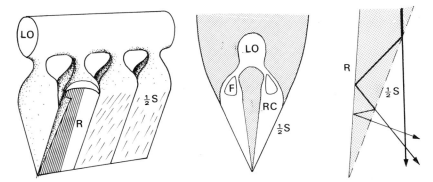

**Figure 10.23.** The organisation of the ventral photophores of *Argyropelecus*. Left: stereo-gram of system showing dorsal light pipe (LO) connected to ventral reflecting chambers (R) backed with a curved reflector and half-silvered on the external surface (½S). Middle: transverse section through ventral part of body showing light organ (LO), filters (F), and reflecting chambers (RC) with half-silvered outer surface (½S). Right: reflections of rays emitted from light organ into reflecting chamber giving rise to appropriate distribution of radiance for ventral camouflage. After Denton (1970) *Phil. Trans. Roy. Soc. Lond.*, *B*, **258**, 285.

(at sea on board 'RRS Discovery') in a chamber in which they could be rotated whilst the light emitted was measured by a photomultiplier. They found that the angular distribution of the light emitted was remarkably close to that of ambient light in the sea, as also was its spectral characteristic (Figure 10.24) so making it virtually certain that *Argyro-pelecus* is hard to see from below!

Obviously, it is not enough for the fish to match the angular distribution and wavelength of the light it emits to the ambient light – it also has to match it in *intensity*. Hatchetfishes and many myctophids have small photophores which are arranged to shine into the eye (a curious arrangement at first sight): these act as reference sources to permit matching of ambient light intensity (Figure 10.24). Provided this little photophore is functionally coupled to the ventral photophores, they will be enabled to match down-welling light and so camouflage the fish.

Hatchetfishes produce light about equivalent in intensity to the down-welling light at depths of around 600 m (where the fish are found during the day). Much nearer the surface, fish cannot produce sufficient light for ventral camouflage, and so rely instead on transparency, or on special morphological modifications.

### 10.8.5 Yellow lenses

The adaptations of fishes are so remarkable that it should come as no surprise to find that a few fishes have devised a means of overcoming the

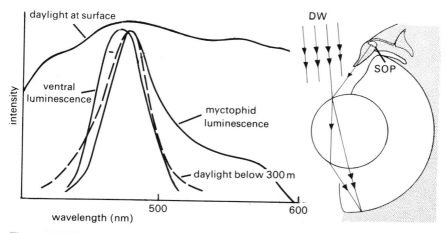

**Figure 10.24** Ventral illumination camouflage. Left: The remarkably accurate spectral match of the light emitted by *Argyropelecus* ventral photophores to down-welling daylight. The light from myctophid photophores is less well matched. Right: Comparison of the light emitted by the supraorbital photophore (sop) of the myctophid *Tarletonbeania*, with down-welling daylight. After Denton (unpublished) and Lawry (1974) *Nature*, **247**, 155.

ventral photophore camouflage system, by using filtering lenses. The emission spectra of the light emitted ventrally by *Opisthoproctus* and by *Argyropelecus* are an extraordinarily close match to that of down-welling daylight, and yellow lenses (which *Argyropelecus* itself possesses) are of no help in detecting such fishes from below. But the emission spectra of myctophid photophores are much broader; so examined with an upward-looking eye equipped with a yellow lens, they will appear brighter than the background. In the clear mesopelagic zone, Muntz (1975) has calculated that their camouflage could be pierced at a distance of around 16 m. However, it seems that, usually, fish with yellow lenses employ them to increase the visibility of lateral photophores.

### 10.8.6 Red headlight fishes

An even more remarkable special case is the use of photophores to circumvent the camouflage of the common red and dark brown animals of the mesopelagic zone. These are not visible when illuminated with blue-emitting photophores, like those on the head of the myctophid *Diaphus*. The stomiatoids *Malacosteus* and *Pachystomias*, however, have large red-emitting headlight photophores underneath the eye, and their retinal pigments absorb at around 575 nm (Figure 10.25) so that they can perceive red light. Most deep-sea animals have pigments with an absorption maximum around 450–490 nm and so cannot perceive red light. With their headlight photophores *Malacosteus* and *Pachystomias* can easily see red

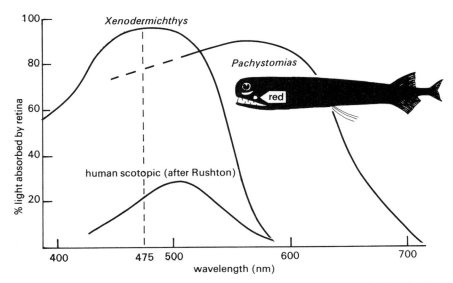

**Figure 10.25** Private wavelength of *Pachystomias*. Curves showing light absorbed by single pass through retina. The dotted line is wavelength maximally transmitted by seawater at 300 m. Note that the alepocephalid *Xenodermichthys* absorbs maximally at the dotted line; human rods are suited to dim light (moonlight is similar but less intense than daylight, Figure 10.19), whereas *Pachystomias* absorbs maximally much further to the red end of the spectrum. *Pachystomias* has a red-reflecting tapetum and thus maximum absorption of the intact eye will be higher than that of the retina alone. After Denton (unpublished).

and brown prey, illuminating them with light of a wavelength that the prey cannot detect! Because red light is less well transmitted in the sea than blue light, *Malacosteus* has a red-reflecting tapetum and increased pigment density in the retina to make up for the inevitable loss of sensitivity.

### 10.9 Olfaction, taste and pheromones

Unlike light and sound, chemical stimuli are persistent and effectively non-directional. Although gradients of concentration may be set up around a chemical source, they are easily disturbed by turbulence caused by wind at the water surface, currents and animals. Chemicals may arouse fishes and under certain circumstances these fish may be able to swim up a gradient of concentration. In particular, some fishes are programmed to swim up-current if they detect a chemical stimulus.

The distinction between smell (olfaction) and taste (gustation) is less clear in water than on land. It is generally assumed that olfaction is distance reception and gustation contact reception but in fishes it is perfectly possible for the taste organs to respond to distant sources of

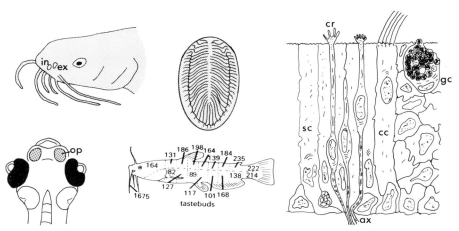

**Figure 10.26** Top left: head of the catfish (*Arius felis*) showing sensory barbels and inhalent (in) and exhalent (ex) nostrils. Top centre: olfactory rosette of the catfish *Silurus glanis*. Lower left: top of head of a fish larva showing olfactory pits (op). Lower centre: distribution of taste buds on the body of the catfish (*Ictalurus*). Right: simplified drawing of a transmission electron microscope section through the olfactory epithelium of a teleost showing ciliated receptor cell (cr), ciliated non-sensory cell (cc), supporting cell (sc), goblet cell (gc) and axon (ax). Redrawn from Zeiske *et al.* (1992) and Bardach *et al.* (1967).

stimuli. In fishes, the taste buds are not only found inside the mouth, as in terrestrial animals, but on the gills, surface of the head and on the fins and barbels (Figure 10.26). They are not usually found on the tongue.

### 10.9.1 The chemoreceptors

Taste buds contain receptor or gustatory cells with apical microvilli and supporting cells (although some controversy exists about different cell types) innervated by cranial nerves. Solitary chemosensory cells also exist. There seems to be a neat division in the gustatory system with the VIIth (facial) nerve innervating the taste buds on the external surface and the IXth (glossopharyngeal) and Xth (vagus) nerves innervating the taste buds in the pharynx. Each system plays its own part in feeding behaviour, the cutaneous taste buds detecting food, the pharyngeal taste buds judging its quality inside the mouth.

The olfactory organ, which is usually paired, contains many thousands of receptor cells arranged in a rosette of very variable form in a chamber with incurrent and excurrent nostrils and with olfactory tracts to the forebrain (Figure 10.26). Lampreys and hagfish have a single chamber and single nostril. In hagfish, there is a nasopharyngeal duct opening into the pharynx but in lampreys this duct ends blindly as a nasopharyngeal pouch. In lungfish also the excurrent nostril opens into the oral cavity. A pressure differential is created between the incurrent and excurrent nostrils by various means in different fishes – by forward movement, respiratory movements or by motile cilia. In some ceratioid angler fish, in *Lophius* and gulper eels and some deep-sea monognathid eels, the female has poorly developed olfactory organs, unlike the male where they are hypertrophied.

The olfactory sensory epithelium consists of receptor cells, either with several apical cilia or numerous microvilli, separated by supporting cells (Figure 10.26). The cilia may have a 9 + 2 or 9 + 0 arrangement of filaments. The receptor cells are present at extremely high density – as many as half a million per square millimetre!

Like the optic cups the olfactory organ develops very early in ontogeny. Initially it occurs as an olfactory pit and this is found in many larval fishes after hatching. Later the pit sinks into the front part of the head in most species, forming a cavity, but it remains open, for example, in the needlefish, *Belone belone*.

The gustatory system responds to a wide range of chemicals but most of the physiological studies have used amino acids. The threshold lies generally between $10^{-5}$ and $10^{-9}$ M although the channel catfish, *Ictalurus punctatus* is especially sensitive, with a threshold near $10^{-11}$ M. The system is also known to respond to bile salts and nucleotides such as AMP and ADP.

The olfactory system also responds down to thresholds of $10^{-7}$–$10^{-9}$ M for amino acids and even lower, $10^{-8}$–$10^{-13}$ M, for bile salts, steroid hormones and prostaglandins. Particularly low thresholds have been claimed for certain alcohols – the eel *Anguilla anguilla* can apparently be trained to respond to about $10^{-18}$ M $\beta$-phenylethyl alcohol, equivalent perhaps to a single alcohol molecule in the olfactory chamber!

### 10.9.2 Feeding and chemoreception

The role of the chemoreceptors can be seen in feeding, predator avoidance, reproduction and homing. It is not surprising that there is a high sensitivity to amino acids, an important component of food. Some species, such as the silurid catfish trail their sensory barbels over the substratum; bullheads, *Ictalurus*, often feed on benthic invertebrates at night or in water of low visibility. The cutaneous taste buds may well trigger pick-up behaviour but the food may be rejected after tasting within the mouth. Other senses such as the lateral line or electroreceptors may also play a part. Tuna, on the other hand, have large olfactory organs although they are typically visual hunters. In experiments, chemical extracts of potential prey cause increased activity, a change of the pattern of swimming and occasional snapping movements. Thus their awareness of food nearby is increased and, after visually-mediated seizure of food they may still reject it on the basis of its taste. Almost certainly, sharks are alerted by the smell of prey (the blood of fish or humans in the water) but orientate to the noise created by splashing and struggling. Probably, *Ictalurus* can make a true gradient search in stagnant water to locate a chemical source, such as liver juice, from 25 body lengths' distance (Figure 10.27). Following odour trails to their source requires the fish to balance the intensity of the odour in its two olfactory organs or to achieve increasing odour concentration when making successive 'sniffs'. The fish should then be orientated towards the source.

There is a great deal of interest in these topics by fishermen and fisheries scientists developing artificial baits. If the most important chemicals for inducing feeding behaviour can be identified they could be incorporated into slow-release bait blocks that might last for weeks or months and greatly reduce the labour of rebaiting lines or traps. Amino acids seem to be the most important but the particular acid or acid combination varies from one fish species to another. Chumming, the attraction of fish to a locality where chemical attractants have been released, has long been established. Usually it is crude, using perforated cans of dog food or fish blood and guts, but fishermen also use live bait, small fish prey species, that provide visual and auditory as well as chemical stimuli.

### 10.9.3 Reproduction and chemoreception

Pheromones, substances secreted into the environment by organisms and received by other individuals of the same species in which they release specific reactions, are often associated with reproductive processes. Some of the earliest work showed that the female estuarine goby (*Bathygobius*

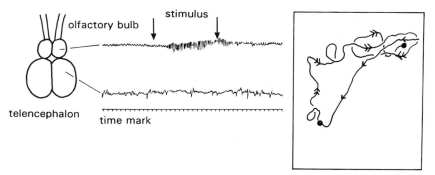

**Figure 10.27** Left: recordings from the olfactory bulb and telencephalon of a charr stimulated with a water sample in which young charr of the same race had been swimming. Redrawn from Døving *et al.* (1973). *Comp. Biochem. Physiol.*, **45A**, 21. Right: paths of intact catfish (*Ictalurus*) (single arrows) towards a point of chemical release in slowly flowing water, and of blinded catfish in still water (double arrows). Redrawn from Bardach *et al.* (1969) *Science*, **155**, 1276.

*soporator*) when gravid produced a chemical in its ovarian fluid that elicited courtship behaviour in the male. Male goldfish use their sense of smell to identify ovulated females. For example, in a maze they preferred water containing ovulated females or ovarian fluid. Several species of oviparous fishes including salmonids, bitterling (*Rhodeus*), loach (*Misgurnus*), ayu (*Plecoglossus*) and eels (*Anguilla*) also show behavioural responses of the male to ovulated females. It may be a more common phenomenon in freshwater where fish are more confined and chemicals less liable to dispersion. Whether hormones, excreted in urine or gradual fluids, are acting as pheromones, is now under examination and there is accumulating evidence that steroids are involved as pheromones in stimulating oocyte maturation and milt production in goldfish. Of particular interest is how species-specific these pheromones are.

### 10.9.4 Homing and chemoreception

The study of homing has been confined almost entirely to salmonids – salmon, trout and charr (p. 309). Many of these are anadromous fish returning to their home stream after a period of maturation elsewhere, often in the open sea. The role of chemoreceptors in helping fish to find their way back, and to identify the home stream precisely, has been investigated over many years using a number of standard techniques. These have included comparing the homing ability of control fish and those with cut olfactory nerves, cauterised olfactory organs or plugged nostrils. Generally fishes treated in this way show high rates of straying and some lose their homing ability completely.

The imprinting hypothesis of Hasler and Wisby proposes that streams have a characteristic odour that the juveniles such as salmon smolts learn by imprinting before they leave for the migratory phase of their life histories. (Of course, imprinting can also apply to other sensory cues.) A further hypothesis by Stabell proposes that the home-stream odour is actually a pheromone released by fish resident in the stream and specific to that fish population. This requires a two-way system of overlapping migrations – downstream of smolts and upstream of adults.

Complex experiments, involving the transplanting of fishes at different life-history stages to foreign rivers, show that there are many inconsistent findings to explain and that is difficult to define an all-embracing mechanism for homing by odour recognition. Generally, transplanted salmon parr and smolts return as adults to the river of release rather than their native river, but the incidence of straying is higher. One of the problems is that no critical period for imprinting has been unequivocally established, making it difficult to plan an appropriate scientific protocol for transplantation and olfactory ablation experiments.

A promising line of work involves the artificial imprinting of salmon with low concentrations of chemicals such as morpholine and phenethyl alcohol. Coho salmon were subjected to one of these chemicals in the hatchery at the smolt stage. At an appropriate time after release, the returning adults were given the choice of streams scented with morpholine or phenethyl alcohol; 95% of the morpholine-treated fish were recaptured in the stream containing morpholine and 92% of the phenethyl alcohol-treated fish were recaptured in the phenethyl alcohol-treated stream.

One of the most interesting new findings is that salmonids can distinguish water which has been occupied by their own siblings. Behaviour experiments were done in various types of flow-through tank where the fish were given a choice of water to swim in, preferring water in which their siblings had swam. Even the electrical response of neurons in the olfactory bulb of Arctic charr (*Salmo alpinus*) is characteristic for the odour of different charr populations (Figure 10.27).

It is becoming increasingly obvious that kin recognition by chemical cues is an attribute of other fish groups, for example the recognition of young by the parents in cichlids and in the threespine stickleback (*Gasterosteus aculeatus*). Recognition of their young may allow the parents to protect them preferentially and even to avoid eating them.

### 10.9.5 Alarm substance

It was first shown that injured minnows (*Phoxinus phoxinus*) produce an alarm substance ('Schreckstoff') from their skin that elicits a fright reaction in conspecifics. This response is found in the Ostariophysi, but not in all species of that group. The substance produced in epidermal 'club' cells is probably hypoxanthine-3(N)-oxide and is not species-specific. This raises the intriguing question of how such a group protective mechanism has evolved since the injured fish derives no benefit.

## Bibliography

Atema, J., Fay, R.R., Popper, A.N. and Tavolga, W.N. (eds) (1988) *Sensory Biology of Aquatic Animals*. Springer, New York.

Blaxter, J.H.S. (1987) Structure and development of the lateral line. *Biological Reviews*, **62**, 471–514.

Blaxter, J.H.S., Gray, J.A.B. and Best, A.C.G. (1983) Structure and development of the free neuromasts and lateral line system of the herring. *Journal of the Marine Biological Association of the United Kingdom*, **63**, 247–260.

Bleckmann, H. (1993) Role of the lateral line in fish behaviour, in *Behaviour of Teleost Fishes*. Pitcher, T. (ed). Chapman & Hall, London, (2nd edn) pp. 201–246.

Coombs, S., Janssen, J. and Webb, J.F. (1988) Diversity of lateral line systems: evolutionary and functional considerations, in *Sensory Biology of Aquatic Animals* (Atema, J., Fay, R.R., Popper, A.N. and Tavolga, W.N. (eds) Springer, New York, pp. 553–593.

Denton, E.J. and Gray, J.A.B. (1979) The analysis of sound by the sprat ear. *Nature*, **282**, 406–407.

Denton, E.J. and Gray, J.A.B. (1993) Stimulation of the acoustico-lateralis system of clupeid fish by external sources and their own movements. *Philosophical Transactions of the Royal Society of London, B*, **341**, 113–127.

Denton, E.J. and Locket, N.A. (1989) Possible wavelength discrimination by multibank retinae in deep-sea fishes. *Journal of the Marine Biological Association of the United Kingdom*, **69**, 409–435.

Denton, E.J. and Nicol, J.A.C. (1964) The chorioidal tapeta of some cartilaginous fishes (Chondrichthyes). *Journal of the Marine Biological Association of the United Kingdom*, **44**, 219–258.

Douglas, R.H. and Djamgoz, M.B.A. (eds) (1990) *The Visual System of Fish*. Chapman & Hall, London.

Engstrom, K. (1963) Cone types and cone arrangements in teleost retinae. *Acta Zoologica (Stockholm)*, **44**, 179–243.

Falcón, J. and Collin, J.P. (1989) Photoreceptors in the pineal of lower invertebrates: functional aspects. *Experientia*, **45**, 909–913.

Hanyu, I. (1978) Salient features in photosensory function of the teleostean pineal organ. *Comparative Biochemistry and Physiology*, **61A**, 49–54.

Hara, T.J. (ed) (1992) *Fish Chemoreception*. Chapman & Hall, London.

Hawkins, A.D. (1973) The sensitivity of fish to sounds. *Oceanography and Marine Biology Annual Review*, **11**, 291–340.

Kramer, B. (1990) Electrocommunication in Teleost fishes. *Zoophysiology*, **29** (Bradshaw, S.D., Burggren, W., Heller, H.C., Ishii, S., Langer, H., Neuweiler, G. and Randall, D.J. eds) Springer-Verlag, Berlin.

Lissmann, H.W. (1963) Electric location by fishes. *Scientific American* Reprint No. 152, 11 pp.

Lythgoe, J.N. (1980) Vision in fish: ecological adaptations, in *Environmental Physiology of Fishes* (Ali, M.A. ed), Plenum Press, New York, pp. 431–445.

Muntz, W.R.A. (1975) Behavioural studies of vision in a fish and possible relationships to the environment, in *Vision in Fishes* (Ali, M.A. ed), Plenum Press, New York, pp. 705–717.

Nicol, J.A.C. (1989) *The Eyes of Fishes*. Clarendon, Oxford.

Puzdrowski, R.C. (1989) Peripheral distribution and central projections of the lateral line nerves in goldfish, *Carassius auratus*. *Brain, Behaviour and Evolution*, **34**, 110–131.

Shand, J. (1993) Changes in the spectral absorption of cone visual pigments during the settlement of the goatfish *Upeneus tragula*: the loss of red sensitivity as a benthic existence begins. *Journal of Comparative Physiology A*, **173**, 115–122.

Stabell, O.B. (1984) Homing and olfaction in salmonids: a critical review with special reference to the Atlantic salmon. *Biological Reviews*, **59**, 333–388.

Walls, G.L. (1963) *The Vertebrate Eye*. Hafner, New York.

Zeiske, E., Theisen, B. and Breucker, H. (1992) Structure, development and evolutionary aspects of the peripheral olfactory system, in *Fish Chemoreception* (Hara, T.J. ed), Chapman & Hall, London, pp. 13–39.

Waltman, B. (1966) Electrical properties and fine structure of the ampullary canals of Lorenzini. *Acta Physiologica Scandinavica*, **66**, suppl. 264.

Webb, J.F. (1989) Gross morphology and evolution of the mechanoreceptive lateral-line system in teleost fishes. *Brain, Behaviour and Evolution*, **33**, 34–53.

# 11   The nervous system

The fish nervous system consists of the brain and spinal cord (the central nervous system or CNS), the motor and sensory nerves linking the central nervous system with receptors and effector organs, and the autonomic nervous system controlling visceral functions. Like all nervous systems, it is based on the electrical activity of its units, the neurons, which depend on their own intrinsic membrane properties, and the changes in these properties brought about by electrical and chemical synapses with other neurons. In the CNS, although almost all studies have concentrated on the connections and properties of neurons, it is important to bear in mind that glial cells of various types (which also sheath the peripheral nerves) make up perhaps half the volume, and that we are only just beginning to understand the way in which they interact with neurons. In special cases, for example the Mauthner neuron (p. 283), the visual system (p. 286) and the electroreceptive system (p. 288), we understand a good deal of the way in which the anatomical arrangements function, but especially in the brain, much more is known of anatomy and histology than of the part that particular anatomical arrangements play in the life of the fish.

## 11.1  Methods for investigating nervous system function

### 11.1.1  Neuroanatomical techniques

*Silver staining.*   The classical method of studying neurons and their connections in the nervous system was staining with different techniques using silver nitrate. In the brain, the Golgi method picked out a few neurons almost completely, whilst reduced silver methods stained almost all neurons and so made it difficult to follow individual neuron processes although tracts between brain nuclei were well shown. At the periphery, reduced silver techniques can give astonishingly complete pictures of the patterns of motor and sensory innervation.

*Degeneration methods.*   Nerve tracts in the brain have been studied after making localised lesions and allowing the animal to survive for a period, so that the axons divorced from their cell bodies degenerate, when they can be selectively stained by special silver methods.

*Enzyme and dye injection.* Silver methods have been supplemented by injection of tracers, such as the enzyme horseradish peroxidase (HRP) or fluorescent dyes such as Lucifer yellow. These may be injected into single neurons via microelectrodes, to spread throughout the neuron, revealing all its processes when suitably treated. Alternatively, larger amounts of HRP and fluorescent dyes like Fast Blue can be injected into regions of the brain when they are taken up by damaged neuron processes, to show connections between the injection site and the cell bodies.

*Immunocytochemistry.* Exquisite pictures of neurons and their connections have been given by immunocytochemical methods, where antisera to substances within neurons like the neurotransmitters γ-aminobutyric acid and neuropeptides such as gonadotrophin releasing hormone show specific classes of neurons in detail (see Figure 11.22, p. 286). To reveal the sites where antisera are bound, the preparations are treated with second antisera (coupled to peroxidase or fluorescent dyes) against the animal in which the original antiserum was raised.

### 11.1.2 Physiological methods

*Extracellular and intracellular recording.* Much information about the functions of different parts of the brain has been gained from extracellular recording of brain electrical activity following specific stimuli, e.g. tectal mapping following particular visual stimuli (see Figure 11.24).

Similarly, intracellular records (technically more difficult) have revealed the operation of specific pathways, such as the jaw-closing reflex (see Figure 11.13).

*Ablation of brain regions.* Removal of specific brain regions, e.g. the cerebellum, and examining the behavioural and physiological effects has given valuable – though limited – information about brain function.

## 11.2 The central nervous system

The CNS is perhaps more daunting than any other system in the fish, for not only does it contain astonishingly large numbers of specialised cells in great variety, but the complications of their connections and interactions are great. Only a few aspects of the CNS will be touched on here.

### 11.2.1 The spinal cord

The fish CNS arises by the inrolling and fusion of neural folds as in other vertebrates, giving rise in most fishes to an essentially circular spinal cord, with a central cruciform grey area surrounded by the fibre tracts of the white (Figure 11.1). White and grey refer to the appearance of sections of the spinal cord after staining the myelin sheaths of nerve fibres with iron haematoxylin. The central zones containing the cell bodies of the cord neurons appear white, contrasting with the greyish or blue-black fibre zones. A small central canal contains the enigmatic Reissner's fibre, passing down the length of the cord from its origin as a continuous secretion of cells in the roof of the brain (p. 203). In hagfishes and lampreys the cord becomes flattened and ribbon-like during ontogeny (Figure 11.1). The functional organisation of the spinal cord is similar to that of other vertebrates, shown schematically in Figure 11.2. The dorsal neurons and fibre tracts of the somatic sensory and visceral sensory divisions lie above the ventral viscero- and somatomotor division composed of the ventro-lateral neuron masses in the grey and the large ventral tracts formed by ascending and descending fibres in the ventral white. In most fishes, large paired Mauthner axons (see p. 283) descend the cord in the ventral white. The similarity of the arrangements of the spinal cord in fishes and other vertebrates is clearly shown by the dorsal horn distribution of neuropeptides (shown by immunocytochemistry) in elasmobranchs and mammals (Figure 11.3). Obviously, the fish spinal cord organisation has been conserved (in evolution) by terrestrial animals, as is also seen by the striking similarities between the embryonic spinal cords of different fishes and amphibians. Most kinds of spinal cord neurons are arranged in segmental patterns, linked to the segmental dorsal and ventral

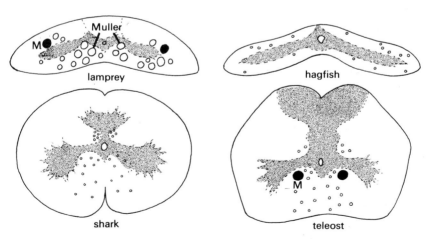

**Figure 11.1** Cross-sections of the spinal cord in different fishes. Note large Mauthner axons (M) in lamprey and teleost, and the equally large Muller axons in the lamprey. After Rovainen *et al.* (1973) *J. Comp. Neurol.*, **149**, 193 and Nieuwenhuys (1964) *Progr. Brain Res.*, **11**, 1.

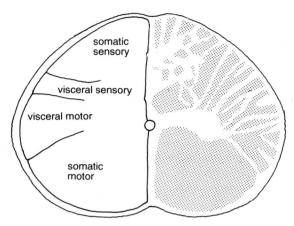

**Figure 11.2** Schematic diagram showing the functional organisation of the spinal cord in the dogfish (*S. acanthias*). Right: distribution of myelinated fibres (stippled). Left: the sensory and motor columns of the spinal cord. Partly after Bertin (1958) in: *Traité de Zoologie*, (P.P. Grassé ed), **13**, fasc.1, 854.

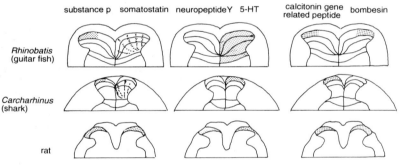

**Figure 11.3** Comparison of the distribution of 5-HT (serotonin) and various neuropeptides shown by immunocytochemistry in the dorsal horn of the spinal cord in two elasmobranchs and in the rat. After Cameron *et al.* (1990) *J. Comp. Neurol.*, **297**, 201–218.

root nerves, but these patterns are somewhat obscured during ontogeny, and it is only in the early stages of a few fishes like the zebra fish (*Brachydanio*), the Australian lungfish (*Neoceratodus*) and the lamprey that the segmental organisation of the spinal cord has been partially worked out (Figure 11.4). It turns out that they are very similar in all, and similar to those in amphibia.

Roberts has modelled the way in which these simple segmental neuron patterns interact to generate rhythmic swimming movements in the embryonic amphibian, and it seems very probable that they operate similarly in fish hatchlings. Indeed, although the adult cord is much more

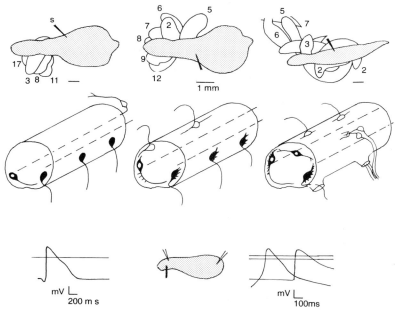

**Figure 11.4** Early movements and spinal cord structure in the embryonic Australian lungfish, *Neoceratodus*. Upper, the gradual development of movements in response to stimulation by a sharp hair (s). In the earliest embryos to show responses (left) the response is head movement to the side away from the stimulus. At a later stage (middle) an initial movement away is followed by a movement towards the stimulated side. Lastly, at a later stage still (right) touch evokes a rapid swim. In each, the outline of the initial position of the embryo is stippled, and the numbers next to the other outlines are milliseconds after stimulation. Middle, spinal cord neurons at stages corresponding to the movement patterns shown above them. Sensory Rohon–Beard cells, clear; motor neurons, dark; internuncial neurons, nuclei shown clear. Bottom, intracellular records of action potentials evoked by touch in the skin of the embryo; right, action potential evoked by stimulation near one recording site travels along the skin to the second electrode (as shown in middle). After Whiting *et al.* (1992), and Bone *et al.* (1989) *Proc. Roy. Soc. B.*, **237**, 127.

complex, with vastly greater numbers of neurons, their interactions are likely to be essentially similar to those in the embryos and hatchlings.

## 11.2.2 Spinal nerves

Except in lampreys (where the dorsal and ventral spinal nerves do not unite), segmental dorsal and ventral roots emerge separately from the spinal cord, and then join to form the mixed sensory and motor spinal nerves. The ventral roots contain mainly the axons of the spinal somatomotor neurons which pass to the myotomal and fin musculature, and in addition, visceromotor autonomic fibres, whilst the axons in the sensory dorsal roots are from the sensory neurons of the dorsal root ganglia, which send their sensory fibres to the periphery (Figure 11.2). In the adult dogfish (*Scyliorhinus*), the ventral roots of the abdominal region contain around 350 fibres, whilst in the much smaller adult zebra fish (*Brachydanio*), each ventral root contains around 100 somatomotor axons. The ventral roots pass out of the cord between the myotomes, and so a single ventral root contains the axons of motoneurons innervating two adjacent myotomes; sometimes in teleosts, axons may cross the two nearest myotomes to innervate muscle fibres in adjacent myotomes, and in the spinal cord itself, motoneuron axons may pass up and down the cord rather than emerging from the nearest ventral root to their cell bodies.

### 11.2.3 Spinal swimming

In sharks, destruction of the brain does not immediately cause paralysis, as it does in lampreys and teleosts. Instead, such 'spinal' sharks continue a stereotyped slow swimming pattern for many hours, provided that they are supported off the bottom and artificially respired. In the spinal dogfish (*Scyliorhinus*), tail beat frequency is around 0.6 Hz, a little slower than in the minimum cruising speed of intact animals, and this slow swimming behaviour can be modified in amplitude and frequency by appropriate stimuli. Spinal dogfish have therefore been much used in investigating such questions as the role of sensory input in modifying the swimming pattern, and the existence of a central pattern generator driving rhythmic swimming. Lampreys have also been used, for although spinal lampreys do not swim, if small amounts of L-DOPA or D-glutamate are added to the experimental baths containing isolated lamprey spinal cords (Figure

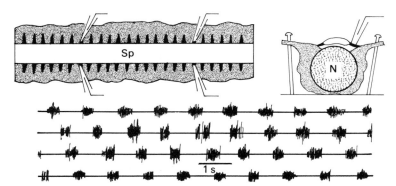

**Figure 11.5** Rhythmic motor activity in isolated spinal cord. Above: experimental arrangement of lamprey spinal cord (Sp) isolated to record from ventral roots (black) with suction electrodes. N: notochord. Below: simultaneous records from four suction electrodes on different right (upper pair) and left (lower pair) ventral roots showing rhythmic motor output in preparation paralysed with curare and bathed in a solution containing 0.6 mM D-glutamate. After Cohen and Wallén (1980).

11.5), rhythmic motor activity similar to that of the intact swimming lamprey is generated. The patterned motor activity of such 'fictive' swimming (as it is called) can be recorded by suction electrodes on the ventral roots, and shows the same phase lags between segments as in normal swimming. It is abolished by the addition of known glutamate inhibitors, suggesting that in the intact lamprey, swimming is evoked by descending glutaminergic pathways synapsing with the local spinal segmental neuron patterns. Unfortunately, this kind of pharmacological 'dissection' of spinal cord pathways has yet to be undertaken in the dogfish, and even the anatomical basis for the pattern-generating circuitry is still unknown. For example, although both intra- and extracellular records of the firing patterns of two types of motoneurons and several types of interneurons have been obtained from spinal dogfish (Figure 11.6), little is known of the functional interactions between them. More progress is likely to be made on embryonic preparations, where the anatomy of the segmental circuitry is simpler than in the adult. An interesting approach here is to attempt to correlate changes in the circuitry as it develops, with changes in locomotor behaviour. For example, in the developing embryos of *Neoceratodus* (Figure 11.4), the first movements are spontaneously generated by the muscle cells themselves (myogenic) before they are reached by the axons of motoneurons. At a slightly later stage, the embryo responds to stimulation by bending away from the stimulus, and later still although it first bends away, this is followed by movements towards the stimulus and just before hatching, these continue as a series of swimming

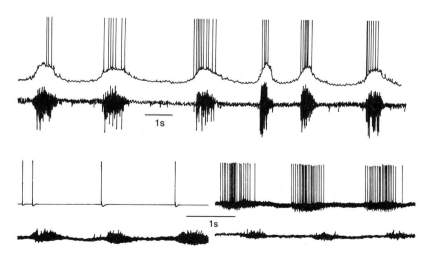

**Figure 11.6** Intracellular records of firing patterns from motoneurons and two types of interneuron in the spinal cord of the dogfish (*Scyliorhinus*). After Mos *et al.* (1991) *Phil. Trans. Roy. Soc. Lond.*, *B*, **330**, 329 and 341.

movements. The first flexures towards the stimulated side of the embryo correlate with the sensory axons of the transient Rohon–Beard column of dorsal sensory neurons (Figure 11.4) reaching the skin, and with the appearance of dorsolateral interneurons. The sustained swimming response correlates with the further development of the ventral inter-neurons and the appearance of dorsal interneurons with bilateral dendritic fields. However, much less is known of the circuitry in *Neoceratodus* (or any other fish) than of the embryonic amphibian spinal cord, where the series of elegant anatomical and behavioural studies made by Roberts and his colleagues on *Xenopus* tadpoles show what should be possible in fish embryos.

Although embryonic and larval fish have very large numbers of neurons in the transient Rohon–Beard system, these all disappear at metamorphosis, and are replaced by the sensory neurons of the dorsal root ganglia. A striking difference between the spinal cord of adult fishes and those of terrestrial vertebrates is the absence of large numbers of proprioceptive fibres from muscle spindles. These seem not to be found in fishes (p. 219). Compared with the better-studied spinal cords of higher vertebrates, those of fishes seem to be characterised by very many more dendrodendritic connections. For example, the motoneurons and interneurons of lampreys or dogfish receive relatively few synaptic inputs on the cell body (soma) itself, but receive synapses on their enormous dendritic fields (Figure 11.7). The functional significance of this striking difference is unclear.

### 11.2.4 Cranial nerves

The cranial nerves (I–XII) were first named by anatomists studying humans, so their names are not necessarily particularly appropriate for fishes. The facial, for example, innervates the snout, but it also innervates the spiracle, which is hardly on the face of the fish! Figure 11.8 shows the distribution of the cranial nerves in the embryonic dogfish; Table 11.1 their functional components, and how they accord with the segmental dorsal sensory–ventral motor arrangement we have seen in the spinal cord. Unlike the spinal nerves, the cranial dorsal and ventral nerves remain separate (as they do in amphioxus and lampreys, hence this may be the primitive condition). Note that the optic nerve (II) is not part of the segmental arrangement of dorsal and ventral root nerves. This will not surprise the reader aware that the eye and optic nerve arise as outpouchings of the brain, hence the IInd nerve is not comparable to the other cranial nerves. The olfactory nerve (I) and nerve 0 (the terminalis nerve, which is a small bundle of fibres entering the brain ventrally just behind the olfactory nerves) may or may not be part of the segmental cranial series of Table 11.1, but however this may be, the olfactory nerve is unique in having its cells of origin, at the *periphery* in the olfactory mucosa.

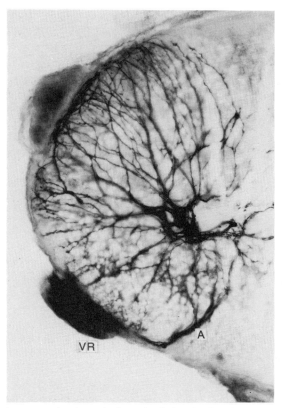

**Figure 11.7** Motoneurons in the spinal cord of the dogfish (*Scyliorhinus*) visualised by antidromic uptake of cobalt lysine injected into a ventral root. The sectioned ventral root (VR) is on left, the axons (A) of the motoneurons pass towards it. Photograph kindly provided by Prof. B.L. Roberts.

## 11.2.5 The brain

To some extent, the brain can be regarded as an enlarged anterior portion of the spinal cord with hypertrophied centres associated with the development of input from the special sense organs. In general plan, the brain is similar in all fishes, though there are very considerable differences between the fish groups in the relative development of the different regions (Figure 11.9). As we should expect, even in fishes of the same kind, the size of the different regions depends on the importance of particular senses, and the requirements of different lifestyles. For example, the cerebellum (concerned with postural control) is relatively larger in more active fishes like the isurid shark or salmon than in the relatively sluggish dogfish or bichir (*Polypterus*). Similarly, where vision is important, the optic lobes are enlarged.

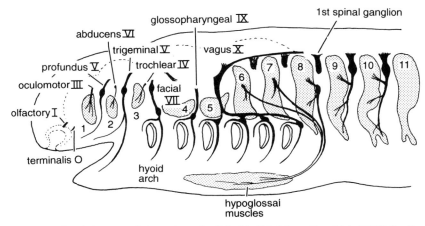

**Figure 11.8** Cranial nerves of the embryo dogfish. Modified from Goodrich (1930) *Studies on the Structure and Development of Vertebrates.*

**Table 11.1** The segmentation and functional components of the cranial nerves in fishes

| Segment | Dorsal root | Ventral root | Components of dorsal root nerves |
|---|---|---|---|
| Premandibular | V (trigeminal) Deep Ophthalmic | III (Oculomotor) | general cutaneous |
| Mandibular | V (trigeminal) Superf. ophthalm., Max. & mandib. | IV (Trochlear) | general cutaneous + visceromotor |
| Hyoid | VII (Facial) VIII (auditory) | VI (Abducens) | lateralis (somatic sensory) + visceral motor + visceral sensory lateralis (somatic sensory) |
| 1st branchial | IX (glosso-pharyngeal) | Absent | somatic sensory + visceral sensory + lateralis (somatic sensory) |
| 2nd branchial | X (vagus) XII (hypoglossal) + XI (accessory) | | lateralis (somatic sensory) + somatic sensory + visceral motor |

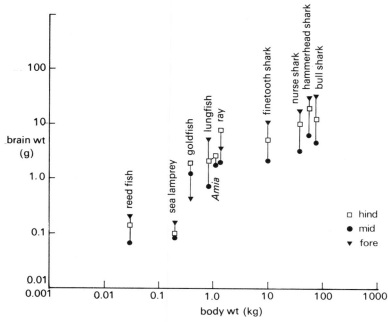

**Figure 11.9** Relative sizes of different brain regions in fishes. Squares, hindbrain; triangles, forebrain; circles, midbrain. Note relatively small brain of lamprey. After Ebbeson and Northcutt (1976) in *Evolution of Brain and Behaviour in Vertebrates* (eds Masterton *et al.*), Lawrence Erlbaum Assoc., NJ. p. 115.

The three main brain regions may be sub-divided as follows:

| | |
|---|---|
| • Forebrain (prosencephalon) | Olfactory lobes or cerebral hemispheres (telencephalon) + between-brain (diencephalon) |
| • Midbrain (mesencephalon) | Optic lobes |
| • Hindbrain (rhombencephalon) | Cerebellum (metencephalon) Medulla oblongata (myelencephalon) |

The medulla oblongata tapers into the spinal cord, where the narrow central canal is the posterior continuation of the brain ventricular system.

## 11.2.6 The development of the fish brain

The neural tube formed by inrolling and fusion becomes progressively subdivided anteriorly by constrictions as development proceeds, so that there are swellings along the anterior end of the tube, the neuromeres, separated by these constrictions. The three most anterior are larger than the remainder and correspond to the definitive telencephalon and diencephalon and to the mesencephalon. Behind these, seven further neuromeres subdivide the major part of the rhombencephalon. Much attention has been paid recently to these neuromeres, because their boundaries coincide with regions where specific regulatory homeobox genes are expressed early in development, so we are beginning to see how genes may control early patterning in the brain. For example, in the zebra fish (*Brachydanio rerio*), the zinc-finger gene *Krox-20* which functions as a transcriptional regulator (it is called zinc-finger since the regulatory protein has a small projection or finger containing zinc atoms) is expressed very early in development in two patches which later develop as the hindbrain rhombomeres 3 and 5 (Figure 11.10). It seems clear that this gene is part of a regulatory network establishing the pattern of sub-division of the early hindbrain. The homologous gene has the same pattern of expression in the mouse (where it was first discovered) as in the zebra fish. Thus it is also evident (as we might perhaps have expected) that regulation is highly conserved in all vertebrates. Using the enzyme acetylcholinesterase as a marker, clusters of primary neurons have been recognised near the centre of each neuromere (Figure 11.10). Their axons grow between the clusters, making the scaffolding seen in Figure 11.10, which can have the same kind of position as the borders of the expression of the different regulatory genes.

There has been remarkable and exciting progress recently in understanding the molecular basis of patterning in the developing brain of the zebra fish, and in view of the highly-conserved nature of brain regulation,

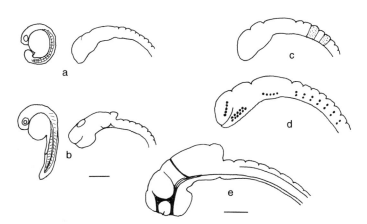

**Figure 11.10** Brain patterning in the development of the zebrafish, *Brachydanio rerio*. (a) At 18 h, about 10 neuromeres are present. (b) At 24 h the embryo is more advanced and the brain begins to take on the adult form. (c) The expression of the *Krox-20* zinc-finger gene in the 18 h embryo (see text). (d) Primary cholinergic neurons at the 18 h stage. (e) The early axon scaffolding at the 24 h stage. All after Kimmel (1993) *Ann. Rev. Neurosci.*, **16**, 707.

there seems little reason to doubt that the results will be applicable to all fishes.

## 11.2.7 Brain size

The absolute size of the brain differs in any single fish during its ontogeny (Figure 11.11) but although the relative size of different brain regions differs (Figure 11.9), the ratio between *total* adult brain weight and body weight is similar for all groups of fishes (except elasmobranchs), and much the same as those for amphibia and reptiles. It is only compared with birds and mammals which have much higher brain:body weight ratios, that fish can be considered animals of little brain. Elasmobranchs are the exception here, for at least some approach the bird and mammal values, and have brain:body weight ratios as much as 400% greater than other fishes. We have no idea why they should show this striking and intriguing difference from other fishes. A number of factors obviously influence brain size, such as neuron number, neuron complexity (i.e. size of dendritic fields), and changes in the complexity of the circuitry linking the neurons. Whilst it is true that elasmobranchs have a very elaborate electroreceptor system, and that both sharks and rays have proprioceptors controlling locomotor movements which are not found in other fishes (p. 219), such specialisations can hardly entirely account for the large size of the elasmobranch brain. It

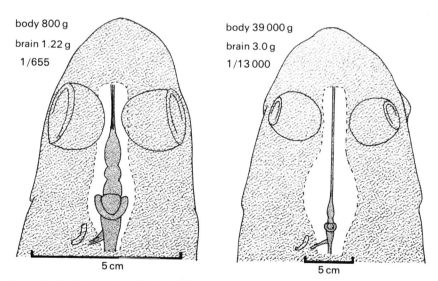

body 800 g
brain 1.22 g
1/655

body 39 000 g
brain 3.0 g
1/13 000

5 cm

5 cm

**Figure 11.11** Changes in relative brain size during growth. Left: dorsal view of head of embryo from oviduct of the coelacanth *Latimeria*, right: similar view of adult. Note vast cranial cavity of adult compared with size of brain, and very large difference in brain/body wt. ratios between embryo and adult. After Anthony and Robineau (1976).

is notable that elasmobranch and mammalian brains share other features as well as size, for example an elaborate mesencephalic system of dopaminergic neurons apparently lacking in actinopterygians, but again, the significance of this is unclear.

### 11.2.8 Brain temperature

In most fishes, the brain (like the rest of the body), is at the same temperature as the ambient water. In swordfish (*Xiphias*) and marlins (*Makaira* and *Tetrapturus*), however, the brain is at least 3–4°C warmer. Carey (1982) found by telemetry experiments from a sensor in the cranial cavity of swordfish that the brain was as much as 10–14°C above ambient. The brain is kept warm by heat generated from special brown liver-like tissue associated with the eye muscles; special retia prevent heat loss.

### 11.2.9 Brain regions and their connections in elasmobranchs

The elasmobranch brain will be described as an example of brain design in fishes, although this does not mean that it can be taken (as it used to be) as a model of the 'primitive' fish brain.

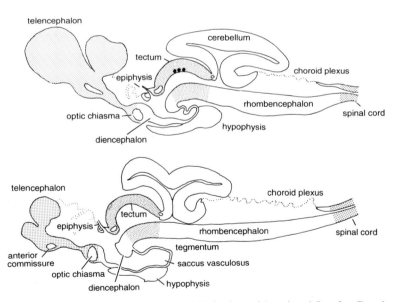

**Figure 11.12** Sagittal sections of the brains of *Scyliorhinus* (above) and *Squalus*. Regular light stipple: telencephalon; random stipple: mesencephalon; darker stipple: spinal cord. The dots in the dorsal mesencephalon next to the ventricle represent the position of mesencephalic Vth neurons seen in Figure 11.13. After Smeets *et al.* (1991).

Figure 11.12 shows the brains of the dogfishes *Squalus* and *Scyliorhinus* in sagittal section. At the front of the brain, all elasmobranchs have large olfactory organs and olfactory bulbs formed by forward bulging out of the forebrain, connected to the forebrain by stalks of various lengths. The size of these reflects the importance of olfaction in feeding behaviour in all sharks, but in some species, like the dogfishes *Mustelus* and *Scyliorhinus* the olfactory bulbs are particularly large, as they are in the hammerhead (*Sphyrna*) compared to the mako (*Isurus*) and sandbar (*Carcharhinus milberti*) sharks. Olfactory fibres enter each bulb, and are gathered in glomeruli connected via interstitial cells to the olfactory tract running to the forebrain. The large telencephalon is not merely an olfactory centre, however, for major parts of it are concerned with vision and probably other modalities also; olfactory activity is restricted to only about 10% of the telencephalon. Indeed, in contrast to earlier ideas that the elasmobranch telencephalon was merely an olfactory processing centre, it is now recognised to have an integrative role, associating different sensory inputs with output motor programmes, like the mammalian limbic system. For example, in the ray, electroreceptive input entering the hindbrain is relayed via the mesencephalon to the diencephalon, and thence to the upper pallial layer of the telencephalon. The telencephalon receives input from the diencephalon and also sends output to it, as well as sending output to the mesencephalic tectum and the motor systems of the brain stem.

The diencephalon is divided into the dorsal epithalamus comprising the epiphysis or pineal organ, and the habenular ganglia, the thalamus itself and the hypothalamus ventrally. There is good evidence that the pineal in dogfish is a sensitive photoreceptor (p. 249), and some evidence that it may be concerned with the paling that dogfish undergo in darkness. Amongst other inputs, the thalamus receives both direct retinal input and secondary visual input from the optic tectum, descending telencephalic fibres, and ascending fibres from the cerebellum and spinal cord. Efferent fibres from the thalamus pass to these regions. The hypothalamus receives fibres from the telencephalon, and taste (gustatory) fibres from the medulla, and sends efferent fibres to the reticular centres of the brainstem. A curious feature is the saccus vasculosus at the hind end of the hypothalamus, which consists of a pigmented folded sac as yet of uncertain function, though perhaps concerned with secretion and resorption of ventricular fluid. In the anterior part of the hypothalamus, the large neurosecretory cells of the supraoptic nucleus send granular neurosecretory material down their axons to the underlying neural lobe of the pituitary (p. 202). Diencephalic functions are not well understood as yet, but it seems that in elasmobranchs as in mammals, the hypothalamus (in association with the telencephalic limbic centres) regulates feeding, escape, attack and sexual behaviours, as well as the homeostatic control of bodily functions such as colour change.

The mesencephalon is large in elasmobranchs (as in all fishes) roofed by the layered tectum. Almost all optic fibres decussate (cross to the opposite side) in the midbrain floor before rising to terminate in the tectum. Although evidence from elasmobranchs is lacking, by analogy with other fishes, maps of the visual field are probably represented in the tectum. As in other vertebrates, there are efferent fibres to the retina, arising from cells in the tectum, and probably also from cells of nerve 0 (the nervus terminalis) in the olfactory bulbs. Experiments in teleosts have shown that retinal efferents are of two types, one probably neurosecretory, but it is still unclear just what they do. Nerve 0 efferents may play a role in the potentiating effect of food extracts on the response to subsequent visual stimuli. The tectum also receives input from other sensory systems, so although often called the optic tectum, it is better simply to call it the tectum. Like the telencephalon, it is evidently an integrative and associative centre. The tegmental floor of the mesencephalon (merging with the rhombencephalon) contains the motor systems of the brainstem; the reticular formation is conspicuous here. For once, there is good experimental evidence for the role of one nucleus in the life of the fish; this is the mesencephalic nucleus in the midline, next to the ventricle. The neurons of this nucleus have large cell bodies (hence are amenable to intracellular recording), and their axons pass to high-threshold touch receptors on the teeth and around the mouth. When the teeth are tapped, the neurons fire a short burst of action potentials (Figure 11.13) to the

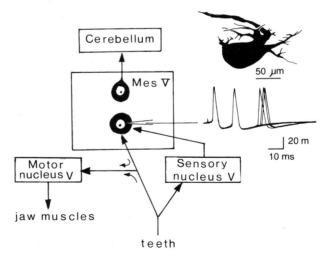

**Figure 11.13** Some features of the mesencephalic Vth neurons involved in the jaw-closing reflex. On the right above, a neuron visualised by the Golgi method, and below it, record of a series of action potentials evoked by tapping the teeth. After Roberts and Witkovsky (1975) *Proc. Roy. Soc Lond. B*, **190**, 473.

motoneurons of the Vth nerve innervating the jaw-closing muscle, and the jaws reflexly snap shut. Interestingly, there is also synaptic input to the cell bodies of these neurons from other brain regions, so snapping at their prey is not simply a reflex driven by touching the teeth.

The cerebellum in elasmobranchs has the same well-ordered and rather complex neural circuitry as in all vertebrates (see Figure 11.14). Mossy fibre input drives the granule cells, which then excite the Purkinje cells and stellate cells via the parallel fibre pathway. Apart from the Purkinje cells, these names refer to the appearance of the cells and fibres in histological sections. The Purkinje cells carry the German version of the name of their discoverer, the distinguished Czech physiologist and histologist Purkyně, a friend of Goethe's, who also gave his name to the fast-conducting fibres of the mammalian heart, and founded the world's first physiology department in 1839. In higher vertebrates, climbing fibre input from cells in the olive nucleus of the rhombencephalon also excites the Purkinje cells, but in fishes, climbing fibres twining around the Purkinje cell dendrites have not been seen. Probably there are similar fibres but they only reach the basal dendrites. Lastly, the stellate and Golgi cells are interneurons presumably modulating Purkinje cell activity. The sole output of the cerebellum is from the Purkinje cells. This is inhibitory to the cerebellar nuclei that excite cell

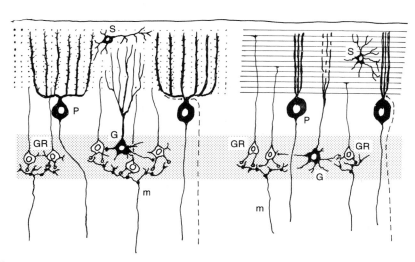

**Figure 11.14** The major cell types of the cerebellum. Left and right of the diagram show sections in planes at right angles to each other; note the remarkable regularity of the parallel fibres and the planar arrangement of the Purkinje dendritic trees. Inhibitory: black. Climbing fibres (dotted) have yet to be definitely demonstrated in any fish, otherwise the pattern is like that of other vertebrates. Purkinje cells (P) provide the only output from the system. Mossy fibre input (m) synapses with the abundant granule cells (GR) whence the parallel fibres of the outer layer arise, modulated by the Golgi (G) and stellate (S) cells. Partly after Eccles *et al.* (1967) *The Cerebellum as a Neuronal Machine*, Springer-Verlag, Berlin.

groups in the brainstem. Here, then, is a complex and specialised circuit whose function has proven rather puzzling, though in higher vertebrates it is obviously somehow concerned with the control of movements, since removal of the cerebellum produces profound motor disturbances.

In *Scyliorhinus*, Roberts and his colleagues have provided very interesting clues to the role of movement control by the cerebellum. They examined fin movements and spinal swimming after various kinds of brain region ablations. About one-third of the Purkinje cells in decerebrate dogfish (after forebrain removal, see Figure 11.15), were found to discharge rhythmically in phase with body movements, and since this activity continued (in curarised fish) when the muscles were paralysed, this activity was presumably driven by spinal cord generator circuits. Here, the role of the cerebellum seems to be to monitor, via this input, the state of spinal cord circuits during movement. Experiments on reflex fin movements gave a clear idea of the role of the Purkinje cell inhibitory output. Dogfish raise their pectoral fins reflexly if these are touched on their upper surface, a rapid upward movement being followed by a longer-lasting tonic phase. Removal of the cerebellum does not abolish this reflex movement, but changes it by increasing the threshold for response and reducing the tonic phase (Figure 11.15). Only a few Purkinje cells in the hind region of the cerebellum are involved, and these increased their rate of firing when the fin moved (after a delay showing that this discharge did not initiate the reflex, but was appropriate to modify it). These Purkinje cells monitor

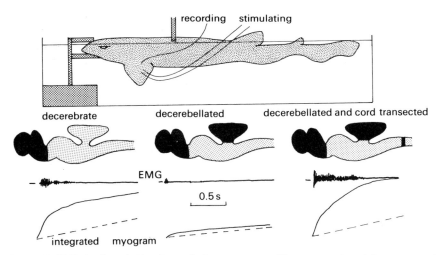

**Figure 11.15** Role of cerebellum in regulating movements. Upper: experimental arrangement; middle: parts of brain removed (black); lower: muscle responses. EMG: electromyogram; below this, the integrated myogram. Note dotted line shows response of integrator in absence of EMG. After Paul and Roberts (1979). *J. Comp. Physiol.*, **134**, 69.

spinal motor circuits, and inhibit a set of cerebellar nuclear neurons which themselves are normally inhibitory to brainstem descending pathways to the same spinal cord circuits. Their inhibitory action is removed, allowing the fin reflex to be expressed fully, whilst at the same time (via other sets of cerebellar nuclear neurons) descending inhibition is increased to other motor systems to prevent their unwanted expression. So the cerebellum is thought to act to control how much of a particular motor pattern originating elsewhere in the nervous system is to be expressed, different regions of the cerebellum dealing with different motor systems. The Purkinje cells of the middle part of the auricles of the cerebellum, for example, give complex responses when the corresponding granule cells are excited by stimulation of the VIIIth nerve, and these are probably concerned with regulating movements controlled by vestibular responses. We should expect on this view that the size and elaboration of the cerebellar cortex would be related to the complexity of movement patterns in different fishes, and this is the case. For example, in Figure 11.16, the active swimmer *Mustelus* has a more elaborate cerebellum than the sluggish *Scyliorhinus*, the very active short fin mako shark (*Isurus*) the most elaborately folded cerebellum.

Lastly, in the medulla oblongata or brainstem, there are the nuclei of origin and the endings of sensory input of all the ten pairs of cranial nerves, except the olfactory (I) and the little nervus terminalis (0). The walls are thickened by the primary nuclei of the acoustico-lateralis sensory systems, and the basal central core of the rhombencephalon contains the reticular formation. The reticular formation is particularly striking in elasmobranchs, since it contains very large neurons with extensive dendritic arborisations, Many of the cells of the reticular formation are driven synaptically by ascending fibres from the spinal cord, others by cutaneous afferents on the head and pharynx, whilst the reticular axons descending the spinal cord influence the intrinsic segmental locomotor circuits of the cord. Cranial

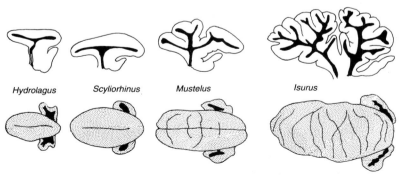

**Figure 11.16** Cerebellar development in different elasmobranchs. After Smeets *et al.* (1991).

outputs from other motor systems of the rhombencephalon control eye movements, and jaw and respiratory movements. Probably, as in mammals, the reticular formation receives and processes information from other parts of the brain, taking part in the control of locomotion, eye movements, and sensorimotor activity, but as yet, there is little experimental evidence for its functions in elasmobranchs.

### 11.2.10  Brains of other fishes

The aim of the remainder of this chapter is to point out some obvious differences from elasmobranch brain morphology shown by other fishes, and then to focus on some of the work (mainly on teleosts) where pathways are known that are involved in specific behaviours.

*Telencephalon.*   The most striking difference between the elasmobranch brain and that of bony fishes and sturgeons is in the telencephalon. In these fishes, the roof of the telencephalon is simply a thin sheet of ependymal cells, and lateral ventricles are absent. In section therefore (Figure 11.17), the teleost telencephalon is quite unlike the thick-roofed telencephalon of elasmobranchs, and indeed, of other classes of vertebrates. This condition of 'eversion' has made comparison with other vertebrates rather difficult (as can be seen in older texts). More recent horseradish peroxidase tracing studies have shown that homologies can be recognised between the tracts

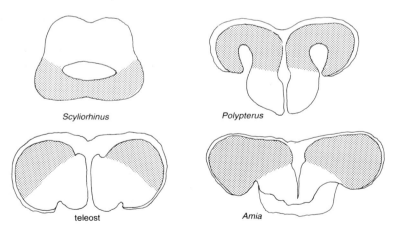

*Scyliorhinus*                    *Polypterus*

teleost                    *Amia*

**Figure 11.17** Transverse sections of the telencephalon in the elasmobranch. *Scyliorhinus* (left) compared with various bony fishes showing the thin roof in the latter. Part of the pallium stippled. After Smeets *et al.* (1991); Kappers *et al.* (1936) *The Comparative Anatomy of the Nervous System of Vertebrates, Including Man* (2 vols.), Macmillan; and Niewenhuys (1966) in *Evolution of the Forebrain: Phylogenesis and Ontogenesis of the Forebrain* (eds Hassler and Stephan), Thieme.

and cell masses of the teleost telencephalon and those of more 'conventional' brains; it now seems clear that despite the differences in adult
gross morphology and embryological development that the 'everted'
telencephalon is functionally organised in a similar way to those of more
conventional brains. As in elasmobranchs, the teleost forebrain is involved
in many non-olfactory behaviours. Lesion and ablation experiments have
shown that the teleost forebrain is concerned with the ability of the fish to
respond to environmental changes and that it plays a role in aggressive and
reproductive behaviour. For example, Siamese fighting fish (*Betta
splendens*) normally build a floating nest of bubbles. After forebrain
removal, they still produce bubbles, but these are not formed into a
cohesive nest.

*Mauthner neurons.*   Another notable difference between the elasmobranch brain and that of other fish groups, is that the important pair of
Mauthner neurons are lacking in elasmobranchs (except in young stages).
These are very large conspicuous multipolar neurons lying in the medulla
at the level of the VIIIth nerve. A large lateral dendrite passes towards the
entry of the VIIIth, covered with synapses from the VIIIth fibres. Input
also comes from the opposite Mauthner cell and neurons associated with a
special axon cap where the axon issues from the cell (inhibitory) and from
the main nucleus of the Vth nerve and from cerebello-tegmental and tecto-
bulbar pathways. The Mauthner axons decussate in the Mauthner chiasma
and descend the cord as the largest fibres within it (Figure 11.18). There
they send collaterals to spinal motoneurons, and drive the rapid C-start
escape reaction (p. 56). The decussation at the Maunthner chiasma means
that when the Mauthner cell on one side is stimulated to fire, the C-start
moves the head (the most vulnerable part) away from the stimulus. The
operation of the system is seen in Figure 11.19. In some teleosts, such as
eels, pipefish, seahorses and angler fishes, a C-start escape response is
inappropriate, and the Mauthner system has disappeared.
   Mauthner neurons are a good example of a 'hard-wired' central system,
which drives a fundamental escape reaction appearing early in ontogeny.
Hard-wired escape systems are much commoner in invertebrates, perhaps
because stereotyped 'invariant' responses are less suited to the more
complex vertebrate CNS. However this may be, curiously enough, escape
reactions can be evoked also without Mauthner activity (just as in squid,
escape jetting can be driven by smaller fibres as well as by the giant axon
system). The interactions of these two ways of evoking escape movements
are not yet well understood.

*Swimming and feeding centres.*   In several teleost fishes, electrical
stimulation of a localised region in the midbrain tegmentum elicits
coordinated swimming movements (Figure 11.20), suggesting that there

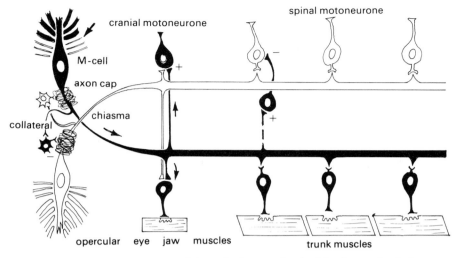

**Figure 11.18** The Mauthner system. Schematic diagram showing Mauthner cells, their axons crossing at the Mauthner chiasma before descending the cord and connecting with contralateral motoneurones.

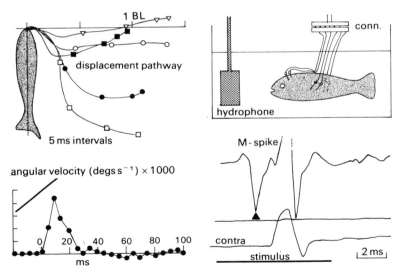

**Figure 11.19** Responses of the Mauthner system. Upper left: displacement pathway of head of goldfish. Points indicate position of head at 5 ms intervals during 5 Mauthner responses. Lower left: angular velocity during Mauthner response of goldfish. Right: experimental arrangement, and record obtained from extracellular electrodes in hindbrain and in white zones of the myotomes of either side. Lower right: upper line, electrode near M-cell, arrow indicates M-spike (followed by a large movement artefact); middle line, ipsilateral muscle; lower line, contralateral muscle (activated). Solid bar below: duration of sound stimulus. Time-marker: 2 ms. After Zottoli (1978) p. 13, and Eaton and Bombardieri (1978) p. 221, in *Neurobiology of the Mauthner Cell* (Faber and Korn eds), Raven Press, NY.

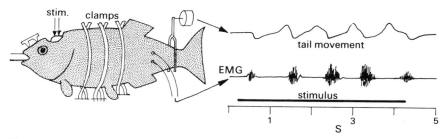

**Figure 11.20** Experimental arrangement and records obtained by stimulation of the swimming centre in the teleost brain. Upper right: tail movement, lower right: electromyogram from caudal muscles. Solid bar: stimulus. After Kashin *et al.* (1974) *Brain Res.*, **82**, 41.

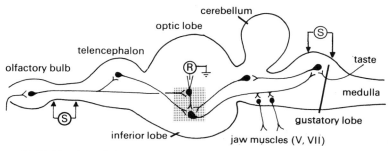

**Figure 11.21** The hypothalamic feeding centre and its connections. S: stimulation. R: recording. After Demski (1981) *Brain Mechanisms of Behaviour in Lower Vertebrates* (Laming ed.) Cambridge Univ. Press, p. 225.

may be a mesencephalic locomotor region (MLR) as in some mammals. Here again there seems to be a 'hard-wired' arrangement, like that of the Mauthner system and the mesencephalic V nucleus (p. 278) involved in the jaw-closing reflex. Similarly, electrical stimulation of areas near the lateral recess of the third ventricle in the inferior lobe of the hypothalamus evoke low-threshold and complete feeding responses in sunfish (*Lepomis*), goldfish and *Tilapia*. Figure 11.21 shows a schematic reconstruction of the pathways to the hypothalamic feeding area (HFA) in the goldfish. Since stimulation of the inferior lobe of the hypothalamus in sharks also leads to feeding responses, it seems that the control system for feeding may be the same in both groups. Similar centres may well exist for similar stereotyped activities like some aspects of reproductive behaviour such as nest-building.

*Modulator neurons and behaviour.* The properties of the ion channels in neuron membranes which are responsible for their electrical activity, are modulated by neurotransmitters and by hormones, and this modulation

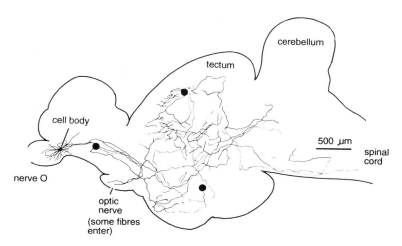

**Figure 11.22** The extensive connections of a single cell in the nucleus of the nervus terminalis (nerve 0) of the teleost *Colisia* visualised by intracellular injection of the tracer neurobiotin. Bold dots (●) indicate commissural connections. After Oka (1992) *Neuroscience Letters,*
**42**, 11.

seems likely to be the basis for long-lasting behavioural changes, such as arousal or motivational state. Recent electrophysiological and immuno-cytochemical work on the small teleost *Colisia lalia* has identified modulator neurons with endogenous beating activity, and an astonishingly wide efferent projection (Figure 11.22). These neurons form a small ventral group at the entry of nerve 0 to the telencephalon, and project to almost all regions of the brain except the cerebellum, as well as to the spinal cord. They contain gonadotrophin releasing hormone, which is known to act as a neuromodulator. In higher vertebrates, similar modulator neurons with endogenous rhythmic discharge patterns, and wide efferent projections, are known from other brain regions and contain serotonin (5-hydroxytryptamine), histamine, noradrenaline and dopamine.

Plainly, if these modulator neurons are themselves affected by environmental or hormonal factors, changes in their rhythmic activity will have very general effects on central neuronal activity. Here we are at an exciting stage in understanding the longer-term aspects of fish behaviour, and although the *Colisia* gonadotrophin releasing hormone modulator system is at present the only modulator system at all well characterised in the fish brain, many more such systems will likely be found in other fishes, and the way in which they modulate behaviour be unravelled.

*Visual processing.* In most teleosts, the eyes are large and vision is important, hence the tectum (where most optic fibres terminate) is usually large, and multilayered, with retinotopic maps of the visual field, and a complex set of neuronal types somewhat resembling those of the cerebellum (Figure 11.23). The most numerous cells have long dendritic rami arranged at right angles to the tectal surface, and from the responses of visual cells in the perch (*Perca*) to light stimuli, using the apparatus shown in Figure 11.24, it has been suggested that these cells are related to functions dependent on the conservation of fine detail and position. Other cells with dendrites spread out in a plane essentially parallel to the tectal surface seem to be involved in generalised functions such as the response to moving objects of any kind. The receptive fields of the cells differ. Those with well-defined boundaries are probably involved in dealing with location and contour of objects, whilst those with irregular and poorly-defined boundaries respond to brightness, contrast and velocity. Yet, how can teleosts generalise shapes of different sizes and positions so that they can be taught to learn to respond, for example, to triangles rather than to squares? This remains a challenging enigma.

The importance of the input to the tectum from the eyes overshadows the fact that even after the eyes (and the pineal photoreceptors) have been removed, minnows (*Phoxinus*) can be trained to respond to changes of illumination; there are neurons sensitive to light elsewhere in the CNS. In lampreys, light-sensitive neurons occur in the caudal spinal cord.

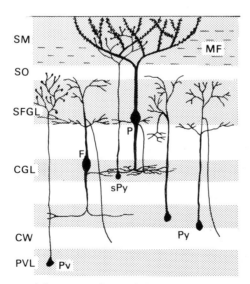

**Figure 11.23** Cell types of the tectum. Some of the cells seen in the tectum impregnated by the Golgi technique. P, pyramidal; F, fusiform; Py, pyriform; SO, marginal layer; SFGL, superficial fibrous and grey layer; MF, marginal fibres; CGL, central grey layer; CW, central white layer; PVL, periventricular layer. Note some similarity with circuit arrangements of cerebellum (Figure 11.26). After Vanegas (1981) *Brain Mechanisms of Behaviour in Lower Vertebrates* (Laming ed), Cambridge Univ. Press, p. 113.

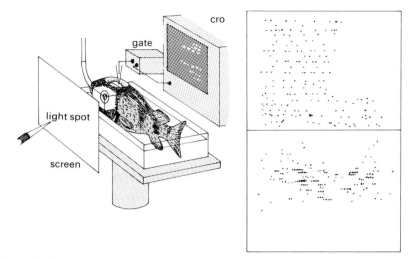

**Figure 11.24** Responses of visual cells of *Perca* to different light stimuli. Left: experimental arrangement (note water-filled box around eye to minimize refractive errors). A light spot is scanned across the screen, and responses of tectal cells (right) are picked up, gated, and displayed on oscilloscope screen (cro). Right upper: cell shows rostral inhibitory field; right lower: cell responds in a more localised pattern by bursts of activity. After Guthrie and Banks (1978).

*The cerebellum in electrolocating teleosts.* Although electroreception has not been evolved by *Amia* and *Lepisosteus*, nor by the majority of teleosts, all other fishes (as well as some amphibians) have modified part of the acoustico-lateralis receptor system for electroreception (p. 229). In these electroreceptive fishes, the lateral line lobes of the hindbrain have a special dorsal electroreceptor nucleus, separate from the acoustico-lateralis nuclei, which projects to the cerebellum. Presumably in most fishes, the system is used chiefly for prey detection (as it is in the platypus, *Ornithorhynchus*).

It is remarkable that two groups of electroreceptive freshwater fishes (the South American Gymnotidae and the African Mormyridae) have in addition to the electroreceptors, independently evolved weak electric organs whose discharges enable them to gain a sophisticated view of their surroundings, and to communicate with each other (p. 233).

No less remarkable are the arrangements in the brain for processing the input from the different kinds of electroreceptors and regulating the electric organ discharges. In the gymnotids, the cerebellum and lateral-line lobes are not especially large, but in the mormyrids, the cerebellum is so greatly enlarged that the valvulae overflow out of the mesencephalic ventricle and cover almost all of the brain (Figure 11.25). Different areas of this huge cerebellum are linked to the input of the three different types

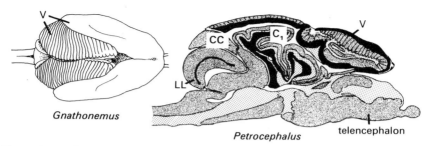

**Figure 11.25** The mormyrid cerebellum. Left: dorsal view of valvulae (V) completely covering brain in *Gnathonemus*. Right: near-sagittal section of brain of *Petrocephalus* showing corpus cerebelli (CC), lateral line lobe (LL), and valvula (V). $C_1$ is the first central lobe, whose neurons are seen in Figure 11.26. After Niewenhuys and Nicholson (1969). *Neurobiology of Cerebellar Evolution and Development* (Llinas ed), Chicago. p. 107.

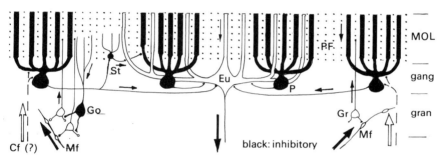

**Figure 11.26** Cell types in lobe $C_1$ of the mormyrid cerebellum (*Gnathonemus*). Cells with inhibitory outputs, black. Note sole output is via axons of eurydendroid cells (Eu), on to which project Purkinje cell (P) axons. Climbing fibre input to the bases of the Purkinje cells is probable but not yet definitely established. Cf, climbing fibres; gang, ganglionic layer; Go, Golgi cells; Gr, granule cells; gran, granular layer; MF, mossy fibres; MOL, molecular layer; St, stellate cells. Apart from the eurydendroid cells, the arrangement is similar to that of a normal cerebellum in a less specialized fish. After Niewenhuys *et al.* (1974). *Z. Anat. Entw. Gesch.*, **144**, 31.

of electroreceptor (p. 230), each of which is involved in a different aspect of electroreception (the ampullary organs, low-frequency electrolocation; mormyromast organs, active electrolocation; the Knollenorgans, communication). The very remarkable hypertrophy of the mormyrid cerebellum must mean that these fishes have an enormous capacity for processing electroreceptive information, lacking in gymnotids, and that there is almost certainly an accurate point-to-point brain map of electroreceptor information. The cell architecture and circuitry of the electric lobes of the cerebellum (Figure 11.26) is similar to that of non-electric fishes (seen in Figure 11.15), but the dendrites of the Purkinje cells are arranged in an extraordinarily regular manner between sheets of

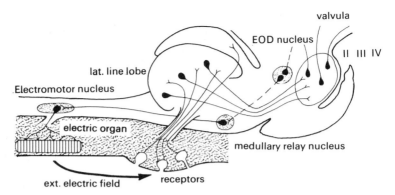

**Figure 11.27** Relations between electric organ discharge and electroreceptors in a mormyrid. After Niewenhuys and Nicholson (1969) *Neurobiology of Cerebellar Evolution and Development* (Llinas ed), p. 107.

parallel fibres, in a much more ordered arrangement. Furthermore, instead of Purkinje axons being the inhibitory output from the cerebellum, they end on enormous eurydendroid cells and it is the axons of these cells which are the sole (excitatory) output from the electric lobes. Figure 11.27 illustrates a much simplified scheme of the relations between the electroreceptor input and electric organ discharge in these curious fishes. There must be analogous arrangements in the brain of rays (which also possess weak electric organs, and so far as is known only one kind of electroreceptor) but these have not been examined.

## 11.3 The autonomic nervous system

Autonomic nerve fibres innervate smooth (involuntary) muscle, for example in the spleen, gut and urogenital tract, and around blood vessels, the heart and glandular chromaffin tissue. Also, in teleosts (but not elasmobranchs), they innervate the pigment cells of the skin. Autonomic fibres reach almost all parts of the body, and regulate visceral functions to maintain homeostasis. In all cases, these efferent pathways from central (viscero-motor) neurons do not go *direct* to the target organ, but synapse first with a peripheral ganglion cell which then innervates the organ, as seen in Figure 11.28. So the fibres from the CNS are called pre-ganglionic fibres, and those from the peripheral ganglion cells themselves, post-ganglionic fibres (see Figure 11.28). The early workers on the fish autonomic system adopted the mammalian divisions of the system into the sympathetic with thoracic and lumbar connections to the spinal cord, the parasympathetic, with pathways in the cranial and sacral nerves, and the enteric intrinsic neurons of the gut plexuses. It seems more sensible in fishes (where the distinction between sacral and lumbar, for example, is less clear than in mammals), simply to divide the autonomic into the cranial autonomic and the spinal autonomic, and the enteric system of the gut. The organisation of these in fishes is best known in elasmobranchs and teleosts, from the early work of J.Z. Young; recent knowledge of its function in elasmobranchs has come largely from work by the same author some 60 years later!

Figure 11.29 shows the arrangement of the autonomic system in the dogfish (*Scyliorhinus*). Cranial autonomic fibres pass out of the IIIrd nerve to the ciliary ganglion close to the IIIrd nerve, whence post-ganglionic fibres innervate the iris, the retractor lentis accommodatory muscle (p. 237) and optic blood vessels. Interestingly (in contrast to teleosts) the iris sphincter is itself light-sensitive, and contracts in response to light; the autonomic innervation of the iris contracts the radial muscles which expands the pupil. Cranial autonomic fibres also emerge from the VIIth, IXth and Xth nerves and pass to the smooth muscle of the pharynx and gut, and (in the Xth nerve) the heart. Spinal autonomic fibres run to segmental paravertebral ganglia associated with masses of catecholamine (mainly noradrenaline)-containing chromaffin cells (p. 210), but longitudinal connections between the ganglia are irregular, and there are no distinct sympathetic chains as are seen in teleosts and higher vertebrates. The cranial autonomic fibres passing to the gut in the IXth and Xth nerves, and in the spinal outflow, modulate and control the intrinsic activity of the enteric autonomic system, where immunocytochemical studies have shown intrinsic neurons containing a wide spectrum of neuropeptides and neurotransmitters such as substance P, serotonin, bombesin, gastrin and vasoactive intestinal polypeptide (VIP). In elasmobranchs, VIP inhibits the spontaneous activity of strips of gut muscle set up in perfusion baths, whilst the other neuropeptides are excitatory. Work by Young and his colleagues on the isolated elasmobranch gut (a more 'physiological' situation) have shown how stimulus frequency of the autonomic fibres in the Xth nerve modifies gut contractions, and have also revealed that 'unconventional' neurotransmitters such as ATP are involved in the enteric autonomic system.

In teleosts and other bony fishes, there are two major differences from the elasmobranch arrangement of Figures 11.28 and 11.29. First, as in higher vertebrates, the spinal autonomic ganglia are linked to the spinal nerves not only by branches carrying the pre-ganglionic fibres from the spinal cord, but also by branches carrying post-ganglionic fibres. These run with the spinal nerves to the skin, where they innervate melanophores (Figure 11.28). Secondly, from the IIIrd nerve, down the body, there is a chain of autonomic ganglia linked by connectives (Figure 11.30). These cranial autonomic ganglia are, however, essentially part of the spinal division of the autonomic system, for they only receive pre-ganglionic

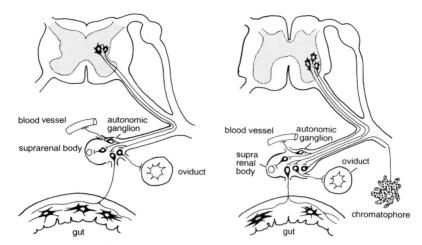

**Figure 11.28** Comparison of autonomic pathways in an elasmobranch (left) and teleost. Note recurrent axons supplying teleost skin chromatophores. After Nichol (1952).

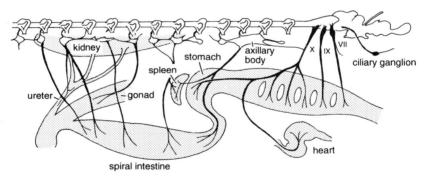

**Figure 11.29** Schematic diagram of the autonomic system in *Scyliorhinus*. After Nilsson and Holmgren (1988) in *Physiology of Elasmobranch Fishes* (ed Shuttleworth). Springer-Verlag, Berlin.

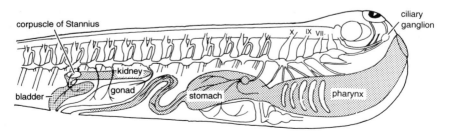

**Figure 11.30** The autonomic system of the teleost *Uranoscopus*. After Young (1933) *Quart. J. Micr. Sc.*, **74**, 491.

fibres from the trunk region (via the autonomic chain), not from their own segments—a similar arrangement to that seen in terrestrial vertebrates. The cranial autonomic system itself is represented by the pre-ganglionic fibres passing in the IIIrd nerve to the ciliary ganglion. These constrict the iris sphincter, pupil expansion being controlled by spinal autonomic fibres passing fowards in the autonomic chain. In contrast to elasmobranchs, where vagal (Xth nerve) fibres seem to play little part in controlling gut movements, the Xth nerve has both excitatory and inhibitory (in different regions) actions on the teleost gut.

## Bibliography

Bullock, T.H. (1982) Electroreception. *Annual Reviews of Neuroscience*, **5**, 121–170.

Cameron, A.A., Plenderleith, M.B. and Snow, P.J. (1990). Organization of the spinal cord in four species of elasmobranch fish: Cytoarchitecture and distribution of serotonin and selected neuropeptides. *Journal of Comparative Neurology*, **297**, 201–218.

Carey, F.G. (1982) A brain heater in the swordfish, *Science*, **216**, 1327–1329.

Douglas, R.H. and Djamgoz, M.B.A. (eds) (1990) *The Visual System of Fish*, Chapman & Hall, London.

Evan, H.M. (1940) *Brain and Body of Fish*, Technical Press, London.

Faber, D.S. and Korn, H. (eds) (1978) *Neurobiology of the Mauthner Cell*, Raven Press, New York.

Goodrich, E.S. (1930) *Studies on the Structure and Development of Vertebrates*, Macmillan, London.

Kimmel, C.B. (1993) Patterning the brain of the zebrafish embryo, *Annual Review of Neuroscience, 1993*, **16**, 707–32.

Laming, P.R. (ed) (1981) *Brain Mechanisms of Behaviour in Lower Vertebrates*, Cambridge Univ. Press, Cambridge.

Metzner, W. and Heiligenberg, W. (1991) The coding of signals in the electric communication of the gymnotiform fish Eigenmannia. *Journal of Comparative Physiology A*, **169**, 135–50.

Nichol, J.A.C. (1952) Autonomic nervous systems in lower chordates. *Biological Reviews*, **27**, 1–49.

Oka, Y. (1992) Gonadotropin-releasing hormone (GnRH) cells of the terminal nerve as a model neuromodulator system, *Neuroscience Letters*, **142**, 119–122.

Roberts, B.L. (1988) The central nervous system, in *Physiology of Elasmobranch Fishes* (Shuttleworth, T.J. ed), Springer-Verlag Berlin, pp. 49–78.

Roberts, A. (1990) How does a nervous system produce behaviour? A case study in neurobiology, *Science Progress, Oxford*, **74**, 31–51.

Roberts, B.L. (1992) Neural mechanisms underlying escape behaviour in fishes, *Reviews in Fish Biology and Fisheries*, **2**, 243–266.

Rovainen, C.M. (1979) Neurobiology of lampreys, *Physiological Reviews*, **59**, 1007–1077.

Smeets, W.J.A.J., Nieuwenhuys, R. and Roberts, B.L. (1983) *The Central Nervous System of Cartilaginous Fishes*. Springer-Verlag, Berlin, Heidelberg and New York, pp. 1–266.

Whiting, H.P., Bannister, L.H., Barwick, R.E. and Bone, Q. (1992) Early locomotor behaviour and the structure of the nervous system in embryos and larvae of the Australian lungfish *Neoceratodus forsteri, Journal of Zoology, London*, **226**, 175–198.

Young, J.Z. (1980) Nervous control of gut movements in *Lophius. Journal of the Marine Biological Association of the United Kingdom*, **60**, 19–30.

Young, J.Z. (1981) *The Life of Vertebrates* 3rd edn, Oxford University Press, Oxford.

# 12 Behaviour

Students of fish behaviour divide into two camps: ethologists analyse behaviour in terms of instinct and learning, breaking down complex behaviour 'patterns' into components, and investigating the neural and sensory basis of behaviour. Behavioural ecologists study behaviour in the context of ecology, looking at behaviour in the wider context of the lifestyle or life history. They investigate metabolic aspects of behaviour, optimal foraging, predator–prey relations, and schooling, migration and spawning as related to the seasons and life history. Frequently there are practical applications in such work if the fish species is exploited by fishermen. Before describing some aspects of behaviour, let us first look at the range of techniques that have been adopted to study this fascinating aspect of fish biology, both in the laboratory and in the field.

## 12.1 Techniques

### 12.1.1 Distribution

One of the simplest techniques to study the behaviour of fish is to record the numbers and positions of fish captured by anglers or commercial fishermen in different seasons. On the assumption that the population is being adequately sampled, analysis should give reliable (and cheap) data on changing distribution. One of the main techniques for investigating fish distribution, the echo-sounder, was first used in the 1930s to find shoals of herring. Since then refinements of the technique allow the echo-sounder to be trained in different directions (sonar) and to use different and higher frequencies to identify 'targets' of different size (Figure 12.1). Later improvements allow the biomass of fish to be estimated by echo-integrators and individuals to be counted as they pass into nets. Special developments for research purposes such as 'sector-scanning' sonar enable the movements of individual fish marked with special sonic tags to be followed for several hours or even days. Surface-living species can be followed by attaching radio transmitters and small aerials; these are especially versatile for locating fish such as basking sharks at the surface at quite long distances.

Tagging of individual fish is a long-established practice (Figure 12.2). Tags may be external 'flags', cylinders or buttons attached by a toggle

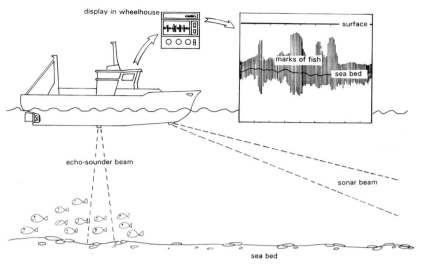

**Figure 12.1** Diagram showing echo-sounding techniques to identify and count fish stocks.

through the muscle. The tags are numbered and carry a message, giving an address and offering a reward if the tagged fish are returned by fishermen or anglers. Internal tagging has been practised on a larger scale to assess not only migrations but also rates of mortality. Fishes like herring or menhaden (its equivalent on the south-east coast of the USA) were caught by purse seine and marked in their thousands by a small steel tag fired into the body cavity by a spring-loaded 'gun'. Return of the tags depended on the fish being processed for meal or oil, the tags being collected on magnets in the fishmeal plant. The most sophisticated tags now depend on electronics. The simplest are attached to the surface of the fish and give a continuous auditory signal that can be picked up by hydrophones working on a triangulation principle. More complicated tags work as transponders and only emit a signal when 'interrogated' by the listening device, e.g. sector-scanning sonar. This increases the life of the battery and makes the tag effective for a longer period.

So-called smart tags (some of which can be inserted into the stomach; Figure 12.2) can telemeter temperature, depth, tail beat frequency, heart beat and jaw movements, giving valuable information about behaviour in natural conditions. Sonic and radio tags (which are only effective at the water surface) have been used to follow the coastal migrations of salmon, the detailed movements of cod in their feeding territories, the far-ranging migrations of sharks and the astonishing vertical migrations of the swordfish.

Other tagging methods include marking fish by feeding or injecting with

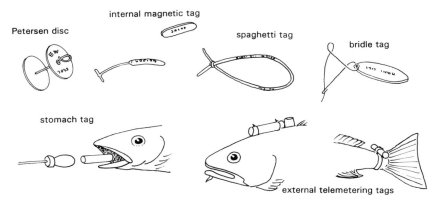

**Figure 12.2** Different types of tag. The top row shows tags attached through the muscles; these depend on the fish being recaptured. The internal tag is collected from magnets in a fish meal plant. The tags in the lower row send information from the fish to hydrophones or other sensors. Redrawn from Hawkins and Urquhart (1983).

tetracycline or other substances that become deposited as fluorescent marks in the bones or otoliths or freeze branding with metal numbers dipped in liquid nitrogen.

### 12.1.2 Behaviour patterns

Tagging gives information about movements over relatively long periods of time. Behaviour in the shorter term can be studied by laboratory actographs which record movements or locomotor patterns of fish. This can be done in a number of ways: by movements of the fish interrupting infra-red light beams (invisible to the fish) and so triggering photo-electric cells, or changing the flow patterns around bead thermistors suspended in the water. Fish counters in salmon ladders also work by the fish interrupting a beam of light.

More detailed analyses of locomotor behaviour have used ciné cameras at high speed, up to 1000 frames s$^{-1}$. However, processing is expensive and takes time. Video tape can be replayed immediately and re-used. Most video cameras operate at 25 or 50 frames s$^{-1}$, but high-speed video cameras operating at up to 400 frames s$^{-1}$ are available, although they require more light, and are much more expensive. Observations of fish behaviour in the dark are also possible by illuminating tanks with infra-red light and using video cameras sensitive to these long wavelengths (Figure 12.3).

### 12.1.3 Sensory capabilities

Classical or operant conditioning techniques are very effective for investigating sensory performance. In classical conditioning fish are trained

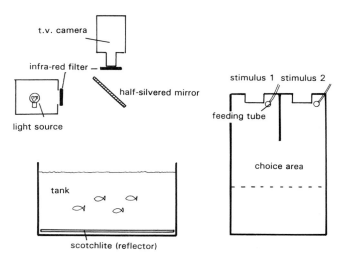

**Figure 12.3** Left: Side view of an experimental tank containing fish viewed by an infra-red sensitive TV camera. The scotchlite reflector is the material used in road signs and gives a very bright background (not visible to the fish which cannot see the infra-red light). Right: Top view of a choice tank in which fish can be trained to discriminate stimuli by being rewarded with food if they select the 'correct' stimuli.

to associate a conditioned stimulus (e.g. a flash of light) with an unconditioned stimulus (e.g. a reward such as food, or a punishment such as an electric shock). After some training, the fish make an appropriate conditioned response such as a feeding movement or a flight reaction when the conditioned stimulus is given but before they receive the unconditioned stimulus. Thus they are shown to have perceived the conditioned stimulus. This makes it possible to measure sensory thresholds by decreasing the intensity of the stimulus in successive trials until the conditioned response disappears. In operant or instrumental conditioning the fish are trained to perform new and more complex tasks. For example, they might be trained to choose between two sources of stimulation, movement to one resulting in a reward of food, movement to the other the absence of reward (Figure 12.3). Differences between the sources can be increased or reduced to test discriminative ability.

The heart beat of fish is highly susceptible to disturbances. Open a door or look into a tank and the occupant's heart starts to race. The most recent classical conditioning techniques involve recording the heart beat or opercular rhythm of fish (Figure 12.4). After being presented with a conditioned stimulus in several trials followed a few seconds later by a mild electric shock, the fish associates the shock with the conditioned stimulus and slows its heart beat or respiratory rhythm *before* the shock, once again indicating that it has perceived the conditioned stimulus. This technique

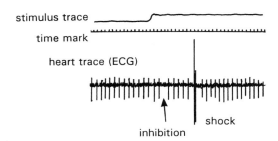

**Figure 12.4** Electrocardiograph of a fish such as cod (lower line) with the heart beating about once per second. The middle line shows time marks and the upper line shows when the stimuli is applied. Note the loss of a heart beat after the stimulus but before the shock is applied (picked up on the ECG electrodes), so demonstrating that the fish has perceived the stimulus.

has been used to investigate all manner of sensory abilities of fish, especially hearing, vision and pressure perception, using sound, light and pressure as the conditioned stimulus. Conditioning techniques lend themselves to establishing thresholds, discrimination of sound frequency or colour of light and the ability to perceive the direction of a source of stimulation. They have been used on a huge range of species from the goldfish to the cod and from the toadfish to the plaice. Conditioning has the great advantage of being a technique under the control of the investigator rather than depending on the whim of the fish.

## 12.2 Instinct and learning

In the past, ethologists have attempted to categorise behaviour into inborn or innate (instinctive) and acquired (learned) categories. Learning implies the transfer of information from other organisms such as parents, siblings, prey or predators and the incorporation of this experience into new or modified behaviour. It is, however, difficult to raise fish in appropriate forms of isolation so as to be sure that no learning has taken place.

### 12.2.1 Primary and secondary orientation

Most fishes have a normal posture or primary orientation which is almost certainly innate; the longitudinal axis is usually kept horizontal and the dorsal surface uppermost. *Chilodus punctatus*, the headstander of tropical aquaria, adopts an oblique posture and a few species such as eel leptocephali, sand eels and seahorses may have a normal vertical posture, while some cichlids may lie on their sides on the bottom to sham death.

Although fish may make oblique movements upwards or downwards, they return to their primary horizontal orientation which is continuously monitored by the eyes and by proprioceptors in the inner ear. Superimposed on the primary orientation is a secondary orientation which determines where a fish remains within its habitat, for example whether it is near the surface, in midwater or near the bottom, pointing up current or down current, and so on. As explained by Fraenkel and Gunn, fish (especially the larvae) sometimes make simple movements towards or away from light (positive and negative photoaxis). They can also move at a fixed angle to a light source, called a *light compass reaction*. A modification of the light compass reaction is the ability to move along a fixed compass bearing using the sun as a reference point (the *sun compass reaction*); by compensating for the movement of the sun across the sky, the bearing can be maintained throughout the day. These are all forms of secondary orientation and we do not know to what extent they are learned.

One of the most common responses to light, demonstrated by many species, is the dorsal light reaction, a form of primary orientation in which the fish remains normal to light coming from above. Von Holst showed that the wrasse (*Crenilabrus rostratus*) orientated at right angles to light and tilted if the light tilted (Figure 12.5). They would, however, only tilt to a limited extent. If the inner ear was removed surgically on both sides, the

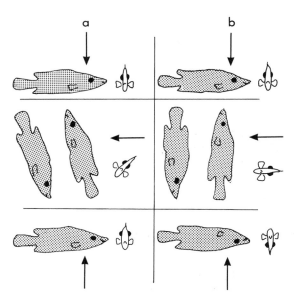

**Figure 12.5** Dorsal light reaction of the wrasse (*Crenilabrus rostratus*) when the light comes from above, the side or below, as shown by the arrows. (a) Intact labyrinth, (b) labyrinths removed. Fish are shown in lateral view and from the front. Without labyrinths the orientation is entirely controlled by the dorsal light reaction. With the labyrinths intact, the fish becomes somewhat tilted when the light is from the side but does not swim upside down when illuminated from below. Redrawn from Fraenkel and Gunn (1961) after von Holst.

fish remained normal to the light and swam upside-down if the light came from below, so showing the interplay between the eye and the inner ear. Obviously this is a simple test – fish do not lose their vertical orientation in the dark, although they may be more prone to swimming off the horizontal and they can cope quite well if the sun rays are entering the water at quite acute angles. In fact, because of refraction, fish see the horizon above the water surface subtended by an angle of about 98° known as Snell's window (p. 234).

## 12.2.2  Reflexes

Reflexes are similar to simple taxes in that they are stereotyped and innate but generally they involve only parts of the body, for example nystagmus (back-and-forth movement) of the eye to a moving light source. One of the best known reflexes in fish is the startle response or C-start, an extreme flexion of the body in response to a potentially harmful stimulus, usually away from the side that the stimulus is applied (p. 56). Much of the research has been done on the larvae of zebrafish (*Brachydanio rerio*) and herring. The resulting escape movement is fast and brief but sufficient to take the fish out of the attack path of a raptorial predator or away from the nematocysts of a predatory medusa. Similar movements seem to be spontaneously generated even before hatching and may assist the larvae in escaping from the chorion. C-starts are usually associated with giant Mauthner neurons or reticulo-spinal cells (p. 283).

## 12.2.3  Complex instinctive behaviour

Instinctive behaviour is often complex, being made up of a number of acts within a hierarchy since a particular act depends on a preceding act and influences a succeeding act. As a case study, let us take one of the best-known examples of reproductive behaviour, that of the three-spined stickleback, *Gasterosteus aculeatus* (Figure 12.6) so well described by Nikko Tinbergen (who might be described as the father of fish ethology and who won the Nobel Prize for Medicine with Konrad Lorenz and Karl von Frisch in 1973). At the beginning of the breeding season the males isolate themselves from the schools and select territories. The eyes develop a blue colour, the back becomes greenish and the belly red. The males protect their territory by a ritual form of aggressive behaviour but rarely fight intruding males. The male then builds a nest of algae in a shallow pit. When this is complete the male goes through a characteristic dance if a female approaches. A receptive female turns towards the male and adopts a semi-upright posture and is then led down to the nest by the male. At the nest the male thrusts its snout into the entrance, the female penetrates the nest, her head on one side, her tail on the other. The male prods the female

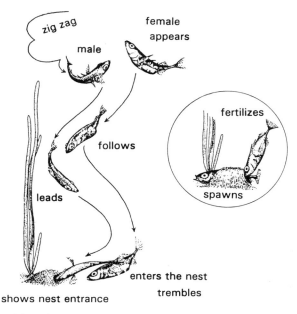

zig zag

male

female
appears

fertilizes

follows

leads

spawns

enters the nest

shows nest entrance          trembles

**Figure 12.6** Courtship and spawning behaviour of the stickleback. Redrawn from Tinbergen
(1951).

which begins to spawn. The female then leaves the nest; the male enters
and fertilises the eggs and subsequently chases the female away. The eggs
are then protected by the male until they hatch. This type of hierarchical
behaviour cannot start half-way but must follow through from a pre-
determined start.

Courtship behaviour associated with holding of territories is widespread
in species that build nests and must distribute themselves at an optimum
density in areas like coral reefs, the seashore and ponds where a suitable
substratum is limited (see p. 191).

### 12.2.4 Learning

We know (p. 297) that fish can establish new relationships by the use of
conditioned response techniques in which naive fish are trained to
associate new stimuli with rewards or punishment (sometimes called,
respectively, positive or negative reinforcement). For example, a rainbow
trout can easily be trained to swim to a feeding station on receipt of a
conditioned stimulus such as transient sound or flash of light. Two hundred
years ago the ill-fated Marie Antoinette called her pet carp to a feeding
station in the Fontainebleau pond by ringing a bell, a practice still
continued but not with the same fish! Goldfish can be trained to pass

through mazes, swim to one of several alternative locations to collect food, to swim rapidly between two flashing lights, and even to control the temperature of their aquarium tank. Rapid swimming between two feeding stations, identified by alternately flashing lights, has been used to assess the swimming speed and stamina of saithe or cod. Thus learning can be used to assess locomotor performance as well as sensory capabilities.

Without doubt, learning plays a major part in survival. Young fish get better at feeding as they gain experience and those that survive predatory attack enhance their survival chances in future encounters.

## 12.3 Complex behaviour patterns

The work of ethologists in trying to analyse behaviour such as feeding, predator avoidance and reproduction shows these to be complex, a basis of instinctive behaviour being modified by learning. Such behaviour is usually intermittent and fish are rarely continuously active. Although much of their lives may be spent searching for food and avoiding predators, movement is wasteful metabolically and may cause the fish to be more conspicuous to their predators. Nevertheless some scombroids and sharks swim continuously, either to maintain respiratory currents across the gills or to maintain their depth using the pectoral fins as aerofoils (see p. 71). Superimposed on short-term, and often irregular, bouts of activity, longer-term rhythmical activity occurs. This may be associated with tidal cycles. For example, inshore fish, feeding in the surf zone, must avoid being stranded (see p. 313). Most frequently, rhythms are associated with day-to-night changes in illumination that control activity, feeding and both vertical and horizontal migrations. On a seasonal basis, especially at higher latitudes, regular migrations are associated with feeding, overwintering and reproduction.

### 12.3.1 Feeding

Much recent interest in the feeding of fish has centred on optimal-foraging theory (OFT, see p. 167). This theory predicts that fish will search for food, select food items if given a choice and cease feeding at an appropriate time so as to maximise the intake of energy for the least energy expenditure. Ultimately, a fish should maximise its fitness, defined as its lifetime reproductive success. OFT is shown to be too simplistic when careful observations are made of the feeding behaviour of social groups of competing fish, of the effect of food being patchy or predators present.

While some species such as the anglerfish, *Lophius piscatorius*, and ceratioid angler fishes are ambush predators, lying in wait for their prey, most species search for their food. If the food is patchy, it can be an

**Figure 12.7** Diagram showing how a school of fish might wheel on to a patch of food.

advantage to search in a social group (Figure 12.7) and, while much searching appears to be random, regular diel vertical migrations may help predators to find their food if it is concentrated in food-rich strata at thermoclines or haloclines.

Detection of food at great distances is difficult underwater. Sighting distances are short but fish may be attracted to the sounds of other fish feeding. Predatory fish biting and chewing their prey (rather than swallowing it whole) are most likely to attract and excite competitors. Herbivorous filter-feeding fish such as the menhaden (*Brevoortia tyrannus*) or anchoveta (*Engraulis ringens*) become more active, and so more 'noisy' during feeding but the main stimulants for other fish are probably the excretory products, for it is well known that feeding is a trigger for excretion and that the chemoreceptors are very sensitive.

In order to home on a food patch by the use of chemical stimuli, fish must swim up a concentration gradient. Such gradients are slow to be established and are easily disturbed. Chemical stimuli are persistent and the source of the stimulus may move away or be eaten. It is likely that fish like sharks and tuna increase their activity on sensing prey chemically but home on the prey using other senses, especially sound. Fish such as bullheads (*Ictalurus*) have highly developed chemosensory organs and search slowly for inactive benthic invertebrates and plants on or in the substratum.

On approaching the food, a series of interlinked behaviours linked with sensory appraisals take place. In a predator–prey situation, the sequence of events shown in Figure 12.8 might take place.

### 12.3.2 Schooling

Although some behaviourists use 'school' and 'shoal' as synonymous terms, it seems sensible to follow Pitcher in using 'school' to describe a group of fish swimming at about the same speed in roughly parallel orientation and maintaining constant nearest-neighbour distance (NND) (Figure 12.9). 'Shoal' describes all social groups of fish so including schools

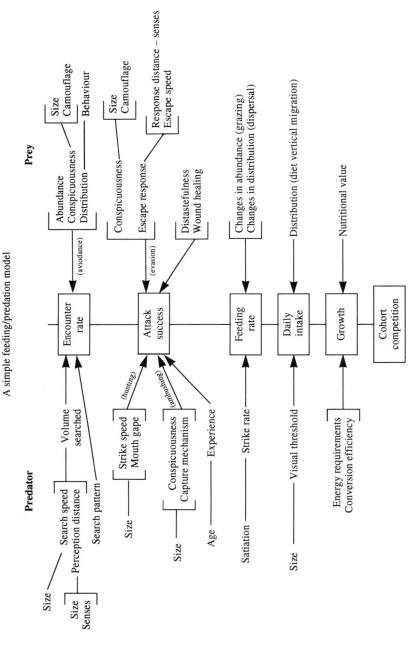

**Figure 12.8** A box model of how feeding takes place. In this instance a fish larva is preying on zooplankton as well as being preyed on itself. The model shows how the 'encounter rate' depends on various factors depending on whether the larva is acting as predator or prey. Encounters may lead to 'attack' to 'feeding' to 'daily intake of food', 'growth' and the possibility of cohorts of prey and predator competing for growth (and survival). The inputs to right and left of the boxes represent sensory and behavioural factors (from Blaxter, 1991).

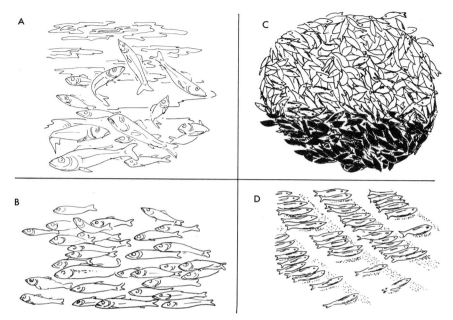

**Figure 12.9** (a) School of herring breaking to feed below surface. (b) School of herring swimming in a tank. (c) 'Pod' *Sebastodes paucispinis*. (d) 'Orderly files' of rainbow trout sheltering behing riffles on the bed of a stream. (a) and (b), redrawn from original underwater photographs; (c) and (d), redrawn from Breder (1959).

as well as aggregations of fish with random orientation and varying NND. One of the more unusual shoals is the pod or ball, where the NND is effectively zero (Figure 12.9) and one of the more deceptive parallel orientations is seen in the ranks of trout sheltering behind riffles in the bed of a stream (Figure 12.9) which is not a social group at all.

The NND determines the extent of packing and in schools is usually of the order of one-half or one body length. Such schools break ranks during feeding and may then themselves be more vulnerable to predators (Figure 12.9). Shoals can comprise anything from a few individuals to many millions, for example in the clupeoids. Large schools are often oblate spheroid in shape, said to minimise the detection envelope under water and so reduce predation.

Schools usually lose their integrity at night. The level of activity falls but it is doubtful whether fish ever sleep in the human sense. Some parrot fishes (scarids) of coral reefs rest immobile and secrete a protective mucous cocoon around themselves at night, and it hard to believe that they are not then asleep. Schools break up at dusk (p. 248) once the illumination has

dropped below a visual threshold of 0.1 mc (0.05 $\mu$W cm$^{-2}$) but they re-
form at dawn. Most schooling species such as the clupeoids and scombroids
follow this pattern, although the mackerel has a very low threshold of 10$^{-7}$
mc. Section of the lateral line nerve in saithe (*Pollachius virens*) causes the
fish to swim closer; this has led to a plausible theory that NND is
determined by the interplay of attraction mediated by visual stimuli and
repulsion by lateral-line stimuli. Although blind saithe were able to school
in a specialised experimental situation, it is likely that both vision and the
lateral line are normally used to maintain school structure and dynamics.
The amazing thing about schooling herring, for example, is how quickly
they move, apparently without colliding, in an impressive display of
coordination. There must be a rapid and effective exchange of information
to achieve this (see p. 229).

Shoaling may play a part in searching for food but its major role is
protective. It is difficult for predators to select individual prey from a shoal
(the confusion effect) and they usually feed on stragglers. Mackerel attack
shoals of sprats in a haphazard way to split off individuals for subsequent
consumption. Shoals of minnows re-form, or become denser, when a
predator is sensed. This may be determined by a 'sentry' effect in which the
fish nearest the predator on the outside of the shoal first respond and
information is rapidly passed across the school, almost certainly initiated
by changes in locomotor behaviour. The possibility of a hydrodynamic
advantage in which schooling fish utilise the vortices created by the fish in
front as a means of increasing the thrust of the tail and reduce drag, still
remains to be validated.

The advantages of shoaling are clear, the disadvantages less so. Shoals
are more conspicuous in conditions of good underwater visibility and from
the air (possibly to avian predators). Limited food has to be shared.
Usually shoals of clupeoids break to feed on particles (but not when filter-
feeding). Particle feeding often leads to intense competition and a feeding
frenzy, and the build-up of metabolites and noise may attract large
predators. Oxygen may become limiting. A large stock of overwintering
herring in a Norwegian fjord, estimated at 2 million tonnes, was recently
reported to be living in very low oxygen levels of 1–2 ml litre$^{-1}$, close to
their minimum tolerance level.

### 12.3.3 Vertical migration

Throughout the oceans, seas and lakes, many species of fish and
invertebrates are found to make diel vertical migrations (with a 24-hour
periodicity) usually towards the surface at dusk and towards the bottom at
dawn. The most plausible general explanation for such a regular event is
that phytoplankton is to be found in the euphotic zone, near the surface.
Herbivores must visit these strata in order to feed but since they can feed in

the dark (unlike most carnivores) the best time to visit the surface layers is at night, while by day they are safer dispersed in deeper water. The carnivores follow the migrations of the herbivores, feeding on them at dusk and dawn when they are in dense concentrations and before the illumination has fallen below the carnivores' visual threshold. Thus, vertical migration is driven by the need to feed and to avoid predators. In particular, it is desirable for many larger species to avoid the surface waters by day where they are vulnerable to avian predators.

Upward vertical migration at dusk seems to be triggered by falling light

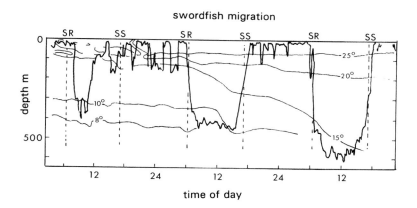

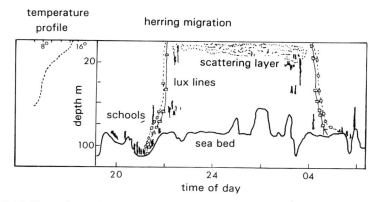

**Figure 12.10** Upper figure: Vertical movements of swordfish depending on the time of day. The depth is given in metres and the isotherms are shown. SR sunrise; SS sunset. Redrawn from Carey and Robison (1981) *Fish. Bull. U.S.*, **79**, 277. Lower figure: Vertical movements of herring in the North Sea over the dusk, night, dawn period. Lines joining points of equal light intensity (isolux lines) are shown. Note how the schools disperse into a scattering layer during the night. Redrawn from Postuma (1957) *Mimeograph Rept. Int. Council Explor. Sea, Copenhagen.*

intensity (Figure 12.10) and downward migration at dawn by increasing light intensity. In high latitudes, in the polar summer and winter, vertical migration is less evident since there is a much reduced diel cycle of light. Vertical migration is also influenced in a predictable way by bright moonlight (which tends to inhibit upward movement) and lunar or solar eclipses (which cause upward movements during the period of darkening).

The amplitude of vertical migration is limited in physoclist fish with closed swimbladders (p. 88). If a physoclist is near neutral buoyancy at a particular depth, rapid downward movements cause it to become negatively buoyant as the swimbladder volume decreases under the increased hydrostatic pressure. If the physoclist moves upwards, the expanding swimbladder not only makes the fish positively buoyant but there is a danger that the swimbladder could burst if the fish moves up too far. The amplitude of movement (usually tens of metres in the cod, *Gadus morhua*, for example) is determined by the rate of gas secretion for a downwardly moving fish and gas resorption for an upwardly moving fish. The 'decompression schedule' for gadid fish is, coincidentally, similar to man, a halving of the pressure followed by a pause. In physostomes like herring, with swimbladders connected to the exterior, no such constraints occur and some oceanic herring may move up or down 100–200 m during the diel cycle, while a swordfish, *Xiphias gladius*, was shown to migrate from the surface to 600 m between night and day (Figure 12.10).

### 12.3.4 Horizontal migration

Daily vertical migrations are modulated by small-scale horizontal migrations involved in feeding and predator avoidance. Reef fish may move on and off the reef with a 24-hour periodicity, feeding by day and hiding at night. Much larger seasonal horizontal migrations occur that are related to spawning and feeding. These are often depicted in the form of oscillatory or triangular movements (Figure 12.11). For example, maturing Atlantic cod migrate to the Norwegian coast to spawn in the spring. After spawning they return to the offshore feeding grounds to recover. Herring in the North Sea move southwards in the early summer. After spawning they tend to drift eastwards, overwintering in the eastern North Sea. In the spring they migrate offshore to the west and north and start to feed and mature for a repeat of the spawning cycle (Figure 12.12). Plaice in the southern North Sea have distinct spawning grounds but wider areas in which they feed and recover after spawning (Figure 12.12).

Some oceanic species of tuna make huge regular trans-ocean migrations, as we know from recaptures of tagged fish. For example, albacore tuna (*Thunnus alalunga*) move from the mid-Pacific to the west coast of the USA or Japan. Bluefin tuna (*T. thynnus*) may migrate between Florida and Norway. Blue sharks (*Prionace glauca*) also migrate across the

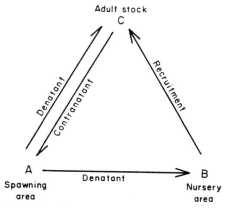

**Figure 12.11** A simple triangular pattern of fish migration (from Harden Jones, 1968).

Atlantic between the eastern USA and the South American and African coasts.

Equally impressive and precise migrations are those of the Atlantic salmon (*Salmo salar*) and eel (*Anguilla*). The anadromous Atlantic salmon migrates from the sea into the upper reaches of rivers to spawn in the autumn and winter. The eggs are laid in gravel beds or redds and hatch after about two months. The young stages take two to three years to reach the smolt stage which returns to the sea. There the smolts grow rapidly on the abundant food, making long migrations as far as the south-west coast of Greenland. After one year, the more precocious fish migrate as grilse back to the home rivers in which they were spawned. More often they remain at sea for two or more years before spawning. The spawners that survive, the kelts, return to the sea and may spawn again. The orientating mechanisms have been extensively studied but are still not fully known. The smolts are certainly imprinted with the odour of the home stream before they start their seaward migration. Their migrations in the ocean may involve residual current systems with possibly some degree of crude orientation using the sun as a reference point (p. 311). This may bring the salmon back to areas where they can identify topographical, auditory or chemical landmarks. These fine-tune the migration to the point where the salmon recognises the unique characteristics of its home stream, which it left as a smolt some time before.

In the Pacific there are several species of salmon (*Oncorhynchus*) each with rather different migration characteristics. Some stocks of coho or silver salmon (*O. kisutch*) are landlocked. Chum salmon (*O. keta*) and pink or humpback salmon (*O. gorbuscha*) spawn near the mouths of rivers and the young stages soon drift to the sea (Figure 12.12) and sockeye salmon (*O. nerka*) makes long migrations into the far Pacific Ocean. The

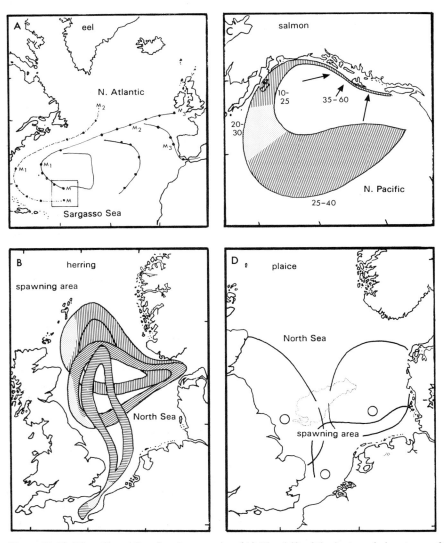

**Figure 12.12** Migration paths of various species. (A) The drift of the leptocephalus stages of the eel across the North Atlantic. (B) The movements of herring of three spawning races around the North Sea. (C) Distribution and movement (arrowed) of Pacific pink salmon (*Oncorhynchus gorbuscha*) in the North Pacific. Lengths in centimetres are given against various shaded areas. (D) Three spawning groups of North Sea plaice with spawning area given (circle) and subsequent area of distribution of the adults. (A) and (D) redrawn from Harden Jones (1968); (B) and (C) redrawn from Tesch (1975).

chinook or king salmon (*O. tshawytscha*) used to make extensive migrations of many hundreds of miles up the Columbia River before its spawning grounds were isolated by dams.

The catadromous European eel (*Anguilla anguilla*) and American eel (*A. rostrata*) migrate upstream as elvers a few centimetres long. They spend some years growing in freshwater habitats, becoming yellow eels which turn silvery *before* they migrate to the sea. The French physiologist, Fontaine, felicitously called this pre-adaptation 'anticipatory'. Although it is not possible to trace their subsequent migrations, there is good circumstantial evidence that they spawn in the Sargasso Sea (Figure 12.12) in deep water. The leaf-like leptocephalus larvae can be traced in plankton catches as they drift with the residual currents eastward and westward to the American and European coasts over a period of one to three years. At the coast they metamorphose to the elver stage and it is likely that they make their way up the estuaries by riding the tides. Observations of elver activity suggest that they stay near the floor of the estuary on ebb tides but move into mid-water on the flood tide, so making net movements upriver.

Riding the tides was fairly recently discovered by acoustic telemetry, and also occurs in fish such as plaice, cod and dogfish in the open sea. In the southern North Sea, plaice migrating for spawning ride the southwardly flowing flood tide and rest on the ebb; after spawning they return northwards by riding the ebb tide and resting on the flood. Other mechanisms may exist to allow fish to orientate their migrations. The active mechanism involved in riding the tide may be superimposed on a passive drift associated with residual current systems, the so-called *denatant* part of a migration cycle. *Contranatant* movement is best seen in salmonids swimming upstream to spawn in the upper (non-tidal) reaches of rivers.

Of the visual mechanisms that may be involved, the most studied is the sun compass reaction. Fish such as coho salmon, centrarchid sunfish (*Lepomis gibbosus*) and the cichlid *Aequidens portalegrensis*, can be trained in circular tanks to escape on a particular compass bearing regardless of the time of day. This implies that they can not only move at an angle to the sun, but allow for the movement of the sun across the sky (Figure 12.13). More simply, changes in activity during different times of day coupled with swimming towards the sun could also lead to orientation in different directions (Figure 12.14) but there is no evidence that this mechanism is used by any fish. These mechanisms are not equivalent to coordinate navigation in which a fish might be able to monitor latitude and longitude by the altitude of the sun at midday and the time of sunrise related to some fixed meridian. Neither coordinate, nor inertial navigation (in which an organism could determine its position by remembering changes of course and distances moved) have yet been demonstrated in fish.

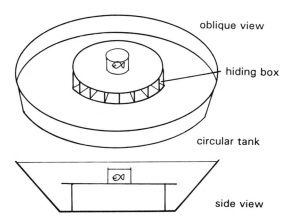

**Figure 12.13** Oblique view (above) and side view (below) of an arena tank with a central circle of refuge boxes to observe the ability of blue-gill sunfish (*Lepomis macrochirus*) to hide in a particular box using the position of the sun as a reference point. The fish was released from a central container at the beginning of each trial. Redrawn from Hasler (1971).

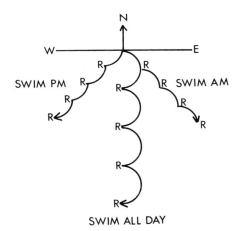

**Figure 12.14** Diagram showing how a fish swimming towards the sun in the morning (AM) during the daylight hours (All Day) and in the afternoon (PM) and resting (R) at other times would make a net movement in a southeasterly, southerly and southwesterly direction (in the northern hemisphere).

Other visual mechanisms that have been postulated include the perception of the direction of passing waves which would give the direction of the wind, and the perception of the plane of the polarised light which is characteristic of the altitude and azimuth of the sun even if obscured by cloud. Olfactory mechanisms operate at relatively close range to a source of stimulation before the chemical stimulus becomes too dilute or chemical gradients are disturbed. The imprinting of salmonids is the best established olfactory mechanism and has been supported by neurophysiological evidence. In coho and chinook salmon, infusion of home-stream water into the nasal cavity causes a marked increase in the activity of the brain recorded by electroencephalograph (p. 259). There is evidence accumulating that skin mucus is at least one of the determinants of home-stream odour. Obviously this mechanism would only operate if non-migrants were present upstream to scent the water.

Whether long-distance migrants, such as eels, could orientate using the Earth's magnetic field remains unclear, though it seems probable. Magnetic material has been found in migratory species like the eel, salmon, mackerel and herring, but also in the non-migratory carp and perch. The yellowfin tuna (*Thunnus albacares*) has the mineral magnetite in the ethmoid sinus. This species, European and American eels, several salmonid species and the stingray (*Urolophus halleri*) have been shown in experiments to detect weak magnetic fields and to alter their orientation in response to manipulation of the Earth's magnetic field. It remains to be shown whether this faculty is employed as an orientation mechanism during migration under natural conditions.

## 12.3.5 Tidal rhythms

Although selective tidal transport described above is a form of tidal rhythm, this behaviour is more prevalent in intertidal areas where there is a strong danger of being stranded as the tide recedes. Many intertidal fishes like blennies or gobies remain in rock pools at low water or, like the rockling, *Ciliata mustela*, can survive under seaweed. In many cases, they return to the same place each time the tide goes out. A reverse process takes place in the mudskippers which are active on the surface of muddy shores at low tide but remain in their underwater burrows at high tide. Often fishes like gobies and flatfish follow the tide in taking food organisms disturbed at the margin of the sea.

Remaining close to the shoreline may reduce predation pressure since larger predators are less at home in very shallow water. For the same reason, the grunion (*Leuresthes tenuis*) lays its eggs in the sand at high water of spring tide and the eggs hatch two weeks later when the tide again reaches the same height. Some *Fundulus* and *Menidia* species also spawn intertidally.

Tidal rhythmicity (Figure 12.15) is known to be endogenous in several species including the plaice, rock goby (*Gobius paganellus*) and shanny (*Lipophrys pholis*), and seems to be entrained by changes of hydrostatic pressure (a so-called *Zeitgeber*). There may be some advantage in *anticipating* tidal movements if stranding is to be avoided, hence the advantage of an endogenous rhythm.

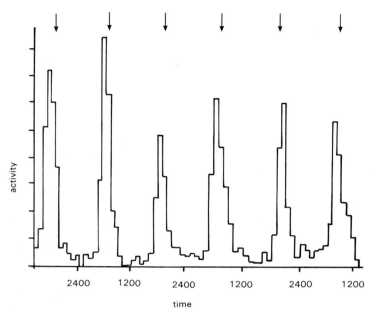

**Figure 12.15** Activity of the blenny (*Blennius pholis*) over three days. The arrows show the predicted time of high water. The fish were removed from the sea and kept in constant conditions of light, temperature and tidal height showing that the activity rhythm is endogenous. Redrawn from Gibson (1965).

## 12.3.6 Symbiosis

The term symbiosis, or living together for mutual benefit, might well be applied to shoaling, an intraspecific behaviour. It is, however, more usually applied interspecifically. One of the best-known examples is cleaning symbiosis in which cleaner species such as shrimps or other fish, remove ectoparasites from the host fish such as sharks and rays. Curiously, this behaviour is rare in freshwater except for a few cichlids and centrarchids. It is especially important in reef fishes but has recently been utilised as a form of biological control in the salmon farming industry. Labrids such as the goldsinny wrasse (*Ctenolabrus rupestris*) are regularly put in salmon cages in seawater to remove sea lice from the skin of the host salmon. Some marine gobies, such as *Psilogobius mainlandi*, live in the burrows of shrimps. The shrimp maintains the burrow while the goby gives warning of danger and may provide particles of food for its symbiont. This exchange of food for protection also applies in the associations between fish and anemones. The best-known associations are between pomacentrids such as *Amphiprion* and corals or anemones but young whiting (*Merlangius merlangus*) and stromateids are frequently found in association with medusae such as *Cyanea* in the pelagic zones of the sea. The stromateid *Nomeus gronovii* seems to have a rather equivocal relationship with the Portuguese man-of-war (*Physalia*), sometimes living apparently unscathed among the stinging tentacles but at other times being observed as a partly digested corpse.

## 12.3.7 Cannibalism

Cannibalism is widespread in both elasmobranchs and teleosts. It has been reported in 36 teleost families and since a substantial size difference is normally required for a predator to take prey, the most common form involves older fish eating younger fish (i.e. intercohort rather than intracohort cannibalism). A time of particular vulnerability is during pelagic spawning when fish such as anchovy may filter-feed (or snap at) their own eggs (or those of adjacent spawners) present at high density. Species such as cichlids, which show parental care, usually have well-developed behaviour that inhibits feeding on their own young, e.g. in mouth-brooding (p. 186).

Cannibalism can be severe in aquaculture if a sufficient size differential is allowed to build up within a cohort. Larger individuals are only likely to eat smaller individuals if they are less than half their own length so that sorting for size can eliminate cannibalism. In natural populations, cannibalism could be a powerful density-dependent mechanism controlling populations size, although this has yet to be demonstrated. An intriguing idea was put forward recently by Nellen that high fecundity in a species has evolved to ensure a food supply for a developing cohort, the fast-growing individuals eventually feeding on the slow-growing individuals or the adults on their young. This hypothesis, of course, requires that cohorts grow up in proximity and do not disperse widely. It is an intriguing idea but hard to test.

## Bibliography

Bardach, J.E., Magnuson, J.J., May, R.C. and Reinhart, J.M. (eds) (1980) *Fish Behaviour and its use in the Capture and Culture of Fishes*, ICLARM, Manila.
Blaxter, J.H.S. (1991) Sensory systems and behaviour of larval fish, in *Marine Biology: its Accomplishment and Future Prospect* (Mauchline, J. and Nemoto, T. eds), Hokusen-Sha, Tokyo. pp. 15–38.
Breder, C.M. Jr. (1959) Studies on social groupings of fishes. *Bulletin of the American Museum of Natural History*, **117** (6), 393–482.
Carey, F.G. and Robison, B.H. (1981) Daily patterns in the activities of swordfish *Xiphias gladius* observed by acoustic telemetry. *Fishery Bulletin of the United States*, **79**, 277–292.
Fraenkel, G.S. and Gunn, D.L. (1961) *The Orientation of Animals*, Dover, New York.
Gibson, R.N. (1965) Rhythmic activity in littoral fish. *Nature, London*, **207**, 544–545.
Gibson, R.N. (1992) Tidally-synchronised behaviour in marine fishes, in *Rhythms in fishes* (Ali, M.A. ed), *NATO ASI series A*, **236**, 63–81.

Harden Jones, F.R.H. (1968) *Fish Migration*, Edward Arnold, London.

Hasler, A.D. (1971) Orientation and fish migration, in *Fish Physiology* Vol. VI. (Hoar, W.S. and Randall, D.J. eds), Academic Press, New York. pp. 429–510.

Hawkins, A.D. and Urquhart, G.G. (1983) Tracking fish at sea, in *Experimental Biology at Sea* (MacDonald, A.G. and Priede, I.G. eds), Academic Press, London.

Kruzhalov, N.B. (1990) Attraction and repellant reactions to amino acids by crucian carp *Carassius auratus*. *Journal of Ichthyology*, **30**, 165–170.

Losey, G.S. (1978) The symbiotic behaviour of fishes. In *The Behaviour of Fishes and other Aquatic Animals* (Mostofsky, D.I. ed), Academic Press, New York. pp. 103–166.

McCleave, J.D., Arnold, G.P., Dodson, J.J. and Neill, W.H. (eds) (1984) *Mechanisms of Migration in Fishes*, Plenum Press, New York.

Neilson, J.D. and Perry, R.I. (1990) Diel vertical migrations of marine fishes: an obligate or facultative process? *Advances in Marine Biology*, **26**, 115–168.

Neilson, L.A. (1992) *Methods of Marking Fish and Shellfish*. Amer. Fish. Soc. Special Public. 23. Bethesda, Maryland.

Pitcher, T.J. (ed) (1993) *Behaviour of Teleost Fishes* 2nd edn, Chapman & Hall, London.

Rankin, J.C., Pitcher, T.J. and Duggan, R. (eds) (1983) *Control Processes in Fish Physiology*, Croom Helm, London.

Smith, C. and Reay, P. (1991) Cannibalism in teleost fish. *Reviews in Fish Biology and Fisheries*, **1**, 41–64.

Tesch, F.W. (1975) 8. Orientation in space. 8.2. Fishes, in *Marine Ecology*, **II**, Pt 2. (Kinne, O. ed) John Wiley, London. pp. 657–707.

Tinbergen, N. (1951) *The Study of Instinct*. Oxford University Press, Oxford.

# 13   Fisheries and aquaculture

Being supported by the water and so with less need for a supporting skeleton, fishes have a higher ratio of muscle to bone than land animals and are thus a very valuable source of protein. The world catch of fin fish, mainly teleosts, from all sources has steadily risen over the last two decades (Figure 13.1) to about 85 million tonnes per annum at present, probably close to the best estimate of the maximum annual sustainable yield. This harvest is worth perhaps US $100 billion (about £67 billion) per annum at the fish market and many times that amount by the time it reaches the consumer. The major part of the catch is taken in the high seas fisheries from wild stocks that man has tried to manage, with singular lack of success. Most of the historically great stocks are being, or have been, overfished. The increase in yield has only been maintained by searching out new stocks and new species in the world's oceans. Freshwater fisheries are probably fully exploited at about 13 million tonnes per annum and,

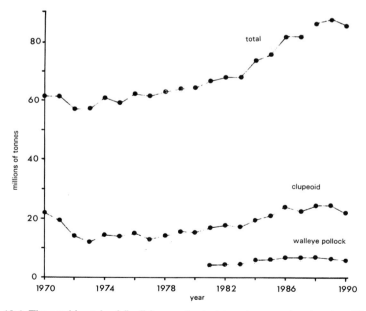

**Figure 13.1** The world catch of fin fish over the last twenty years showing, in addition, the catch of all clupeoids and of the walleye pollock. From *FAO Yearbooks*.

while aquaculture at about perhaps 8–10 million tonnes per annum has assumed much greater importance in the last three decades, it is unlikely to compete with hunting techniques on wild stocks in terms of weight. In many countries, however, the value of aquaculture production may exceed that of the fisheries on wild stocks. In China in 1989, these two sources of production were, in fact, equal in terms of weight.

## 13.1 Productive areas and species

The biggest fisheries occur over the continental shelf (less than 200-m depth) where nutrient turnover occurs in winter, or in areas of upwelling of deep water as on the coast of Peru and Chile. This nutrient-rich water from the deeper layers then replenishes the nutrients already used up in the primary production of the phytoplankton at the base of the food chains, ultimately supporting the reproduction and growth of fish stocks. The yield from the open ocean is low, except for migratory species like tuna, and the biomass in deep oceanic water is very low except close to the ocean bed.

## 13.2 Species

The most productive group of fish (in terms of weight, if not of value) are the clupeoids – anchovies, herring, menhadens, sardines and sprats – which feed mainly near the base of the food chain on zooplankton or phytoplankton, often by filter-feeding (p. 158). In the 1960s and 1970s, the annual world catch of the Peruvian anchoveta (*Engraulis ringens*) was as high as 10 million tonnes but it fluctuated on a yearly basis depending on El Niño, a weather pattern which in some years influences the direction of the prevailing wind and so the extent of the upwelling. In recent years, the total clupeoid catch has been near 20 million tonnes, about one-quarter of the world catch, the yield being maintained by the wealth of species and stocks with high biomass.

A large and important group in terms of weight and value are the gadoids – cod, haddock, hake, pollock, saithe and whiting – which tend to feed higher up the food chain. A single species, the Alaska or walleye pollock (*Theragra chalcogramma*) from the North Pacific and Bering Sea now supports one of the most important world fisheries, yielding about 6 million tonnes per annum. The flatfish are another valuable group of mainly benthic feeders. Fish such as brill, flounder, halibut, lemon sole, plaice, sole and turbot are caught on the sea bed of the continental shelf in many parts of the world, especially in more temperate areas. Tuna and mackerel are pelagic predatory schooling species usually supporting migratory fisheries of high value and in some cases substantial weight.

Over the continental slope there are large stocks of blue whiting (*Micromesistius poutassou*) and on the ocean floor rat-tails (macrourids) and morids, related to the cod, are fairly abundant. It has been found that the orange roughy (*Hoplostethus atlanticus*), until recently the object of a considerable fishery off Australasia, is present in possibly fishable quantities to the west of the British Isles.

There are other important commercial and sport fisheries for migrating species like the salmon in the North Pacific and North Atlantic and for sturgeon (*Acipenser*) and eels. Salmon and sturgeon move into freshwater to spawn (although some populations are landlocked) and the eels into the deep part of the Sargasso Sea in the Central Atlantic (p. 310). Salmon are now vulnerable to capture during their high seas phase of growth (though this was not traditionally a high seas fishery) as well as in estuaries and rivers. The elasmobranch fishes – sharks, dogfish, rays and skates – have much less commercial value but are important in sport fisheries throughout the world.

In freshwater, the fisheries are supported by the relatives of the salmon – rainbow, brown and steelhead trout, charr (*Salvelinus*) and whitefish (*Coregonus*) – and by a huge group of fish like carp, perch, roach and ruffe. In the brackish waters of the Black Sea and Caspian Sea, sprats comprise an important fishery.

## 13.3 The fisheries

Two opposing trends have occurred in the high seas fishing fleets. Some countries, like the UK, have contracted towards their coasts as the traditional distant or middle-water grounds were closed by changes in fishery limits, e.g. those of Iceland and Norway. The Common Fisheries Policy of the EC has, however, opened up near-water fisheries to other countries of the Community. The pressure on the near-water stocks has thus been enormous. Other countries like Japan, Russia and Eastern European countries have sent fishing fleets, consisting of mother or factory ships and smaller catchers, to new grounds, for example in the South-West Atlantic where they remain for many months. Littoral, estuarine and freshwater fisheries are important in many parts of the world where they can be prosecuted by small boats. In the Great Lakes of North America and the inlands seas of the former Soviet Union, serious pollution problems have influenced the fisheries which have the potential for a good yield in optimum conditions.

The gill nets and small beam trawls, used by sailing boats and lightly-powered vessels of the last century, have gradually been superseded by much larger beam trawls and large otter trawls fishing on the sea bed, and by pelagic trawls and purse seines fishing in midwater (Figure 13.2). Gill

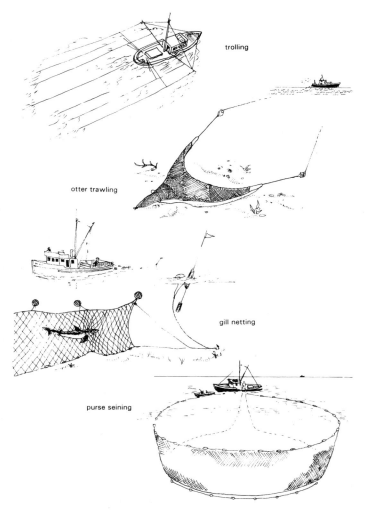

trolling

otter trawling

gill netting

purse seining

**Figure 13.2** Various types of fishing gear. Based on drawings in Rounsefell (1975).

(entangling) nets and lines are of relatively minor importance in terms of weight of fish caught.

## 13.4 Aquaculture

In 1990 the world-wide production of fish from aquaculture was 8.4 million tonnes, of which 6.6 million tonnes was produced in developing countries. About 85% of production is of herbivorous fish especially cyprinids but also tilapias, milkfish and mullets. China produced just over 4 million tonnes, mainly of cyprinids like the carp. It is estimated that the likely maximum output of aquaculture could eventually reach 25 million tonnes p.a. but is never likely to exceed ⅓ to ¼ of the yield from the freshwater and marine fisheries. This represents about 4% of terrestrial production of protein.

Aquaculture has been practised for many centuries in Africa and the Far East but also in some parts of Europe where carp (*Cyprinus carpio*) were reared in the ponds of monasteries. Technologies for new species have been developed more recently. The channel catfish (*Ictalurus punctatus*) is now farmed on a large scale in the southern USA. Other important species are eels, especially in Japan, the cichlid tilapias, with a worldwide production of at least 250 000 tonnes per annum and the milkfish (*Chanos chanos*). More recently, there has been an explosion of interest in seawater cage farming of rainbow trout and salmon, especially in Norway, Scotland, Canada and Denmark (Figure 13.3). Total annual production is now approaching 250 000 tonnes and has outstripped the requirements of the market, resulting in a fall in price. The culture of high-value marine species such as bass, bream, halibut, sole and sturgeon is now being developed, mainly by a breakthrough in rearing of the young stages (not a problem with salmonids). This technology is only beginning to reach an economically-viable phase.

Much of aquaculture is based on herbivores like the grass-eating carp (*Ctenopharyngodon idella*) and milkfish and on omnivores like common carp and tilapias that feed low in the food web. Most high-priced species are, however, fed on fish-based diets. This makes high-conversion rates essential since cheaper protein is being upgraded to more expensive protein with little room for saving if the price of the final product falls. Even if the food is based on offal or other waste protein, it is unlikely that

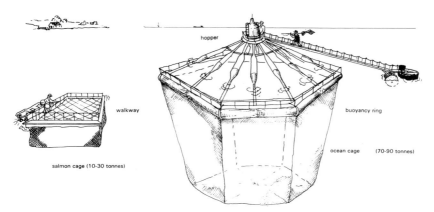

**Figure 13.3** Two types of cage used in salmon farming.

the yield of low-value carnivorous fish from aquaculture will be much increased. Genetic manipulation in aquaculture is becoming commonplace. By chromosome manipulation (gynogenesis), appropriate hybridization and hormonal treatment it is now possible to produce unisex populations. In salmon and rainbow trout, for example, it is advantageous to rear all-female populations; this is because the males often mature early at a small size, become aggressive with a correspondingly lower growth rate and have unpalatable flesh. In the tiliapias, the males are larger than the females and all-male populations are optimal. Hybrids of salmonids, bass (*Morone*), sturgeon and other species, often combine the best qualities of the parents used. Although they are likely to be sterile, hybrids may be useful in aquaculture or in stocking sport fisheries. Sterility in non-hybrids can be induced by pressure or cold shock of newly fertilised eggs which induces triploidy. Triploid salmon, for example, are sterile and triploid females are especially useful since they do not mature sexually and maintain their growth until ready for harvesting. Another advantage of using sterile salmon is that escaped fish cannot interbreed with wild stocks. Interbreeding has been a matter of some concern to anglers and managers over the past few years, not only with salmon, but also with tilapias such as *Oreochromis mossambicus*.

## 13.5 Ranching

The release into the wild of species reared in shore facilities is analogous to ranching cattle as long as there is some guarantee of the fish being available for later harvesting. Since the cost of feed is the greatest financial burden in aquaculture, ranching is likely to be viable even with substantial losses through straying and predation.

Salmon smolts can be imprinted with the characteristics of their home area (p. 260) and after one or more years in the sea, travelling perhaps many hundreds of miles, can be expected to return for harvesting. At present, salmon ranching is only being practised actively in Iceland. Sturgeons are also reared and allowed to migrate into the sea in the hope of an enhanced return later in life. In Norway, shore-reared cod are being released into the fjords and, in Denmark plaice and turbot are being released into sheltered bays. Some, at least, remain in the neighbourhood and, if the carrying capacity of the area proves adequate, can potentially be harvested later at a marketable size. Present trials suggest that stock enhancement is of limited value.

## 13.6 Management

Over the past decades, a huge amount of research on fishes has been driven by the need to manage the wild stocks to obtain the *maximum sustainable yield*. The yield from a stock depends on recruitment of young fish and growth on the plus side, and fishing and natural mortality on the minus side. The first three parameters can be measured in a satisfactory manner but usually a guess has to be made at the rate of natural mortality. Monitoring the fish stocks is done by experimental fishing using research boats but also by recording the catches from commercial vessels and sampling their catches at the fish market.

What is meant by a *stock* of fish? It seems that some commercial species may be divided into stocks that should be considered independent of other stocks of the same species for management purposes. These stocks usually separate out during spawning but often mix outside the spawning season, a classic case being the North Sea herring. For many years, fisheries scientists have attempted to distinguish between stocks using identifying characteristics such as growth patterns determined from the scales and otoliths (p. 193), and counts of meristic characters such as vertebrae and fin rays. Stocks of fish can sometimes be identified by parasites that are specific to particular areas of the sea. Genetic techniques allow stocks to be identified by the electrophoretic isozyme pattern obtained from tissues like the muscles and eye lens. Some cod and salmon released into the Norwegian fjords and rivers for restocking can be identified by a special isozyme pattern. Genetic fingerprinting makes it theoretically possible to identify individual fish or particular parental lines in the future, although such techniques are unlikely to become quick or cheap.

In the last century the fishing fleets were not powerful enough to deplete stocks below a sustainable level. With the introduction of steam-, and then diesel-powered, engines and winches, the boats could travel further and faster and use more effective gear. Fish-finding techniques such as sonar have also made fishing practice more efficient. Many, if not most, fish stocks now show signs of overfishing. Overfished stocks are characterised by reduced yields for a given fishing effort and by the absence of larger fish. Early management techniques consisted of increasing the mesh sizes of nets to allow small fish to escape and by forbidding the landing of undersized fish, the idea being to allow the smaller fish to grow to a more optimal size before being harvested. These measures were not successful, mainly because of the difficulty of agreeing mesh sizes with the industry—it was a classical instance of too little, too late. *Total allowable catches* were then introduced in which the annual sustainable catch for any stock was estimated by fishery biologists and divided up into quotas amongst the interested fishing nations. This measure, in use currently in many fisheries, is only partially successful because there is a lag in collecting statistics, difficulty in closing the fishery when the quota has been reached and losses of the protected species in other mixed-species fisheries that are still operating. Discards of fish that cannot be landed because the quota has been reached (or of undersized fish) are an appalling source of loss of future marketable fish. The main problem at the social level is that the quota can be taken very quickly, leaving the fishermen with nothing to do. The obvious solution is to reduce the size of the fleets in many countries and allow the remaining boats a larger share of the national quota, but this has severe social implications.

## Bibliography

Cloud, J.G. and Thorgaard, G.H. (eds) (1993) *Genetic Conservation of Salmonid Fishes NATO ASI ser. A, Life Sciences*, **248**, Plenum Press, London.

Jolly, C.M. and Clonts, H.A. (1993) *Economics of Aquaculture*. Haworth Press, New York.

L'Abee-Lund, J.H. and Jensen, A.J. (1993) Otoliths as natural tags in the systematics of salmonids. *Environmental Biology of Fishes*, **36**, 389–393.

Mullin, M.A. (1993) *Webs and scales. Physical and Ecological Processes in Marine Fish Recruitment*. Univ. Washington Press, Seattle and London. pp. 135.

Pitcher, T.J. and Hart, P.J.B. (1983) *Fisheries Ecology*. Chapman & Hall, London.

Pullin, R.V.S. (1991) Cichlids in aquaculture. In *Cichlid Fishes, Behaviour, Ecology and Evolution* (Keenleyside, M.H.A. ed), Chapman & Hall, London. pp. 280–309.

Pullin, R.S.V., Rosenthal, H. and MacLean, J.L. (1993) *Environment and Aquaculture in Developing Countries*. ICLARM, Manila, The Philippines. (Proceedings of Conference 31, 359 pp.)

Purdom, C.E. (1993) *Genetics and Fish Breeding*. Chapman & Hall, London.

Rounsefell, G.A. (1975) *Ecology, Utilization and Management of Marine Fisheries*. C.V. Mosby, St. Louis.

Scott, A.P. and Sumpter, J.P. (1983) The control of trout reproduction: basic and applied research on hormones. In *Control Processes in Fish Physiology* (Rankin, J.C., Pitcher, T.J., and Duggan, R. ed), Croom Helm, London. pp. 200–220.

Steele, J.H. and Henderson, E.W. (1984) Modelling long-term fluctuations in fish stocks. *Science*, **224**, 985–987.

# Species index

(Species in figures and tables shown in heavy type)

# Subject index